Calculus III

Tunc Geveci

cognella
San Diego, CA

I0047546

Copyright © 2011 by Tunc Geveci. All rights reserved. No part of this publication may be reprinted, reproduced, transmitted, or utilized in any form or by any electronic, mechanical, or other means, now known or hereafter invented, including photocopying, microfilming, and recording, or in any information retrieval system without the written permission of University Readers, Inc.

First published in the United States of America in 2011 by Cognella, a division of University Readers, Inc.

Trademark Notice: Product or corporate names may be trademarks or registered trademarks, and are used only for identification and explanation without intent to infringe.

15 14 13 12 11   1 2 3 4 5

Printed in the United States of America

ISBN: 978-1-935551-45-4

www.cognella.com 800.200.3908

Contents

11 Vectors   **1**
   11.1 Cartesian Coordinates in 3D and Surfaces 1
   11.2 Vectors in Two and Three Dimensions 6
   11.3 The Dot Product 18
   11.4 The Cross Product 27

12 Functions of Several Variables   **35**
   12.1 Tangent Vectors and Velocity 35
   12.2 Acceleration and Curvature 46
   12.3 Real-Valued Functions of Several Variables 61
   12.4 Partial Derivatives 65
   12.5 Linear Approximations and the Differential 74
   12.6 The Chain Rule 85
   12.7 Directional Derivatives and the Gradient 94
   12.8 Local Maxima and Minima 107
   12.9 Absolute Extrema and Lagrange Multipliers 121

13 Multiple Integrals   **131**
   13.1 Double Integrals over Rectangles 131
   13.2 Double Integrals over Non-Rectangular Regions 138
   13.3 Double Integrals in Polar Coordinates 144
   13.4 Applications of Double Integrals 152
   13.5 Triple Integrals 157
   13.6 Triple Integrals in Cylindrical and Spherical Coordinates 167
   13.7 Change of Variables in Multiple Integrals 179

14 Vector Analysis   **187**
   14.1 Vector Fields, Divergence and Curl 187
   14.2 Line Integrals 194
   14.3 Line Integrals of Conservative Vector Fields 210
   14.4 Parametrized Surfaces and Tangent Planes 220
   14.5 Surface Integrals 239
   14.6 Green's Theorem 259
   14.7 Stokes' Theorem 274
   14.8 Gauss' Theorem 278

K Answers to Some Problems   **285**

L Basic Differentiation and Integration formulas   **309**

Preface

This is the third volume of my calculus series, **Calculus I, Calculus II** and **Calculus III**. This series is designed for the usual three semester calculus sequence that the majority of science and engineering majors in the United States are required to take. Some majors may be required to take only the first two parts of the sequence.

Calculus I covers the usual topics of the first semester: **Limits, continuity, the derivative, the integral and special functions such exponential functions, logarithms, and inverse trigonometric functions.** **Calculus II** covers the material of the second semester: **Further techniques and applications of the integral, improper integrals, linear and separable first-order differential equations, infinite series, parametrized curves and polar coordinates.** **Calculus III** covers topics in **multivariable calculus: Vectors, vector-valued functions, directional derivatives, local linear approximations, multiple integrals, line integrals, surface integrals, and the theorems of Green, Gauss and Stokes.**

An important feature of my book is its **focus on the fundamental concepts, essential functions and formulas of calculus**. Students should not lose sight of the basic concepts and tools of calculus by being bombarded with functions and differentiation or antidifferentiation formulas that are not significant. I have written the examples and designed the exercises accordingly. I believe that "less is more". That approach enables one to demonstrate to the students the beauty and utility of calculus, without cluttering it with ugly expressions. Another important feature of my book is **the use of visualization as an integral part of the exposition**. I believe that the most significant contribution of technology to the teaching of a basic course such as calculus has been the effortless production of graphics of good quality. Numerical experiments are also helpful in explaining the basic ideas of calculus, and I have included such data.

Remarks on some icons: I have indicated the end of a proof by ■, the end of an example by □ and the end of a remark by ◊.

Supplements: An **instructors' solution manual** that contains the solutions of all the problems is available as a PDF file that can be sent to an instructor who has adopted the book. The student who purchases the book can access the **students' solutions manual** that contains the solutions of odd numbered problems via **www.cognella.com**.

Acknowledgments: ScientificWorkPlace enabled me to type the text and the mathematical formulas easily in a seamless manner. **Adobe Acrobat Pro** has enabled me to convert the LaTeX files to pdf files. **Mathematica** has enabled me to import high quality graphics to my documents. I am grateful to the producers and marketers of such software without which I would not have had the patience to write and rewrite the material in these volumes. I would also like to acknowledge my gratitude to two wonderful mathematicians who have influenced me most by demonstrating the beauty of Mathematics and teaching me to write clearly and precisely: **Errett Bishop and Stefan Warschawski**.

Last, but not the least, I am grateful to **Simla** for her encouragement and patience while I spent hours in front a computer screen.

Tunc Geveci (tgeveci@math.sdsu.edu)
San Diego, January 2011

Chapter 11

Vectors

In this chapter we will introduce the concept of a **vector** and discuss the relevant **algebraic operations.** We will also discuss the **dot product** that is related to angles between vectors and the **cross product** that produces a vector that is orthogonal to a pair of vectors. These concepts and operations will be needed when we develop the calculus of functions of several variables.

11.1 Cartesian Coordinates in 3D and Surfaces

Cartesian Coordinates in Three Dimensions

Our starting point is the familiar **Cartesian coordinate plane**. Let's designate the axes as the x and y axes. Picture the xy-plane as a plane in the three-dimensional space. The third axis is placed so that it is perpendicular to the xy-plane and its origin coincides with the origin of the xy-plane. The positive direction is determined by the right-hand rule. Let us label the third axis as the z-axis.

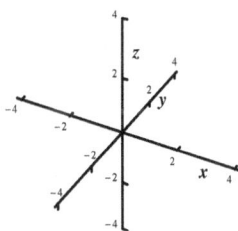

Figure 1

We will associate an **ordered triple** (x, y, z) with each point P in space as follows: If P is the points at which the three axes intersect, we will call P the **origin** and denote it by O. The ordered triple $(0, 0, 0)$ is associated with O. Let P be a point other than the origin. Consider the line that passes through P and is perpendicular to the xy-plane. Let Q be the intersection

1

of that line with the xy-plane. We will associate with Q the triple $(x, y, 0)$, where x and y are determined as Cartesian coordinates in the xy-plane. Consider the plane that passes through P and is parallel to the xy-plane. If z is the point at which that plane intersects the third axis, we will associate the ordered triple (x, y, z) with the point P. We are speaking of an **"ordered triple"**, since the order of the numbers x,y and z matters. We will identify P with the triple (x, y, z), and refer to "the point (x, y, z)", just as we identify a point in the Cartesian coordinate plane with the corresponding order pair of numbers. Thus, we have described the **Cartesian coordinate system** in the three-dimensional space. The system is also referred to as a rectangular coordinate system.

The set of all ordered triples (x, y, z) of real real numbers will be denoted by $\mathbb{R}^3$. Thus, $\mathbb{R}^3$ can be identified with the three-dimensional space that is equipped with the Cartesian coordinate system, just as the set of all ordered pairs (x, y) if real numbers can be denoted as $\mathbb{R}^2$ and identified with the set of points in the Cartesian coordinate plane.

The xy-plane consists points of the form $(x, y, 0)$, the xz-plane consists of points of the form $(x, 0, z)$, and the yz-plane consists of points of the form $(0, y, z)$. We will refer to these planes as **the coordinate planes.**

The first octant consists of points (x, y, z) such that $x \geq 0, y \geq 0$ and $z \geq 0$.

Definition 1 The (Euclidean) **distance** between $P_1 = (x_1, y_1, z_1)$ and $P_2 = (x_2, y_2, z_2)$ is

$$dist\,(P_1, P_2) = \sqrt{(x_2 - x_1)^2 + (y_2 - y_1)^2 + (z_2 - z_1)^2}.$$

Surfaces

Definition 2 Let $F(x, y, z)$ be an expression in the variables x, y, z. and let C be a constant. The set of points $(x, y, z) \in \mathbb{R}^3$ such that $F(x, y, z) = C$ is a **surface** in $\mathbb{R}^3$. We say that the surface is the **graph of the equation** $F(x, y, z) = C$.

Example 1 Let a, b, c, d be given constants. The graph of the equation

$$ax + by + cz = d$$

is a **plane**.

For example, the equation

$$x + y + z = 1$$

describes a plane that intersects the coordinate axes at the points $(1, 0, 0)$, $(0, 1, 0)$ and $(0, 0, 1)$.

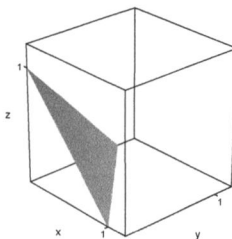

Figure 2

The equation $z = 2$ describes a plane that is parallel to the the xy-plane.

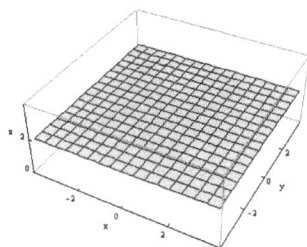

Figure 3

Definition 3 Given a point $P_0 = (x_0, y_0, z_0)$ and a positive number r, the set of points $P = (x, y, z)$ such that

$$(x - x_0)^2 + (y - y_0)^2 + (z - z_0)^2 = r^2$$

is the **sphere of radius r centered at P_0**.

Indeed, P is such a point if

$$dist\,(P, P_0) = \sqrt{(x - x_0)^2 + (y - y_0)^2 + (z - z_0)^2} = r.$$

For example, the sphere of radius 2 centered at $(3, 3, 1)$ is the graph of the equation

$$(x - 3)^2 + (y - 3)^2 + (z - 1)^2 = 4.$$

Figure 4 displays that sphere.

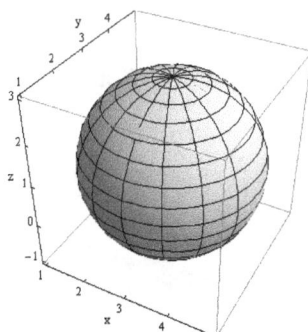

Figure 4

Example 2 Figure 5 shows the surface that is the graph of the equation

$$z = x^2 + y^2.$$

This surface is referred to as a paraboloid.

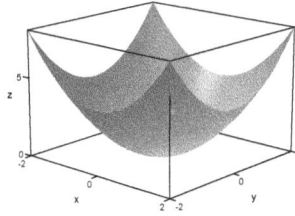

Figure 5

Since $z = x^2 + y^2 \geq 0$ for each $(x, y) \in \mathbb{R}^2$, the surface is above the xy-plane. If $(x, y) = (0, 0)$, then $z = 0$ so that $(0, 0, 0)$ is on the surface. If $c > 0$, the intersection of the surface with the plane $z = c$ is a circle whose projection onto the xy-plane is the circle

$$x^2 + y^2 = c^2$$

that is centered at $(0, 0)$ and has radius c. As c increases, these concentric circles expand. The intersection of the surface with a plane of the form $x = c$ is a parabola whose projection onto the yz-plane is the graph of the equation

$$z = c^2 + y^2.$$

Similarly, the intersection of the surface with a plane of the form $y = c$ is a parabola whose projection onto the xz-plane is the graph of the equation

$$z = x + c^2.$$

$\square$

Example 3 Figure 6 shows the surface that is described by the equation

$$x^2 - y^2 + z^2 = 1$$

Such a surface is referred to as a **hyperboloid of one sheet.**

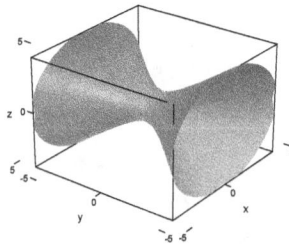

Figure 6

Since

$$x^2 + z^2 = 1 + y^2,$$

the intersection of the surface with a plane of the form $y = c$ is a circle that projects onto the xz-plane as the circle

$$x^2 + y^2 = 1 + c^2$$

that is centered at $(0, 0)$ and has radius $\sqrt{1 + c^2}$.

The intersection of the surface with a plane of the $z = c$ is a hyperbola that projects onto the xy-plane as the graph of the equation

$$x^2 - y^2 = 1 - c^2.$$

Similarly, the intersection of the surface with a plane of the form $x = c$ is a a hyperbola that projects onto the yz-plane as the graph of the equation

$$-y^2 + z^2 = 1 - c^2.$$

□

Example 4 Figure 7 shows the surface that is described by the equation

$$x^2 - y^2 - z^2 = 1$$

Such a surface is referred to as a **hyperboloid of two sheets.**

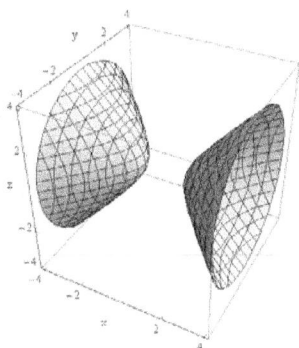

Figure 7

Since

$$y^2 + z^2 = x^2 - 1,$$

the surface does not intersect a plane of the form $x = c$ if $-1 < c < 1$. If $c < -1$ or $c > 1$, the intersection of the surface with a plane of the form $x = c$ is a circle that projects onto the yz-plane as the circle

$$y^2 + z^2 = c^2 - 1$$

that is centered at $(0, 0)$ and has radius $\sqrt{c^2 - 1}$.

The intersection of the surface with a plane of the $z = c$ is a hyperbola that projects onto the xy-plane as the graph of the equation

$$x^2 - y^2 = 1 + c^2.$$

Similarly, the intersection of the surface with a plane of the form $y = c$ is a a hyperbola that projects onto the xz-plane as the graph of the equation

$$x^2 - z^2 = 1 + c^2.$$

□

Problems

[C] In problems 1-10, make use of a graphing device to plot the surface that is the graph of the given equation. In each case identify the curves that are the intersections of the surface with planes of the form

$$x = \text{constant}, \ y = \text{constant}, \ z = \text{constant}.$$

1.
$$2x + y + 4z = 8$$

2.
$$z = 4x - 3y$$

3.
$$z = 4x^2 + 9y^2$$

4.
$$x = y^2 + 4z^2$$

5.
$$x^2 + 2y^2 + 4z^2 = 4$$

6.
$$4x^2 + y^2 + 9z^2 = 4$$

7.
$$x^2 - 9y^2 - 4z^2 = 1$$

8.
$$4y^2 - x^2 - 2z^2 = 1$$

9.
$$y^2 - x^2 + z^2 = 1$$

10.
$$x^2 + y^2 - z^2 = 4$$

11.2   Vectors in Two and Three Dimensions

In this section we will introduce the concept of a vector. Roughly speaking, a vector is an entity such as velocity or force that has a magnitude and direction. Most probably you have encountered vectors intuitively in a Physics course already. Now we will try to express the idea more precisely.

Two Dimensional Vectors

Algebraically, **a two-dimensional vector is simply an ordered pair**

$$\mathbf{v} = (v_1, v_2).$$

Even though we use the same the notation for a vector and a point in the plane, the context will clarify whether the ordered pair refers to a vector or a point. As you will soon see, the distinction will be possible due to the way we visualize vectors and define certain operations involving vectors.

We will denote vectors by boldface letters in print. The vector $\mathbf{v}$ can be denoted as $\vec{v}$ in writing. The number v_1 is **the first component $\mathbf{v}$** and v_2 is **the second component** of $\mathbf{v}$. Two vectors $\mathbf{v} = (v_1, v_2)$ and $\mathbf{w} = (w_1, w_2)$ are declared to be **equal** if and only if

$$v_1 = w_1 \text{ and } v_2 = w_2.$$

In this case we write $\mathbf{v} = \mathbf{w}$. Thus, two vectors are equal iff their corresponding components are equal. We define **the length of $\mathbf{v} = (v_1, v_2)$** as

$$\|\mathbf{v}\| = \sqrt{v_1^2 + v_2^2}.$$

The 0-vector $\mathbf{0}$ is the ordered pair $(0, 0)$. If $\mathbf{v} = (v_1, v_2)$ is not the 0-vector, we can associate with $\mathbf{v}$ a **directed line segment** that begins at a point $P_1 = (x_1, y_1)$ and ends at a point $P_2 = (x_2, y_2)$ such that

$$x_2 - x_1 = v_1 \text{ and } y_2 - y_1 = v_2.$$

We denote such a directed line segment as $\overrightarrow{P_1 P_2}$ and visualize it as an **arrow** along the line segment $\overrightarrow{P_1 P_2}$ with its tip at P_2. We will refer to $\overrightarrow{P_1 P_2}$ as **a representation of the vector $\mathbf{v}$**, or as **the vector $\mathbf{v}$ located at** P_1. There are infinitely many representations of the vector $\mathbf{v}$, but **any two representations of $\mathbf{v}$ have the same length $\|\mathbf{v}\|$ and direction**. The 0-vector cannot be represented by a directed line segment.

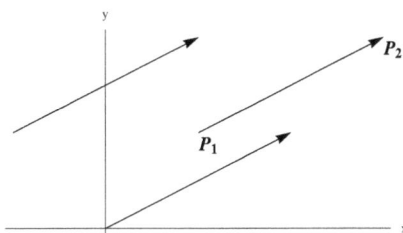

Figure 1: Representations of a vector have the same length and direction

Conversely, given the points $P_1 = (x_1, y_1)$ and $P_2 = (x_2, y_2)$, **the vector**

$$\mathbf{v} = (x_2 - x_1, y_2 - y_1)$$

is the vector that is associated with the directed line segment $\overrightarrow{P_1 P_2}$. Usually, we will simply refer to $\mathbf{v}$ as the vector $\overrightarrow{P_1 P_2}$.

We can associate the origin with the 0-vector. Every nonzero vector can be represented uniquely by a directed line segment $\overrightarrow{OP}$ that is located at the origin. We will refer to $\overrightarrow{OP}$ as **the position vector of the point** P. Thus, if $P = (a, b)$ then the position vector of P is designated by the same ordered pair (a, b). The context will clarify whether (a, b) refers to the point P or the position vector of P.

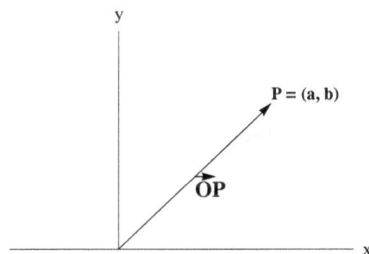

Figure 2

Example 1

a) Let $P_1 = (3, 2)$ and $P_2 = (2, 4)$. Determine the vector $\mathbf{v}$ that is associated with $\overrightarrow{P_1 P_2}$.

b) Determine Q_2 such that $\overrightarrow{Q_1 Q_2}$ is the representation of v that is located at $Q_1 = (-1, 1)$. Sketch the directed line segments $\overrightarrow{P_1 P_2}$.and $\overrightarrow{Q_1 Q_2}$.

Solution

a) We have

$$\mathbf{v} = (2 - 3, 4 - 2) = (-1, 2).$$

b) If $Q_2 = (x_2, y_2)$ the

$$x_2 - (-1) = -1 \text{ and } y_2 - 1 = 2,$$

so that $x_2 = -2$ and $y_2 = 3$. Therefore, $Q_2 = (-2, 3)$. $\square$

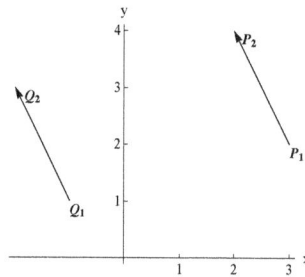

Figure 3

Aside from the fact that we visualize an ordered pair as a point and or a vector differently, the distinction results from the fact that we endow vectors with two algebraic operations.

Definition 1 Given vectors $\mathbf{v} = (v_1, v_2)$ and $\mathbf{w} = (w_1, w_2)$ we define **the sum v + w** by setting

$$\mathbf{v} + \mathbf{w} = (v_1 + w_1, v_2 + w_2).$$

Thus, we add vectors by adding the corresponding components.

Within the context of vectors, real numbers are referred to as **scalars**.

Definition 2 The **scalar multiplication of the vector** $\mathbf{v} = (v_1, v_2)$ **by the scalar** c is denoted as $c\mathbf{v}$ and we set

$$c\mathbf{v} = (cv_1, cv_2).$$

Thus, we multiply each component of $\mathbf{v}$ by the real number c.

Note that

$$||c\mathbf{v}|| = |c|\,||\mathbf{v}||.$$

Indeed, if $v = (v_1, v_2)$ then

$$||c\mathbf{v}|| = \sqrt{(cv_1)^2 + (cv_2)^2} \;\; = \;\; \sqrt{c^2\left(v_1^2 + v_2^2\right)}$$
$$= \;\; |c|\,\sqrt{v_1^2 + v_2^2} = |c|\,||\mathbf{v}||.$$

Geometrically, the addition of the vectors $\mathbf{v}$ and $\mathbf{w}$ can be represented by the **parallelogram picture**: If we represent $\mathbf{v}$ and $\mathbf{w}$ by arrows emanating from the same point, the sum $v + \mathbf{w}$ can be represented by the arrow along the diagonal of the parallelogram that is determined by $\mathbf{v}$ and $\mathbf{w}$.

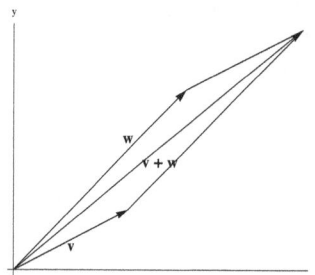

Figure 4: The parallelogram picture for the sum of $\mathbf{v}$ and $\mathbf{w}$

We can also represent $\mathbf{v} + \mathbf{w}$ by the **triangle picture**: If we have any representation of $\mathbf{v}$, we can attach $\mathbf{w}$ to the tip of that representation. The sum is represented by the arrow from the point at which $\mathbf{v}$ is located to the tip of the arrow that represents $\mathbf{w}$.

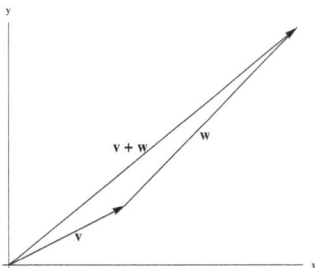

Figure 5: The triangle picture for $\mathbf{v} + \mathbf{w}$

Example 2 Let $\mathbf{v} = (1, 2)$ and $\mathbf{w} = (-2, 3)$.

a) Determine $\mathbf{v} + \mathbf{w}$
b) Sketch the parallelogram picture for the addition of $\mathbf{v}$ and $\mathbf{w}$.

Solution

a) We have
$$\mathbf{v} + \mathbf{w} = (1 - 2, 2 + 3) = (-1, 5)$$

b) Figure 6 illustrates the parallelogram picture for $\mathbf{v} + \mathbf{w}$. $\square$

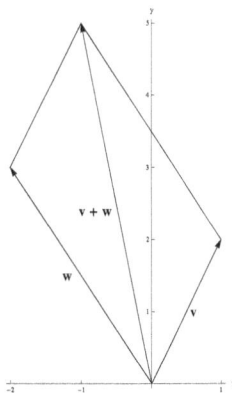

Figure 6

Remark 1 If the vectors $\mathbf{v}$ and $\mathbf{w}$ are located at the point P_0 and represent forces that act on object at P_0, the sum $\mathbf{v} + \mathbf{w}$ is the resultant force that acts on that object. If $\mathbf{v}$ represents the velocity of an object in still water and $\mathbf{w}$ is the velocity of a current, the sum $\mathbf{v} + \mathbf{w}$ is the net velocity of the object.$\Diamond$

If $c = 0$, the scalar multiple $c\mathbf{v} = \mathbf{0}$ for any vector $\mathbf{v}$. The scalar multiple is also the 0-vector if $\mathbf{v} = \mathbf{0}$. If $\mathbf{v} \neq \mathbf{0}$ and $c > 0$, we can represent $c\mathbf{v}$ by an arrow along a representation of $\mathbf{v}$ whose length is $c\|\mathbf{v}\|$. If $c < 0$ then $c\mathbf{v}$ can be represented by an arrow along the line determined by v that points in the opposite direction. The length of the arrow is $|c|\,\|\mathbf{v}\| = -c\,\|\mathbf{v}\|$.

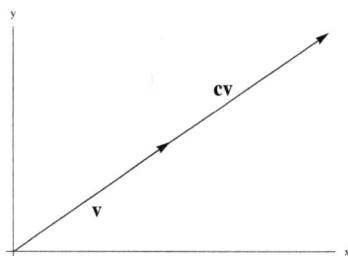

Figure 7: $c\mathbf{v}$ if $c > 0$

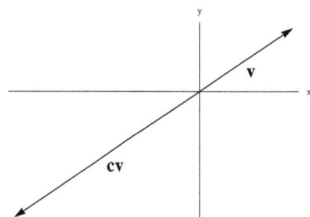

Figure 8: $c\mathbf{v}$ if $c < 0$

Example 3 Let $v = (1, 2)$.

a) Determine $3\mathbf{v}$ and $-3\mathbf{v}$ and sketch their representations.
b) Confirm that $||3\mathbf{v}|| = ||-3\mathbf{v}|| = 3\,||\mathbf{v}||$.

Solution

a) We have

$$3\mathbf{v} = 3\,(1, 2) = (3, 6) \ \text{ and } \ -3\mathbf{v} = -3\,(1, 2) = (-3, -6)\,.$$

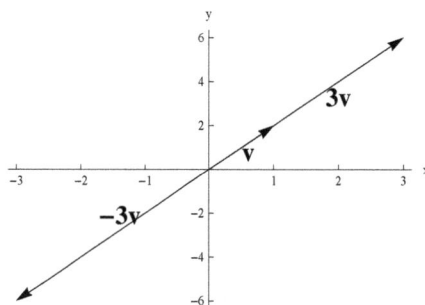

Figure 9

b) We have

$$||3\mathbf{v}|| = \sqrt{3^2 + 6^2} = \sqrt{45} = 3\sqrt{5},$$

and

$$3\,||\mathbf{v}|| = 3\sqrt{1^1 + 2^2} = 3\sqrt{5}.$$

Thus, $||3\mathbf{v}|| = 3\,||\mathbf{v}||$.

We have

$$\|-3\mathbf{v}\| = \sqrt{(-3)^2 + (-6^2)} = \sqrt{3^3 + 6^2} = \sqrt{45} = 3\sqrt{5} = 3\|\mathbf{v}\|.$$

$\square$

If $\mathbf{v} = (v_1, v_2)$ we set

$$-\mathbf{v} = (-1)\mathbf{v} = (-v_1, -v_2).$$

Thus, $\mathbf{v} - \mathbf{v} = 0$. We can represent $-\mathbf{v}$ by an arrow that points in the opposite direction that is determined by (a representation of) $\mathbf{v}$ and has the same length.

If $\mathbf{v} = (v_1, v_2)$ and $\mathbf{w} = (w_1, w_2)$ then

$$\mathbf{v} - \mathbf{w} = \mathbf{v} + (-\mathbf{w}) = (v_1 - w_1, v_2 - w_2).$$

We can represent $\mathbf{v} - \mathbf{w}$ by an arrow from the tip of $\mathbf{w}$ to the tip $\mathbf{v}$.

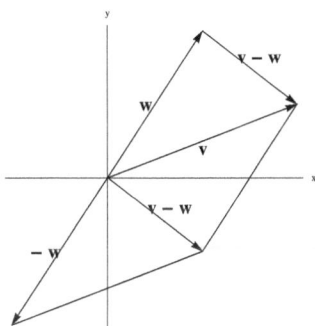

Figure 10: Representations of $\mathbf{v} - \mathbf{w}$

Example 4 Let $\mathbf{v} = (4, 5)$ and $\mathbf{w} = (3, 2)$. Determine $\mathbf{v} - \mathbf{w}$ and sketch a representation of the operation.

Solution

We have

$$\mathbf{v} - \mathbf{w} = (4 - 3, 5 - 2) = (1, 3).$$

Figure 11 illustrates $\mathbf{v} - \mathbf{w}$. $\square$

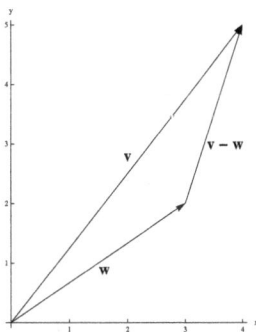

Figure 11

Definition 3 If $\mathbf{v}$ and $\mathbf{w}$ are vectors, c and d are real numbers (scalars), the vector $c\mathbf{v} + d\mathbf{w}$ is a **linear combination of** v and $\mathbf{w}$.

Assume that c and d are real numbers (scalars), and $\mathbf{u}, \mathbf{v}, \mathbf{w}$ are vectors. The following rules apply are valid for addition and scalar multiplication:

$$
\begin{aligned}
\mathbf{v} + \mathbf{w} &= \mathbf{w} + \mathbf{v}, \\
\mathbf{u} + (\mathbf{v} + \mathbf{w}) &= (\mathbf{u} + \mathbf{v}) + \mathbf{w}, \\
\mathbf{v} + \mathbf{0} &= \mathbf{v}, \\
\mathbf{v} + (-\mathbf{v}) &= \mathbf{0}, \\
c\,(\mathbf{v} + \mathbf{w}) &= c\mathbf{v} + c\mathbf{w}, \\
(c + d)\,\mathbf{v} &= c\mathbf{v} + d\mathbf{v}, \\
(cd)\,\mathbf{v} &= c\,(d\mathbf{v}), \\
1\mathbf{v} &= \mathbf{v}.
\end{aligned}
$$

You can confirm the validity of these rules easily (exercise).

Definition 4 A **unit vector** is a vector of unit length.

Thus, $\mathbf{u} = (u_1, u_2)$ is a unit vector iff

$$
\|\mathbf{u}\| = \sqrt{u_1^2 + u_2^2} = 1.
$$

Definition 5 The **normalization of the nonzero vector** $\mathbf{v}$ or **the unit vector along** $\mathbf{v}$ is

$$
\mathbf{u} = \frac{1}{\|\mathbf{v}\|}\mathbf{v}.
$$

Note that $\mathbf{u}$ is in the same direction as $\mathbf{v}$ and has unit length:

$$
\|\mathbf{u}\| = \left\| \frac{1}{\|\mathbf{v}\|}\mathbf{v} \right\| = \frac{1}{\|\mathbf{v}\|}\|\mathbf{v}\| = 1.
$$

The normalization of $\mathbf{v}$ can be written as

$$
\frac{\mathbf{v}}{\|\mathbf{v}\|}.
$$

Example 5 Let $\mathbf{v} = (2, -3)$. Determine the unit vector along $\mathbf{v}$.

Solution

We have

$$
\|\mathbf{v}\| = \sqrt{2^2 + (-3)^2} = \sqrt{13}.
$$

Therefore, the unit vector along $\mathbf{v}$ is

$$
\mathbf{u} = \frac{1}{\|\mathbf{v}\|}\mathbf{v} = \frac{1}{\sqrt{13}}\,(2, -3) = \left(\frac{2}{\sqrt{13}}, -\frac{3}{\sqrt{13}} \right).
$$

$\square$

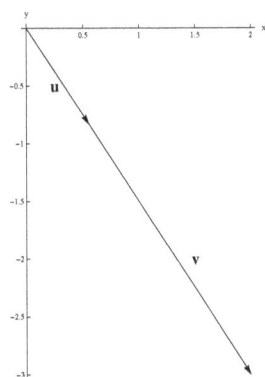

Figure 12

Standard Basis Vectors

We set

$$\mathbf{i} = (1, 0) \text{ and } \mathbf{j} = (0, 1).$$

Thus, we can represent $\mathbf{i}$ as the position vector of the point $(1, 0)$ and $\mathbf{j}$ as the position vector of the point $(0, 1)$. The vector $\mathbf{i}$ points in the positive direction of the horizontal axis and $\mathbf{j}$ points in the positive direction of the vertical axis. We will refer to $\mathbf{i}$ and $\mathbf{j}$ as **the standard basis vectors.**

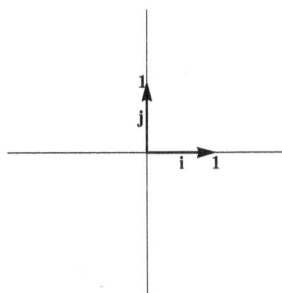

Figure 13: Standard basis vectors

If $\mathbf{v} = (v_1, v_2)$ then

$$\mathbf{v} = v_1 (1, 0) + v_2 (0, 1) = v_1 \mathbf{i} + v_2 \mathbf{j}.$$

Thus, we can express any vector as a sum of scalar multiples of the standard basis vectors $\mathbf{i}$ and $\mathbf{j}$. We will refer to the expression $v_1 \mathbf{i} + v_2 \mathbf{j}$ for the vector $\mathbf{v}$ as the **ij**representation of $\mathbf{v}$. We will favor the **ij**representation when we wish to emphasize a geometric representation of a vector. If $\mathbf{v} = v_1 \mathbf{i} + v_2 \mathbf{j}$ and $\mathbf{w} = w_1 \mathbf{i} + w_2 \mathbf{j}$ then

$$\mathbf{v} + \mathbf{w} = (v_1 + w_1) \mathbf{i} + (v_2 + w_2) \mathbf{j},$$

and

$$c\mathbf{v} = cv_1 \mathbf{i} + cv_2 \mathbf{j}$$

for any real number c.

Example 6 Let $\mathbf{v} = (4, 1)$ and $\mathbf{w} = (-2, 3)$.

a) Express $\mathbf{v}$ and $\mathbf{w}$ in terms of the standard basis vectors.

b) Perform the operations to determine $3\mathbf{v}$, $\mathbf{v} + \mathbf{w}$ and $\mathbf{v} - \mathbf{w}$ by making use the expressions for $\mathbf{v}$ and $\mathbf{w}$ in terms of the standard basis vectors.

Solution

a) We have

$$\mathbf{v} = (4, 1) = 4\mathbf{i} + \mathbf{j} \text{ and } \mathbf{w} = (-2, 3) = -2\mathbf{i} + 3\mathbf{j}.$$

b)

$$
\begin{aligned}
3\mathbf{v} &= 3\left(4\mathbf{i} + \mathbf{j}\right) = 12\mathbf{i} + 3\mathbf{j}, \\
\mathbf{v} + \mathbf{w} &= \left(4\mathbf{i} + \mathbf{j}\right) + \left(-2\mathbf{i} + 3\mathbf{j}\right) = 2\mathbf{i} + 4\mathbf{j}, \\
\mathbf{v} - \mathbf{w} &= \left(4\mathbf{i} + \mathbf{j}\right) - \left(-2\mathbf{i} + 3\mathbf{j}\right) = 6\mathbf{i} - 2\mathbf{j}.
\end{aligned}
$$

$\square$

Three Dimensional Vectors

The concept of a three dimensional vector is a straightforward analog of the concept of a two dimensional vector, and the operations are similar.

Algebraically, a 3D-vector $\mathbf{v}$ is an ordered triple (v_1, v_2, v_3) where the components v_1, v_2 and v_3 are real numbers (scalars). If $\mathbf{v} = (v_1, v_2, v_3)$ and $\mathbf{w} = (w_1, w_2, w_3)$ we declare that $\mathbf{v} = \mathbf{w}$ iff the corresponding components are equal:

$$v_1 = w_1, \ v_2 = w_2, \ v_3 = w_3.$$

The 0-vector $\mathbf{0}$ is $(0, 0, 0)$.

The length of $\mathbf{v} = (v_1, v_2, v_3)$ is

$$||\mathbf{v}|| = \sqrt{v_1^2 + v_2^2 + v_3^2}.$$

A (geometric) representation of $\mathbf{v} = (v_1, v_2, v_3)$ is a directed line segment $\overrightarrow{P_1 P_2}$, where $P_1 = (x_1, y_1, z_1)$, $P_2 = (x_2, y_2, z_2)$ are points in $\mathbb{R}^3$ such that

$$v_1 = x_2 - x_1, \ v_2 = y_2 - y_1 \text{ and } v_3 = z_2 - z_1.$$

We may refer to $\overrightarrow{P_1 P_2}$ as the vector v that is located at P_1 or as v that is attached to P_1. Representations of the same vector have the same length and direction.

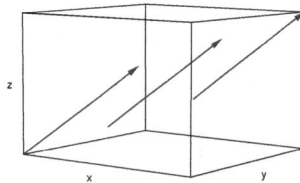

Figure 14: Representations of a vector in 3D

If $P = (x, y, z)$ is a point in $\mathbb{R}^3$, the position vector of P is the directed line segment $\overrightarrow{OP}$. The sum of $\mathbf{v} = (v_1, v_2, v_3)$ and $\mathbf{w} = (w_1, w_2, w_3)$ is the vector

$$\mathbf{v} + \mathbf{w} = (v_1 + w_1, v_2 + w_2, v_3 + w_3).$$

We can visualize addition via the parallelogram picture or the triangle picture, as in the 2D case.

If c is a real number (scalar), the scalar multiple $c\mathbf{v}$ of $\mathbf{v} = (v_1, v_2 v_3)$ is the vector

$$c\mathbf{v} = (cv_1, cv_2, cv_3).$$

Thus $c\mathbf{v}$ is along $\mathbf{v}$, points in the same direction as $\mathbf{v}$ if $c > 0$ and in the opposite direction if $c < 0$. We have

$$||cv|| = |c|\, ||v||.$$

Addition and scalar multiplication obey the rules that were listed for two dimensional vectors. Just as the two dimensional case, the normalization of the nonzero vector $\mathbf{v}$ is

$$\mathbf{u} = \frac{1}{||\mathbf{v}||}\mathbf{v},$$

which can be written as

$$\frac{\mathbf{v}}{||\mathbf{v}||}.$$

Example 7 Let $\mathbf{v} = (2, -2, 3)$ and $\mathbf{w} = (-1, 1, 2)$. Determine $\mathbf{v} + \mathbf{w}$, $4\mathbf{v} + 3\mathbf{w}$ and the normalization of $\mathbf{v}$.

Solution

$$\mathbf{v} + \mathbf{w} = (2, -2, 3) + (-1, 1, 2) = (1, -1, 5).$$

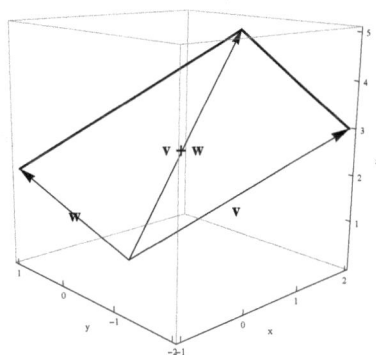

Figure 15

We also have

$$2\mathbf{v} = 2(2, -2, 3) = (4, -4, 6),$$

$$\begin{aligned}
4\mathbf{v} + 3\mathbf{w} &= 4(2, -2, 3) + 3(-1, 1, 2)\\
&= (8, -8, 12) + (-3, 3, 6) = (5, -5, 18).
\end{aligned}$$

The length of $\mathbf{v}$ is

$$||\mathbf{v}|| = \sqrt{2^2 + (-2)^2 + 3^2} = \sqrt{17}.$$

Therefore, the normalization of **v** is

$$\mathbf{u} = \frac{1}{\|\mathbf{v}\|}\mathbf{v} = \frac{1}{\sqrt{17}}\,(2,-2,3) = \left(\frac{2}{\sqrt{17}}, -\frac{2}{\sqrt{17}}, \frac{3}{\sqrt{17}}\right).$$

In the 3D case, the standard basis vectors are

$$\mathbf{i} = (1,0,0)\,,\ \mathbf{j} = (0,1,0)\ \text{and}\ \mathbf{k} = (0,0,1)\,.$$

If we label the Cartesian coordinate axes as the x, y and z axes, then **i** is a unit vector in the direction of the positive x-axis, **j** is a unit vector in the direction of the positive y-axis, and **k** is a unit vector in the direction of the positive z-axis.

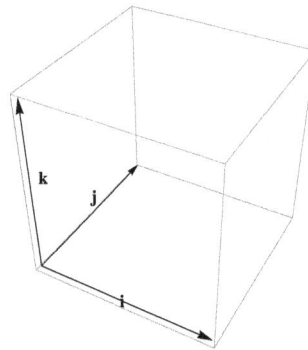

Figure 16: Standard basis vectors in 3D

If $\mathbf{v} = (v_1, v_2, v_3)$ then

$$\mathbf{v} = v_1\mathbf{i} + v_2\mathbf{j} + v_3\mathbf{k}.$$

We may refer to the above expression for v as the **ijk**representation of **v**.

If $\mathbf{v} = v_1 i + v_2 j + v_3 k$, $\mathbf{w} = w_1 i + w_2 j + w_3 k$.and $c \in \mathbb{R}$ then

$$\mathbf{v} + \mathbf{w} = (v_1 + w_1)\,\mathbf{i} + (v_2 + w_2)\,\mathbf{j} + (v_3 + w_3)\,\mathbf{k}$$

and

$$c\mathbf{v} = cv_1\mathbf{i} + cv_2\mathbf{j} + cv_3\mathbf{k}.$$

Example 8 Let

$$\mathbf{v} = 2\mathbf{i} - 3\mathbf{j} + 5\mathbf{k} \text{ and } \mathbf{w} = -3\mathbf{i} + 2\mathbf{j} - 4\mathbf{k}.$$

Determine $-3\mathbf{v}$ and $\mathbf{v} - \mathbf{w}$.

Solution

$$-3\mathbf{v} = -3\,(2\mathbf{i} - 3\mathbf{j} + 5\mathbf{k}) = -6\mathbf{i} + 9\mathbf{j} - 15\mathbf{k},$$

$$\mathbf{v} - \mathbf{w} = (2\mathbf{i} - 3\mathbf{j} + 5\mathbf{k}) - (-3\mathbf{i} + 2\mathbf{j} - 4\mathbf{k}) = 5\mathbf{i} - 5\mathbf{j} + 9\mathbf{k}.$$

□

Problems

In problems 1-4,

a) Determine the vector **v** that is associated with the directed line segment $\overrightarrow{P_1P_2}$.

b) Determine Q_2 such that $\overrightarrow{Q_1Q_2}$ is the representation of v that is located at Q_1. Sketch the directed line segments $\overrightarrow{P_1P_2}$.and $\overrightarrow{Q_1Q_2}$.

1.

$$P_1 = (1,2),\ P_2 = (3,5),\ Q_1 = (2,1)$$

2.

$$P_1 = (-2,3),\ P_2 = (-4,2),\ Q_1 = (1,3)$$

3.

$$P_1 = (2,-3),\ P_2 = (4,2),\ Q_1 = (3,2)$$

4.

$$P_1 = (-2,-1),\ P_2 = (-4,2),\ Q_1 = (-1,2)$$

In problems 5-8

a) Determine $\mathbf{v} + \mathbf{w}$,

b) Sketch the parallelogram picture for the addition of **v** and **w** (represent **v** and **w** by directed line segments attached to the origin):

5. $\mathbf{v} = (2,1)$, $\mathbf{w} = (3,4)$
6. $\mathbf{v} = (2,-3)$, $\mathbf{w} = (3,2)$

7. $\mathbf{v} = (-2,-1)$, $\mathbf{w} = (2,4)$
8. $\mathbf{v} = (2,3)$, $\mathbf{w} = (-1,-5)$

In problems 9 and 10

a) Determine $\mathbf{v} + \mathbf{w}$,

b) Sketch the triangle picture for the addition of **v** and **w**.(represent **v** by a directed line segment attached to the origin):

9. $\mathbf{v} = (2,4)$, $\mathbf{w} = (-1,2)$

10. $\mathbf{v} = (-4,2)$, $\mathbf{w} = (3,1)$

In problems 11 and 12

a) Determine $\mathbf{v} - \mathbf{w}$,

b) Sketch the triangle picture for $\mathbf{v} - \mathbf{w}$.(represent **v** and **w** by directed line segments attached to the origin):

11. $\mathbf{v} = (2,4)$, $\mathbf{w} = (-2,2)$

12. $\mathbf{v} = (-4,2)$, $\mathbf{w} = (3,1)$

In problems 13 and 14, determine the given linear combination of **v** and **w**.

13. $2\mathbf{v} - 3\mathbf{w}$ if $\mathbf{v} = (3,-1)$ and $\mathbf{w} = (-2,5)$

14. $-4\mathbf{v} + 5\mathbf{w}$ if $\mathbf{v} = (2,4)$ and $\mathbf{w} = (1,-4)$

In problems 15 and 16

a) Determine **u,** the unit vector along **v** (the normalization of **v**),

b) Sketch **u** and **v**.(represent **v** and **u** by directed line segments attached to the origin):

15. $\mathbf{v} = (3, 4)$ 16. $\mathbf{v} = (-2, 2)$

In problems 17-20
a) Express the vectors $\mathbf{v}$ and $\mathbf{w}$ in terms of the standard basis vectors,
b) Determine the given linear combination of $\mathbf{v}$ and $\mathbf{w}$ by using their representations in terms of the standard basis vectors.
17. $\mathbf{v} = (3, 2)$, $\mathbf{w} = (-2, 4)$, $2\mathbf{v} - 3\mathbf{w}$
18. $\mathbf{v} = (-4, 1)$, $\mathbf{w} = (4, 3)$, $2\mathbf{v} + \mathbf{w}$
19. $\mathbf{v} = (-2, 3, 6)$, $\mathbf{w} = (4, -2, 1)$, $-\mathbf{v} + 4\mathbf{w}$
20. $\mathbf{v} = (-1, -3, 5)$, $\mathbf{w} = (7, 2, -2)$, $3\mathbf{v} - 2\mathbf{w}$

11.3   The Dot Product

In this section we will introduce the dot product. This is an operation that assigns a real number to pairs of vectors and enables us to measure the angle between them.

The Definition of the Dot Product

Definition 1 If $\mathbf{v} = (v_1, v_2)$ and $\mathbf{w} = (w_1, w_2)$ are two dimensional vectors, **the dot product** $\mathbf{v} \cdot \mathbf{w}$ is the real number that is defined as

$$\mathbf{v} \cdot \mathbf{w} = v_1 w_1 + v_2 w_2.$$

If $\mathbf{v} = (v_1, v_2, v_3)$ and $\mathbf{w} = (w_1, w_2, w_3)$ are three dimensional vectors, the dot product $\mathbf{v} \cdot \mathbf{w}$ is the real number that is defined as

$$\mathbf{v} \cdot \mathbf{w} = v_1 w_1 + v_2 w_2 + v_3 w_3.$$

Note that we can express **the length of a vector** in terms of the product:

$$\|\mathbf{v}\| = \sqrt{\mathbf{v} \cdot \mathbf{v}}$$

Example 1 Determine $\mathbf{v} \cdot \mathbf{w}$ if

a) $\mathbf{v} = (2, 5)$ and $\mathbf{w} = (4, 3)$,
b) $\mathbf{v} = 2\mathbf{i} - 3\mathbf{j} + \mathbf{k}$ and $\mathbf{w} = -\mathbf{i} + 2\mathbf{j} + 4\mathbf{k}$.

Solution

a)
$$\mathbf{v} \cdot \mathbf{w} = (2, 5) \cdot (4, 3) = 8 + 15 = 23.$$

b)
$$\mathbf{v} \cdot \mathbf{w} = (2\mathbf{i} - 3\mathbf{j} + \mathbf{k}) \cdot (-\mathbf{i} + 2\mathbf{j} + 4\mathbf{k}) = -2 - 6 + 4 = -4.$$

□

The basic properties on the dot product: If $\mathbf{u}$, $\mathbf{v}$ and $\mathbf{w}$ are vectors in $\mathbb{R}^n$ ($n = 2$ or $n = 3$) and c is a real number (scalar) then

$$\begin{aligned}
\mathbf{v} \cdot \mathbf{w} &= \mathbf{w} \cdot \mathbf{v}, \\
\mathbf{u} \cdot (\mathbf{v} + \mathbf{w}) &= \mathbf{u} \cdot \mathbf{v} + \mathbf{u} \cdot \mathbf{w}, \\
(c\mathbf{v}) \cdot \mathbf{w} &= c(\mathbf{v} \cdot \mathbf{w}) = \mathbf{v} \cdot (c\mathbf{w}), \\
\mathbf{0} \cdot \mathbf{v} &= 0.
\end{aligned}$$

These properties can be verified easily (exercise).

Theorem 1 (The Cauchy-Schwarz Inequality) For each v and w in $\mathbb{R}^n$ ($n = 2$ or $n = 3$) we have

$$|\mathbf{v} \cdot \mathbf{w}| \leq ||\mathbf{v}||\, ||\mathbf{w}||$$

Proof

Let $t \in \mathbb{R}$. Set

$$
\begin{aligned}
f(t) = ||\mathbf{v} - t\mathbf{w}||^2 &= (\mathbf{v} - t\mathbf{w}) \cdot (\mathbf{v} - t\mathbf{w}) \\
&= \mathbf{v} \cdot \mathbf{v} - t\mathbf{v} \cdot \mathbf{w} - t\mathbf{w} \cdot \mathbf{v} + t^2 \mathbf{w} \cdot \mathbf{w} \\
&= ||\mathbf{v}||^2 - 2t\mathbf{v} \cdot \mathbf{w} + \mathbf{t}^2 ||\mathbf{w}||^2 \\
&= ||\mathbf{w}||^2 t^2 - 2\mathbf{v} \cdot \mathbf{w}t + ||\mathbf{v}||^2
\end{aligned}
$$

Since

$$f(t) = ||\mathbf{v} - t\mathbf{w}||^2 \geq 0 \text{ for each } t \in \mathbb{R},$$

the discriminant of the quadratic expression $||\mathbf{w}||^2 t^2 - 2\mathbf{v} \cdot \mathbf{w}t + ||\mathbf{v}||^2$ is nonpositive. Thus,

$$(-2\mathbf{v} \cdot \mathbf{w})^2 - 4 ||\mathbf{w}||^2 ||\mathbf{v}||^2 \leq 0,$$

so that

$$4(\mathbf{v} \cdot \mathbf{w})^2 \leq 4 ||\mathbf{w}||^2 ||\mathbf{v}||^2 \Rightarrow (\mathbf{v} \cdot \mathbf{w})^2 \leq ||\mathbf{v}||^2 ||\mathbf{w}||^2 \Leftrightarrow |\mathbf{v} \cdot \mathbf{w}| \leq ||\mathbf{v}||\, ||\mathbf{w}||,$$

as claimed. ∎

The following fact is referred to as **the triangle inequality** for the obvious reason: The length of one side of a triangle is less than the sum of the lengths of other two sides.

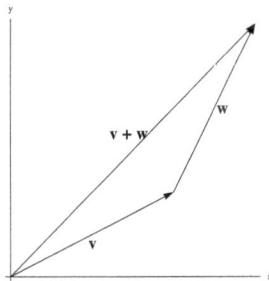

Figure 1: $||\mathbf{v} + \mathbf{w}|| < ||\mathbf{v}|| + ||\mathbf{w}||$

Theorem 2 (The Triangle Inequality) For each v and w in $\mathbb{R}^n$ ($n = 2$ or $n = 3$) we have

$$||\mathbf{v} + \mathbf{w}|| \leq ||\mathbf{v}|| + ||\mathbf{w}||.$$

Proof

We have

$$
\begin{aligned}
||\mathbf{v} + \mathbf{w}||^2 &= (\mathbf{v} + \mathbf{w}) \cdot (\mathbf{v} + \mathbf{w}) \\
&= \mathbf{v} \cdot \mathbf{v} + \mathbf{v} \cdot \mathbf{w} + \mathbf{w} \cdot \mathbf{v} + \mathbf{w} \cdot \mathbf{w} \\
&= ||\mathbf{v}||^2 + 2\mathbf{v} \cdot \mathbf{w} + ||\mathbf{w}||^2
\end{aligned}
$$

By the Cauchy-Schwarz Inequality,

$$2\mathbf{v} \cdot \mathbf{w} \leq \mathbf{2} \left\|\mathbf{v}\right\| \left\|\mathbf{w}\right\|.$$

Therefore,

$$\left\|\mathbf{v} + \mathbf{w}\right\|^2 \leq \left\|\mathbf{v}\right\|^2 + \mathbf{2} \left\|\mathbf{v}\right\| \left\|\mathbf{w}\right\| + \left\|\mathbf{w}\right\|^2 = \left(\left\|\mathbf{v}\right\| + \left\|\mathbf{w}\right\|\right)^2.$$

Thus,

$$\left\|\mathbf{v} + \mathbf{w}\right\| \leq \left\|\mathbf{v}\right\| + \left\|\mathbf{w}\right\|$$

■

Assume that $\mathbf{v}$ and $\mathbf{w}$ are nonzero vectors in $\mathbb{R}^n$ ($n = 2$ or $n = 3$). By the Cauchy-Schwarz inequality,

$$\left|\mathbf{v} \cdot \mathbf{w}\right| \leq \left\|\mathbf{v}\right\| \left\|\mathbf{w}\right\|,$$

so that

$$-\left\|\mathbf{v}\right\| \left\|\mathbf{w}\right\| \leq \mathbf{v} \cdot \mathbf{w} \leq \left\|\mathbf{v}\right\| \left\|\mathbf{w}\right\|.$$

Thus

$$-1 \leq \frac{\mathbf{v} \cdot \mathbf{w}}{\left\|\mathbf{v}\right\| \left\|\mathbf{w}\right\|} \leq 1.$$

Therefore there exists a unique angle $\theta \in [0, \pi]$ (in radians) such that

$$\cos\left(\theta\right) = \frac{\mathbf{v} \cdot \mathbf{w}}{\left\|\mathbf{v}\right\| \left\|\mathbf{w}\right\|} \Leftrightarrow \theta = \arccos\left(\frac{\mathbf{v} \cdot \mathbf{w}}{\left\|\mathbf{v}\right\| \left\|\mathbf{w}\right\|}\right).$$

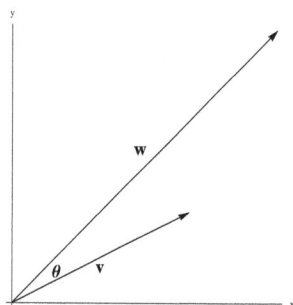

Figure 2: The angle between two vectors

Definition 2 If $\mathbf{v}$ and $\mathbf{w}$ are nonzero vectors in $\mathbb{R}^n$ ($n = 2$ or $n = 3$).the angle $\theta \in [0, \pi]$ such that

$$\cos\left(\theta\right) = \frac{\mathbf{v} \cdot \mathbf{w}}{\left\|\mathbf{v}\right\| \left\|\mathbf{w}\right\|} \Leftrightarrow \theta = \arccos\left(\frac{\mathbf{v} \cdot \mathbf{w}}{\left\|\mathbf{v}\right\| \left\|\mathbf{w}\right\|}\right)$$

is the angle between the vectors $\mathbf{v}$ and $\mathbf{w}$.

Remark 1 It can be shown that the above definition is consistent with the law of cosines:

$$\left\|\mathbf{v} - \mathbf{w}\right\|^2 = \left\|\mathbf{v}\right\|^2 - 2 \left\|\mathbf{v}\right\| \left\|\mathbf{w}\right\| \cos\left(\theta\right) + \left\|\mathbf{w}\right\|^2.$$

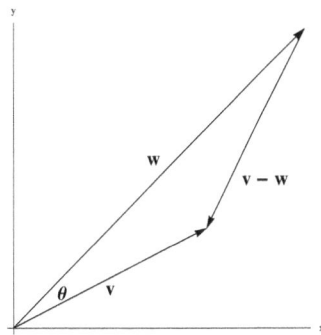

Figure 3

Indeed,

$$||\mathbf{v} - \mathbf{w}||^2 = (\mathbf{v} - \mathbf{w}) \cdot (\mathbf{v} - \mathbf{w}) = ||\mathbf{v}||^2 - 2\mathbf{v} \cdot \mathbf{w} + ||\mathbf{w}||^2.$$

Thus,

$$\begin{aligned} ||\mathbf{v}||^2 - 2\mathbf{v} \cdot \mathbf{w} + ||\mathbf{w}||^2 &= ||\mathbf{v} - \mathbf{w}||^2 \\ &= ||\mathbf{v}||^2 - 2||\mathbf{v}||\,||\mathbf{w}||\cos(\theta) + ||\mathbf{w}||^2. \end{aligned}$$

Therefore,

$$\mathbf{v} \cdot \mathbf{w} = ||\mathbf{v}||\,||\mathbf{w}||\cos(\theta).$$

$\Diamond$

Example 2 Let
$$\mathbf{v} = (1,2) \text{ and } \mathbf{w} = \left(\sqrt{3} + 2, 2\sqrt{3} - 1\right).$$

Determine the angle between v and w.

Solution

We have
$$||\mathbf{v}|| = \sqrt{5},\ ||\mathbf{w}|| = \sqrt{\left(\sqrt{3} + 2\right)^2 + \left(2\sqrt{3} - 1\right)^2} = 2\sqrt{5},$$

and
$$\mathbf{v} \cdot \mathbf{w} = (1,2) \cdot \left(\sqrt{3} + 2, 2\sqrt{3} - 1\right) = \sqrt{3} + 2 + 4\sqrt{3} - 2 = 5\sqrt{3}.$$

Therefore,

$$\theta = \arccos\left(\frac{\mathbf{v} \cdot \mathbf{w}}{||\mathbf{v}||\,||\mathbf{w}||}\right) = \arccos\left(\frac{5\sqrt{3}}{\left(\sqrt{5}\right)\left(2\sqrt{5}\right)}\right) = \arccos\left(\frac{\sqrt{3}}{2}\right) = \frac{\pi}{6}.$$

Figure 4 illustrates the angle θ. $\square$

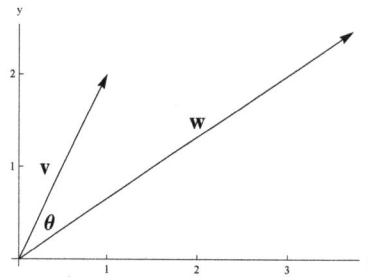

Figure 4

Definition 3 The vectors $\mathbf{v}$ and $\mathbf{w}$ in $\mathbb{R}^n$ ($n = 2$ or $n = 3$) are **orthogonal** (or **perpendicular**) if $\mathbf{v} \cdot \mathbf{w} = 0$.

Example 3 Let

$$\mathbf{v} = (2, 1) \text{ and } \mathbf{w} = (-1, 2).$$

Show that $\mathbf{v}$ and $\mathbf{w}$ are orthogonal.

Solution

We have

$$\mathbf{v} \cdot \mathbf{w} = (2, 1) \cdot (-1, 2) = -2 + 2 = 0.$$

Therefore $\mathbf{v}$ and $\mathbf{w}$ are orthogonal.

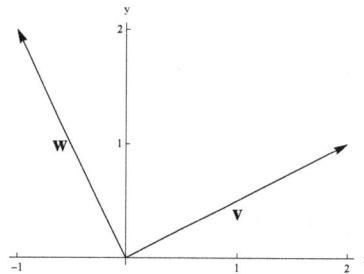

Figure 5: Orthogonal vectors

.$\square$

Remark 2 The 0-vector is orthogonal to any vector since $\mathbf{0} \cdot \mathbf{v} = 0$ for any $\mathbf{v} \in \mathbb{R}^n$. $\Diamond$

Remark 3 The standard basis vectors are mutually orthogonal. Therefore, if $\mathbf{v} = v_1\mathbf{i} + v_2\mathbf{j}$ then

$$\mathbf{v} \cdot \mathbf{i} = (v_1\mathbf{i} + v_2\mathbf{j}) \cdot \mathbf{i} = v_1(\mathbf{i} \cdot \mathbf{i}) + v_2(\mathbf{i} \cdot \mathbf{j}) = v_1 \|\mathbf{i}\|^2 + 0 = v_1,$$

and

$$\mathbf{v} \cdot \mathbf{j} = (v_1\mathbf{i} + v_2\mathbf{j}) \cdot \mathbf{j} = v_1(\mathbf{i} \cdot \mathbf{j}) + v_2(\mathbf{j} \cdot \mathbf{j}) = 0 + v_2 \|\mathbf{j}\|^2 = v_2.$$

Thus,

$$\mathbf{v} = (\mathbf{v} \cdot \mathbf{i})\mathbf{i} + (\mathbf{v} \cdot \mathbf{j})\mathbf{j}.$$

Similarly, if $\mathbf{v} \in \mathbb{R}^3$ then

$$\mathbf{v} = (\mathbf{v} \cdot \mathbf{i})\mathbf{i} + (\mathbf{v} \cdot \mathbf{j})\mathbf{j} + (\mathbf{v} \cdot \mathbf{k})\mathbf{k},$$

where $\mathbf{i}$, $\mathbf{j}$ and $\mathbf{k}$ are the standard basis vectors in $\mathbb{R}^3$. $\Diamond$

Let $\mathbf{v}$ be a nonzero vector in $\mathbb{R}^2$ and let $\mathbf{u}$ be the unit vector along $\mathbf{v}$. Thus,

$$\mathbf{u} = \frac{1}{||\mathbf{v}||}\mathbf{v}.$$

Let α be the angle between $\mathbf{v}$ and the positive x-axis and let β be the angle between $\mathbf{v}$ and the positive y-axis. We have

$$\cos\left(\alpha\right) = \frac{\mathbf{v}\cdot\mathbf{i}}{||\mathbf{v}||\,||\mathbf{i}||} = \frac{\mathbf{v}\cdot\mathbf{i}}{||\mathbf{v}||} = \mathbf{u}\cdot\mathbf{i},$$

and

$$\cos\left(\beta\right) = \frac{\mathbf{v}\cdot\mathbf{j}}{||\mathbf{v}||\,||\mathbf{j}||} = \frac{\mathbf{v}\cdot\mathbf{j}}{||\mathbf{v}||} = \mathbf{u}\cdot\mathbf{j}$$

Thus

$$\mathbf{u} = \left(\mathbf{u}\cdot\mathbf{i}\right)\mathbf{i} + \left(\mathbf{u}\cdot\mathbf{j}\right)\mathbf{j} = \cos\left(\alpha\right)\mathbf{i} + \cos\left(\beta\right)\mathbf{j}.$$

Similarly, if $\mathbf{v}$ is a nonzero vector in $\mathbb{R}^3$ and $\mathbf{u}$ is the unit vector along $\mathbf{v}$, then

$$\mathbf{u} = \left(\mathbf{u}\cdot\mathbf{i}\right)\mathbf{i} + \left(\mathbf{u}\cdot\mathbf{j}\right)\mathbf{j} = \cos\left(\alpha\right)\mathbf{i} + \cos\left(\beta\right)\mathbf{j} + \cos\left(\gamma\right)\mathbf{k},$$

where α, β and γ are the angles between $\mathbf{v}$ and the positive directions of the x, y and z axes, respectively.

Definition 4 Given a nonzero vector $\mathbf{v} \in \mathbb{R}^2$ the direction cosines of $\mathbf{v}$ are the components of the unit vector along $\mathbf{v}$. The **direction cosines** of a nonzero vector $\mathbf{v}$ in $\mathbb{R}^3$ are the components of the unit vector along $\mathbf{v}$.

Example 4 Let $\mathbf{v} = (2, 4, 3)$. Find the direction cosines of $\mathbf{v}$.

Solution

All we need to do is to determine the unit vector $\mathbf{u}$ along $\mathbf{v}$. The length of $\mathbf{v}$ is

$$||\mathbf{v}|| = \sqrt{2^2 + 4^2 + 3^2} = \sqrt{29}$$

Therefore,

$$\mathbf{u} = \frac{1}{||\mathbf{v}||}\mathbf{v} = \frac{1}{\sqrt{29}}(2, 4, 3) = \left(\frac{2}{\sqrt{29}}, \frac{4}{\sqrt{29}}, \frac{3}{\sqrt{29}}\right).$$

Thus, the direction cosines of $\mathbf{v}$ are

$$\frac{2}{\sqrt{29}}, \quad \frac{4}{\sqrt{29}} \quad \text{and} \quad \frac{3}{\sqrt{29}}.$$

$\square$

Projections

Let $\mathbf{w}$ be a nonzero vector. In a number of applications it is useful to express a given vector $\mathbf{v}$ as

$$\mathbf{v} = \mathbf{v}_1 + \mathbf{v}_2,$$

where $\mathbf{v}_1$ is along $\mathbf{w}$, i..e, a scalar multiple of $\mathbf{w}$, and $\mathbf{v}_2$ is orthogonal to $\mathbf{w}$.

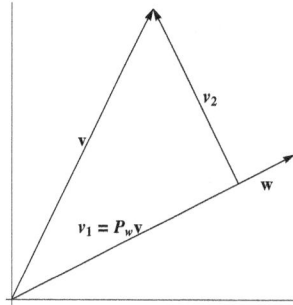

Figure 6

Thus,
$$\mathbf{v} = \alpha\mathbf{w} + \mathbf{v}_2,$$
where $\mathbf{v}_2 \cdot \mathbf{w} = 0$. Therefore,
$$(\mathbf{v} - \alpha\mathbf{w}) \cdot \mathbf{w} = 0.$$
Thus,
$$\mathbf{v} \cdot \mathbf{w} - \alpha \|\mathbf{w}\|^2 = 0 \Rightarrow \alpha = \frac{\mathbf{v} \cdot \mathbf{w}}{\|\mathbf{w}\|^2}.$$
Therefore,
$$\mathbf{v} = \left(\frac{\mathbf{v} \cdot \mathbf{w}}{\|\mathbf{w}\|^2}\right)\mathbf{w} + \mathbf{v}_2,$$
where
$$\left(\frac{\mathbf{v} \cdot \mathbf{w}}{\|\mathbf{w}\|^2}\right)\mathbf{w}$$
is along $\mathbf{w}$ and $\mathbf{v}_2$ is orthogonal to w. Note that
$$\left(\frac{\mathbf{v} \cdot \mathbf{w}}{\|\mathbf{w}\|^2}\right)\mathbf{w} = \left(\frac{\mathbf{v} \cdot \mathbf{w}}{\|\mathbf{w}\|}\right)\frac{\mathbf{w}}{\|\mathbf{w}\|} = \left(\mathbf{v} \cdot \frac{\mathbf{w}}{\|\mathbf{w}\|}\right)\frac{\mathbf{w}}{\|\mathbf{w}\|}.$$
The vector
$$\mathbf{u} = \frac{\mathbf{w}}{\|\mathbf{w}\|}$$
is the unit vector along $\mathbf{w}$. Thus, we can express $\mathbf{v}$ as
$$\mathbf{v} = (\mathbf{v} \cdot \mathbf{u})\,\mathbf{u} + \mathbf{v}_2,$$
where $\mathbf{u}$ is **the unit vector along $\mathbf{w}$** and $\mathbf{v}_2$ **is orthogonal to $\mathbf{w}$.**

Definition 5 If $\mathbf{w}$ is a nonzero vector, **the projection of $\mathbf{v}$** along $\mathbf{w}$ is
$$\boldsymbol{P}_{\mathbf{w}}\mathbf{v} = (\mathbf{v} \cdot \mathbf{u})\,\mathbf{u},$$
where
$$\mathbf{u} = \frac{\mathbf{w}}{\|\mathbf{w}\|}$$
is **the unit vector along $\mathbf{w}$. The component of $\mathbf{v}$ along $\mathbf{w}$** is
$$comp_{\mathbf{w}}\mathbf{v} = \mathbf{v} \cdot \mathbf{u},$$
so that
$$\boldsymbol{P}_{\mathbf{w}}\mathbf{v} = (comp_{\mathbf{w}}\mathbf{v})\,\mathbf{u}$$

By the discussion that preceded Definition 4, the vector $\mathbf{v} - \boldsymbol{P}_{\mathbf{w}}\mathbf{v}$ is orthogonal to $\mathbf{w}$.

Remark 4 If θ is the angle between v and w, then

$$comp_w\mathbf{v} = \mathbf{v} \cdot \mathbf{u} = \frac{\mathbf{v} \cdot \mathbf{w}}{||\mathbf{w}||} = \frac{||\mathbf{v}||\,||\mathbf{w}||\cos(\theta)}{||\mathbf{w}||} = ||\mathbf{v}||\cos(\theta),$$

and

$$\boldsymbol{P}_{\mathbf{w}}\mathbf{v} = (\mathbf{v} \cdot \mathbf{u})\,\mathbf{u} = ||\mathbf{v}||\cos(\theta)\,\mathbf{u}.$$

$\Diamond$

Remark 5 We have noted that a vector $\mathbf{v} \in \mathbb{R}^3$ can be expressed as

$$\mathbf{v} = (\mathbf{v} \cdot \mathbf{i})\,\mathbf{i} + (\mathbf{v} \cdot \mathbf{j})\,\mathbf{j} + (\mathbf{v} \cdot \mathbf{k})\,\mathbf{k}.$$

Thus,

$$\mathbf{v} = (comp_\mathbf{i}\mathbf{v})\,\mathbf{i} + (comp_\mathbf{j}\mathbf{v})\,\mathbf{j} + (comp_\mathbf{k}\mathbf{v})\,\mathbf{k}$$

$\Diamond$

Example 5 Let $\mathbf{v} = (1,2)$ and $\mathbf{w} = (2,1)$.
a) Determine the component of $\mathbf{v}$ along $\mathbf{w}$ and the projection of $\mathbf{v}$ along $\mathbf{w}$.
b) Express $\mathbf{v}$ as $\boldsymbol{P}_{\mathbf{w}}\mathbf{v} + \mathbf{v}_2$ where $\mathbf{v}_2$ is orhogonal to $\boldsymbol{P}_{\mathbf{w}}\mathbf{v}$

Solution

a) The unit vector along $\mathbf{w}$ is

$$\mathbf{u} = \frac{1}{||\mathbf{w}||}\mathbf{w} = \frac{1}{\sqrt{5}}(2,1).$$

Therefore,

$$comp_w\mathbf{v} = \mathbf{v} \cdot \mathbf{u} = (1,2) \cdot \left(\frac{1}{\sqrt{5}}(2,1)\right) = \frac{1}{\sqrt{5}}(2+2) = \frac{4}{\sqrt{5}},$$

and

$$\boldsymbol{P}_{\mathbf{w}}\mathbf{v} = (\mathbf{v} \cdot \mathbf{u})\,\mathbf{u} = \frac{4}{\sqrt{5}}\left(\frac{1}{\sqrt{5}}(2,1)\right) = \frac{4}{5}(2,1) = \left(\frac{8}{5},\frac{4}{5}\right).$$

b)

$$\mathbf{v}_2 = \mathbf{v} - \boldsymbol{P}_{\mathbf{w}}\mathbf{v} = (1,2) - \left(\frac{8}{5},\frac{4}{5}\right) = \left(-\frac{3}{5},\frac{6}{5}\right)$$

$\square$

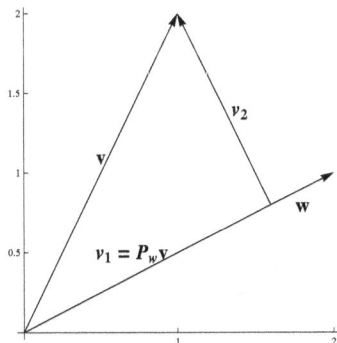

Figure 7

The idea of projection is related to the definition of **work** in Physics. Assume that an object that is moving along a line is subjected to the **constant force** (vector) **F**. Let **w** be the vector that represents **the displacement** of the object. The work done by **F** is defined as

$$W = \mathbf{F} \cdot \mathbf{w} = ||\mathbf{F}||\,||\mathbf{w}||\cos(\theta) = (\boldsymbol{comp_w}\mathbf{F})\,||\mathbf{w}||$$

(in appropriate units).

Example 6 Let $\mathbf{F} = 2\mathbf{i} + 3\mathbf{j}$ be the force that is acting on an object whose displacement is represented by the vector $\mathbf{w} = 4\mathbf{i} - 2\mathbf{j}$. Compute the work done by **F**.

Solution

The work done by **F** is simply

$$\mathbf{F} \cdot \mathbf{w} = (2\mathbf{i} + 3\mathbf{j}) \cdot (4\mathbf{i} - 2\mathbf{j}) = 8 - 6 = 2.$$

$\square$

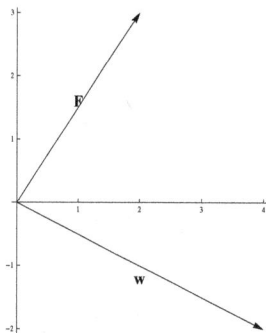

Figure 8

Problems

In problems 1-6,
a) Determine $||\mathbf{v}||$, $||\mathbf{w}||$ and the dot product $\mathbf{v} \cdot \mathbf{w}$,
b) Determine θ, the angle between **v** and **w**. Are the vectors orthogonal to each other?

1.

$$\mathbf{v} = (1, 0),\ \ \mathbf{w} = (1, 1)$$

2.

$$\mathbf{v} = (1, 0),\ \ \mathbf{w} = \left(-1, \sqrt{3}\right)$$

3.

$$\mathbf{v} = (1, 1),\ \ \mathbf{w} = \left(\frac{\sqrt{3}+1}{2}, \frac{\sqrt{3}-1}{2}\right)$$

4.

$$\mathbf{v} = \mathbf{i} + \mathbf{j},\ \ \mathbf{w} = -\mathbf{i} + \mathbf{j}$$

5.

$$\mathbf{v} = (0, 1, 1), \ \mathbf{w} = \left(0, \frac{\sqrt{3}+1}{2}, \frac{1-\sqrt{3}}{2}\right)$$

6.

$$\mathbf{v} = \mathbf{i} + \mathbf{j}, \ \mathbf{w} = \mathbf{i} - \mathbf{j}$$

In problems 7-10,
a) Determine $\|\mathbf{v}\|$, $\|\mathbf{w}\|$ and the dot product $\mathbf{v} \cdot \mathbf{w}$,
b) Determine $\cos(\theta)$, where θ is the angle between $\mathbf{v}$ and $\mathbf{w}$,
c) Make use of your computational utility to obtain an approximate value of θ (display 6 significant digits).

7.

$$\mathbf{v} = (1, 2), \ \mathbf{w} = (1, -1)$$

8.

$$\mathbf{v} = 3\mathbf{i} + 2\mathbf{j}, \ \mathbf{w} = -2\mathbf{i} + 4\mathbf{j}$$

9.

$$\mathbf{v} = (-2, -1), \ \mathbf{w} = (-1, 3)$$

10.

$$\mathbf{v} = \mathbf{i} + 3\mathbf{j} - \mathbf{k}, \ \mathbf{w} = 2\mathbf{i} - \mathbf{j} + \mathbf{k}$$

In problems 11-14
a) Determine the unit vector $\mathbf{u}$ along $\mathbf{v}$ (the normalizaton of $\mathbf{v}$).
b) Determine the direction cosines of $\mathbf{v}$.

11.

$$\mathbf{v} = (2, 3)$$

12.

$$\mathbf{v} = -\mathbf{i} + 2\mathbf{j}$$

13.

$$\mathbf{v} = (-3, 4)$$

14.

$$\mathbf{v} = \mathbf{i} - 2\mathbf{j} + \mathbf{k}$$

In problems 15-18, determine
a) the unit vector $\mathbf{u}$ along $\mathbf{w}$,
b) $comp_{\mathbf{w}}\mathbf{v}$, the component of $\mathbf{v}$ along $\mathbf{w}$,
c) $P_{\mathbf{w}}\mathbf{v}$, the projection of $\mathbf{v}$ along $\mathbf{w}$,
d) $\mathbf{v_2}$ such that $\mathbf{v} = P_{\mathbf{w}}\mathbf{v} + \mathbf{v_2}$ and $\mathbf{v_2}$ is orthogonal to $\mathbf{w}$.

15.

$$\mathbf{v} = (3, 4), \ \mathbf{w} = (6, 2)$$

16.

$$\mathbf{v} = (-2, 1), \ \mathbf{w} = (2, 1)$$

17.

$$\mathbf{v} = \mathbf{i} - 2\mathbf{j}, \ \mathbf{w} = 2\mathbf{i} - \mathbf{j}$$

18.

$$\mathbf{v} = (-2, -6), \ \mathbf{w} = (-3, -4)$$

11.4   The Cross Product

In the previous section we introduced the dot product that assigns a scalar to a pair of vectors and enables us to calculate the angle between them. In this section we will introduce the **cross product** that assigns to a pair of vectors another vector that is orthogonal to both vectors.
The cross product has many physical and geometric applications. In this section we will describe a plane in terms of a vector that is orthogonal to that plane by making use of the cross product. We will also introduce the **scalar triple product** that corresponds to the volume of the parallelepiped spanned by three vectors.

The Definition of the Cross Product

Suppose that we wish to construct the **cross product** of the vectors $\mathbf{v}$ and $\mathbf{w}$ in $\mathbb{R}^3$ as another vector in $\mathbb{R}^3$ that is denoted by $\mathbf{v} \times \mathbf{w}$ that has the following properties:

1. $\mathbf{v} \times \mathbf{w}$ is orthogonal to $\mathbf{v}$ and $\mathbf{w}$

2. $\|\mathbf{v} \times \mathbf{w}\|$ is the area of the parallelogram that is spanned by $\mathbf{v}$ and $\mathbf{w}$, i.e., $\|\mathbf{v}\| \, \|\mathbf{w}\| \sin{(\theta)}$, where θ is the angle between $\mathbf{v}$ and $\mathbf{w}$. In particular, $\mathbf{v} \times \mathbf{w} = \mathbf{0}$ if $\mathbf{v}$ and $\mathbf{w}$ are parallel vectors.

3. The direction of $\mathbf{v} \times \mathbf{w}$ is determined by the right-hand rule: If you imagine that a right-handed screw turns from $\mathbf{v}$ to $\mathbf{w}$, the vector $\mathbf{v} \times \mathbf{w}$ points upwards. Thus, $\mathbf{w} \times \mathbf{v} = -\mathbf{v} \times \mathbf{w}$, so that the cross product is anticommutative.

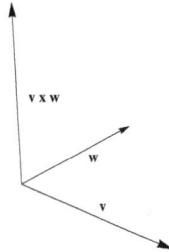

Figure 1

4. The cross product is distributive: If α is a scalar and $\mathbf{u}, \mathbf{v}, \mathbf{w}$ are vectors in $\mathbb{R}^3$ then

$$(\alpha \mathbf{v}) \times \mathbf{w} \;=\; \alpha \, (\mathbf{v} \times \mathbf{w}) = \mathbf{v} \times (\alpha \mathbf{w}),$$
$$\mathbf{v} \times (\mathbf{u} + \mathbf{w}) \;=\; \mathbf{v} \times \mathbf{u} + \mathbf{v} \times w.$$

In particular,

$$\mathbf{i} \times \mathbf{j} \;=\; \mathbf{k}, \; \mathbf{j} \times \mathbf{k} = \mathbf{i}, \; \mathbf{k} \times \mathbf{i} = \mathbf{j},$$
$$\mathbf{j} \times \mathbf{i} \;=\; -\mathbf{k}, \; \mathbf{k} \times \mathbf{j} = -\mathbf{i}, \; \mathbf{i} \times \mathbf{k} = -\mathbf{j}.$$

The above properties lead to the calculation of the product of vectors $\mathbf{v} = v_1 \mathbf{i} + v_2 \mathbf{j} + v_3 \mathbf{k}$ and $\mathbf{w} = w_1 \mathbf{i} + w_2 \mathbf{j} + w_3 \mathbf{k}$:

$$
\begin{aligned}
\mathbf{v} \times \mathbf{w} \;&=\; (v_1 \mathbf{i} + v_2 \mathbf{j} + v_3 \mathbf{k}) \times (w_1 \mathbf{i} + w_2 \mathbf{j} + w_3 \mathbf{k}) \\
&=\; v_1 w_2 \mathbf{k} - v_1 w_3 \mathbf{j} - v_2 w_1 \mathbf{k} + v_2 w_3 \mathbf{i} + v_3 w_1 \mathbf{j} - v_3 w_2 \mathbf{i} \\
&=\; (v_2 w_3 - v_3 w_2) \mathbf{i} - (v_1 w_3 - v_3 w_1) \mathbf{j} + (v_1 w_2 - v_2 w_1) \mathbf{k}. \\
&=\; \begin{vmatrix} v_2 & v_3 \\ w_2 & w_3 \end{vmatrix} \mathbf{i} - \begin{vmatrix} v_1 & v_3 \\ w_1 & w_3 \end{vmatrix} \mathbf{j} + \begin{vmatrix} v_1 & v_2 \\ w_1 & w_2 \end{vmatrix} \mathbf{k} \\
&=\; \begin{vmatrix} \mathbf{i} & \mathbf{j} & \mathbf{k} \\ v_1 & v_2 & v_3 \\ w_1 & w_2 & w_3 \end{vmatrix}.
\end{aligned}
$$

The above expression can be remembered easily by introducing the "symbolic determinant"

$$\begin{vmatrix} \mathbf{i} & \mathbf{j} & \mathbf{k} \\ v_1 & v_2 & v_3 \\ w_1 & w_2 & w_3 \end{vmatrix}.$$

This is a three-by-three array. The first row consists of the standard basis vectors $\mathbf{i}, \mathbf{j}$ and $\mathbf{k}$, the second row is made up of the components of $\mathbf{v}$ and the third row is made up of the components of **w**. The rule for the evaluation of this symbolic determinant is as follows:

$$\begin{vmatrix} \mathbf{i} & \mathbf{j} & \mathbf{k} \\ v_1 & v_2 & v_3 \\ w_1 & w_2 & w_3 \end{vmatrix} = \begin{vmatrix} v_2 & v_3 \\ w_2 & w_3 \end{vmatrix} \mathbf{i} - \begin{vmatrix} v_1 & v_3 \\ w_1 & w_3 \end{vmatrix} \mathbf{j} + \begin{vmatrix} v_1 & v_2 \\ w_1 & w_2 \end{vmatrix} \mathbf{k}.$$

Here the coefficients of **i**, **j** and **k** are two-by-two determinants:

$$\begin{vmatrix} v_2 & v_3 \\ w_2 & w_3 \end{vmatrix} = v_2 w_3 - v_3 w_2,$$

$$\begin{vmatrix} v_1 & v_3 \\ w_1 & w_3 \end{vmatrix} = v_1 w_3 - v_3 w_1,$$

$$\begin{vmatrix} v_1 & v_2 \\ w_1 & w_2 \end{vmatrix} = v_1 w_2 - v_2 w_1.$$

The two-by-two determinant that is the coefficient of **i** is obtained by deleting the first row and the first column of the symbolic determinant. The two-by-two determinant that is the coefficient of **j** is obtained by deleting the first row and the second column of the symbolic determinant. The two-by-two determinant that is the coefficient of **k** is obtained by deleting the first row and the third column of the symbolic determinant.

It can be shown that the cross product that is defined by the above expression has the properties 1-4.

Example 1 Let $\mathbf{v} = \mathbf{i} - 3\mathbf{j} + 2\mathbf{k}$ and $\mathbf{w} = 2\mathbf{i} - \mathbf{j} + 4\mathbf{k}$.

a) Determine $\mathbf{v} \times \mathbf{w}$,
b) Confirm that $\mathbf{v} \times \mathbf{w}$ is orthogonal to **v** and **w**.
c) Compute the area of the parallelogram spanned by **v** and **w**.

Solution

a)

$$\begin{aligned} \mathbf{v} \times \mathbf{w} &= \begin{vmatrix} \mathbf{i} & \mathbf{j} & \mathbf{k} \\ 1 & -3 & 2 \\ 2 & -1 & 4 \end{vmatrix} \\ &= \begin{vmatrix} -3 & 2 \\ -1 & 4 \end{vmatrix} \mathbf{i} - \begin{vmatrix} 1 & 2 \\ 2 & 4 \end{vmatrix} \mathbf{j} + \begin{vmatrix} 1 & -3 \\ 2 & -1 \end{vmatrix} \mathbf{k} \\ &= -10\mathbf{i} + 5\mathbf{k}. \end{aligned}$$

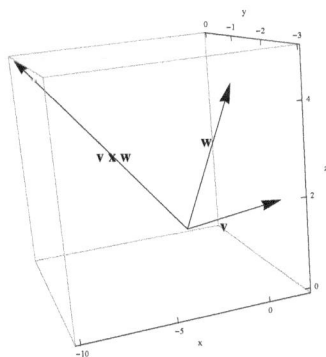

Figure 2

b)

$$\begin{aligned}
(\mathbf{v} \times \mathbf{w}) \cdot \mathbf{v} &= (-10\mathbf{i} + 5\mathbf{k}) \cdot (\mathbf{i} - 3\mathbf{j} + 2\mathbf{k}) = 0, \\
(\mathbf{v} \times \mathbf{w}) \cdot \mathbf{w} &= (-10\mathbf{i} + 5\mathbf{k}) \cdot (2\mathbf{i} - \mathbf{j} + 4\mathbf{k}) = 0.
\end{aligned}$$

c) The area of the parallelogram spanned by $\mathbf{v}$ and $\mathbf{w}$ .is

$$\|\mathbf{v} \times \mathbf{w}\| = \sqrt{10^2 + 5^2} = \sqrt{125}.$$

$\square$

Example 2 Make use of the cross product to compute the area of the parallelogram that is spanned by the vectors $\mathbf{v} = (2, 1)$ and $\mathbf{w} = (1, 2)$.

Solution

The cross product is defined for three dimensional vectors. Therefore, we consider the plane to be embedded in $\mathbb{R}^3$, and consider the vectors $\tilde{\mathbf{v}} = (2, 1, 0)$ and $\tilde{\mathbf{w}} = (1, 2, 0)$ that span the same parallelograms as the vectors $\mathbf{v}$ and $\mathbf{w}$. Thus, we can compute the required area as the magnitude of $\tilde{\mathbf{v}} \times \tilde{\mathbf{w}}$.

We have

$$\begin{aligned}
\tilde{\mathbf{v}} \times \tilde{\mathbf{w}} &= \begin{vmatrix} \mathbf{i} & \mathbf{j} & \mathbf{k} \\ 2 & 1 & 0 \\ 1 & 2 & 0 \end{vmatrix} \\
&= \begin{vmatrix} 1 & 0 \\ 2 & 0 \end{vmatrix} \mathbf{i} - \begin{vmatrix} 2 & 0 \\ 1 & 0 \end{vmatrix} \mathbf{j} + \begin{vmatrix} 2 & 1 \\ 1 & 2 \end{vmatrix} \mathbf{k} = 3\mathbf{k}.
\end{aligned}$$

Therefore, the area of the parallelogram that is spanned by the vectors $\mathbf{v}$ and $\mathbf{w}$ is 3. $\square$

The Scalar Triple Product

The scalar triple product of the vectors $\mathbf{u}, \mathbf{v}, \mathbf{w}$ is

$$\mathbf{u} \cdot (\mathbf{v} \times \mathbf{w}) = \begin{vmatrix} u_1 & u_2 & u_3 \\ v_1 & v_2 & v_3 \\ w_1 & w_2 & w_3 \end{vmatrix}$$

The absolute value of the scalar triple product of u, v, w is the volume of the parallelepiped that is spanned by u, v and w: Since $\mathbf{v} \times \mathbf{w}$ is perpendicular to the parallelogram spanned by $\mathbf{v}$ and $\mathbf{w}$, the height of the parallelepiped is

$$\|\mathbf{u}\| \cos(\theta),$$

where θ is the angle between $\mathbf{u}$ and $\mathbf{v} \times \mathbf{w}$. Thus, the volume of the parallelepiped is

$$(\text{height}) \times (\text{area of the base}) = \|\mathbf{u}\| |\cos(\theta)| \times \|\mathbf{v} \times \mathbf{w}\| = |\mathbf{u} \cdot (\mathbf{v} \times \mathbf{w})|.$$

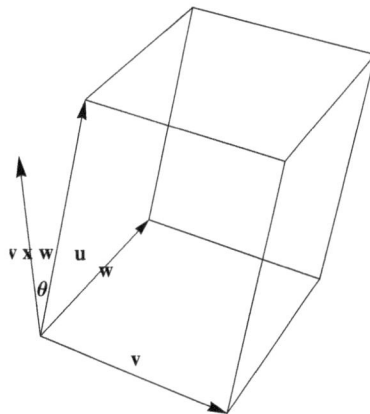

Figure 3

Example 3 Let
$$\mathbf{u} = (1, 0, 4), \ \mathbf{v} = (1, 1, 1) \ \text{and} \ \mathbf{w} = (-1, 1, 2).$$

Compute the volume of the parallelepiped spanned by **u**, **v** and **w**.

Solution

The volume of the parallelepiped spanned by **u**, **v** and **w** .is

$$\mathbf{u} \cdot (\mathbf{v} \times \mathbf{w}) = \begin{vmatrix} 1 & 0 & 4 \\ 1 & 1 & 1 \\ -1 & 1 & 2 \end{vmatrix} = 9$$

(confirm) . $\square$

Planes

Assume that the equation of a plane is given as

$$ax + by + cz = d.$$

If $P_0 = (x_0, y_0, z_0)$ is a given point on the plane and $P = (x, y, z)$ is an arbitrary point on the plane, we have

$$ax + by + cz = d \ \text{and} \ ax_0 + by_0 + cz_0 = d,$$

so that

$$a(x - x_0) + b(y - y_0) + c(z - z_0) = 0.$$

Thus, if we set

$$\mathbf{N} = a\mathbf{i} + b\mathbf{j} + c\mathbf{k},$$

we have

$$\mathbf{N} \cdot \left(\overrightarrow{P_0 P} \right) = 0.$$

We refer to a vector **N** such that $\mathbf{N} \cdot \left(\overrightarrow{P_0 P} \right) = 0$ for each P on the plane as a vector that is **orthogonal to the plane** or a **normal vector** for the plane.

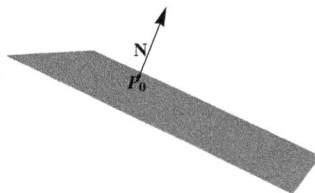

Figure 4

Conversely, assume that a point $P_0 = (x_0, y_0, z_0)$ is a given point on the plane and

$$\mathbf{N} = a\mathbf{i} + b\mathbf{j} + c\mathbf{k}$$

is orthogonal to the plane. If $P = (x, y, z)$ is an arbitrary point on the plane, then

$$\mathbf{N} \cdot \left(\overrightarrow{P_0 P} \right) = 0.$$

Thus,

$$(a\mathbf{i} + b\mathbf{j} + c\mathbf{k}) \cdot ((x - x_0)\,\mathbf{i} + (y - y_0)\,\mathbf{j} + (z - z_0)\,\mathbf{k}) = 0.$$

Therefore,

$$a\,(x - x_0) + b\,(y - y_0) + c\,(z - z_0) = 0$$

so that

$$ax + by + cz = d,$$

where

$$d = ax_0 + by_0 + cz_0.$$

Example 4 Assume that $\mathbf{N} = (1, -1, 2)$ is orthogonal to the plane Π and that $(-2, 3, 2)$ is a point on Π. Find an equation for the plane.

Solution

Set $P_0 = (-2, 3, 2)$ and let $P = (x, y, z)$ be an arbitrary point on the plane. We have

$$\mathbf{N} \cdot \left(\overrightarrow{P_0 P} \right) = 0$$

so that

$$(\mathbf{i} - \mathbf{j} + 2\mathbf{k}) \cdot ((x + 2)\,\mathbf{i} + (y - 3)\,\mathbf{j} + (z - 2)\,\mathbf{k}) = 0.$$

Thus,

$$(x + 2) - (y - 3) + 2\,(z - 2) = 0,$$

i.e.,

$$x - y + 2z = -1$$

$\square$

A plane is also determined by specifying three points on the plane. Assume that $P_0 = (x_0, y_0, z_0)$, $P_1 = (x_1, y_1, z_1)$ and $P_2 = (x_2, y_2, z_2)$ are on the plane. Then

$$\mathbf{N} = \overrightarrow{P_0 P_1} \times \overrightarrow{P_0 P_2}$$

is orthogonal to the plane. We can determine an equation of the plane as before.

Example 5 Find an equation for the plane that contains the points $P_0 = (1, 2, 1)$, $P_1 = (-1, 1, 2)$ and $P_2 = (3, -2, 1)$.

Solution

We have

$$\overrightarrow{P_0 P_1} = (-2, -1, 1) \text{ and } \overrightarrow{P_0 P_2} = (2, -4, 0).$$

Therefore,

$$\mathbf{N} = \overrightarrow{P_0 P_1} \times \overrightarrow{P_0 P_2} = \begin{vmatrix} \mathbf{i} & \mathbf{j} & \mathbf{k} \\ -2 & -1 & 1 \\ 2 & -4 & 0 \end{vmatrix}$$

$$= \begin{vmatrix} -1 & 1 \\ -4 & 0 \end{vmatrix} \mathbf{i} - \begin{vmatrix} -2 & 1 \\ 2 & 0 \end{vmatrix} \mathbf{j} + \begin{vmatrix} -2 & -1 \\ 2 & -4 \end{vmatrix} \mathbf{k}$$

$$= 4\mathbf{i} + 2\mathbf{j} + 10\mathbf{k}.$$

A point $P = (x, y, z)$ is on the plane iff

$$\mathbf{N} \cdot \left(\overrightarrow{P_0 P} \right) = 0,$$

i.e.,

$$(4\mathbf{i} + 2\mathbf{j} + 10\mathbf{k}) \cdot ((x - 1)\mathbf{i} + (y - 2)\mathbf{j} + (z - 1)\mathbf{k}) = 0.$$

Therefore,

$$4(x - 1) + 2(y - 2) + 10(z - 1) = 0,$$

so that

$$4x + 2y + 10z = 18,$$

$\square$

Problems

In problems 1-4 determine $\mathbf{v} \times \mathbf{w}$.

1.

$$\mathbf{v} = (2, 1, 0), \ \mathbf{w} = (1, 3, 0)$$

2.

$$\mathbf{v} = 3\mathbf{j} + 2\mathbf{k}, \ \mathbf{w} = -2\mathbf{j} + \mathbf{k}$$

3.

$$\mathbf{v} = (3, 1, 4), \ \mathbf{w} = (-2, 3, 1)$$

4.

$$\mathbf{v} = -2\mathbf{i} + \mathbf{j} + 4\mathbf{k}, \ \mathbf{w} = \mathbf{i} - 2\mathbf{j} + 5\mathbf{k}$$

In problems 5 and 6, Compute the area of the parallelogram spanned by $\mathbf{v}$ and $\mathbf{w}$ by making use of the cross product.

5.   **6.**

$$\mathbf{v} = (-1, 3),\ \mathbf{w} = (2, 6)$$ $$\mathbf{v} = -\mathbf{i} + 3\mathbf{j} - 2\mathbf{k},\ \mathbf{w} = 4\mathbf{i} - \mathbf{j} - 6\mathbf{k}$$

In problems 7 and 8, compute the volume of the parallelepiped spanned by $\mathbf{u}, \mathbf{v}$ and $\mathbf{w}$ by making use of the Scalar Triple Product.

7.

$$\mathbf{u} = (2, 1, 3),\ \mathbf{v} = (3, 1, 0),\ \mathbf{w} = (1, 0, 4)$$

8.

$$\mathbf{u} = 2\mathbf{i} + 2\mathbf{j} + 5\mathbf{k},\ \mathbf{v} = 3\mathbf{i} - \mathbf{j} - 4\mathbf{k},\ \mathbf{w} = \mathbf{i} - \mathbf{j}$$

In problems 9-12, assume that $\mathbf{N}$ is orthogonal to the plane Π and that P_0 is a point on Π. Find an equation for the plane.

9.   **11.**

$$\mathbf{N} = (3, 2, 1),\ P_0 = (1, 3, 4)$$

$$\mathbf{N} = -\mathbf{i} + \mathbf{j} + 2\mathbf{k},\ P_0 = (2, 1, -3)$$

10.

12

$$\mathbf{N} = (1, -2, 4),\ P_0 = (2, 0, 5)$$

$$\mathbf{N} = \mathbf{i} - 4\mathbf{j} - \mathbf{k},\ P_0 = (3, 1, 2)$$

In problems 13 and 14, find an equation for the plane that contains the points P_0, P_1 and P_2, by making use of the cross product.

13.

$$P_0 = (1, 2, 2),\ P_1 = (0, 2, 1),\ P_2 = (1, -3, 4)$$

14.

$$P_0 = (3, -2, -1),\ P_1 = (0, 4, 1),\ P_2 = (1, 3, -2)$$

Chapter 12

Functions of Several Variables

In this chapter we will discuss the differential calculus of functions of several variables. We will study functions of a single variable that take values in two or three dimensions. The images of such functions are **parametrized curves** that may model the trajectory of an object. We will calculate their **velocity** and **acceleration**. Geometrically, we will calculate **tangents** to curves and their **curvature**. Then we will take up the differential calculus **real-valued functions of several variables**. We will study their **rates of change** in different directions and their **maxima and minima**. In the process we will generalize the ideas of **local linear approximation** and the **differential** from functions of a single variable to functions of several variables.

12.1 Parametrized Curves, Tangent Vectors and Velocity

Parametrized Curves

Definition 1 Assume that σ **is a function from an interval J to the plane $\mathbb{R}^2$.** If $\sigma(t) = (x(t), y(t))$, then $x(t)$ and $y(t)$ are the **component functions of σ.** The letter t is **the parameter.** The image of σ, i.e., the set of points $C = \{(x(t), y(t)) : t \in J\}$ is said to be **the curve that is parametrized by the function σ.** We say that the function σ is a parametric representation of the curve C.

We will use the notation $\sigma : J \to \mathbb{R}^2$ to indicate that σ is a function from J into $\mathbb{R}^2$. We can identify the point $\sigma(t) = (x(t), y(t))$ with its position vector, and regard σ as a **vector-valued function** whose values are two-dimensional vectors. We may use the notation $\sigma(t) = x(t)\mathbf{i} + y(t)\mathbf{j}$. It is useful to imagine that $\sigma(t)$ is the position of a particle in motion at the fictitious time t. Figure 1 illustrates a parametrized curve and the "motion of the particle" is indicated by arrows. In many applications, the parameter does refer to time.

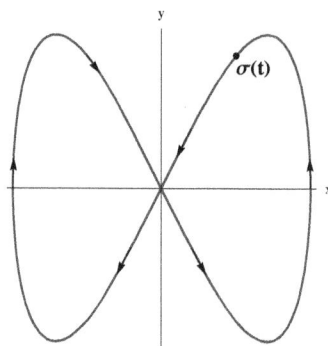

Figure 1: A parametrized curve in the plane

Example 1 Let $\boldsymbol{\sigma}(t) = (\cos(t), \sin(t))$, where $t \in [0, 2\pi]$.

The curve that is parametrized by the function $\boldsymbol{\sigma}$ is the unit circle, since

$$\cos^2(t) + \sin^2(t) = 1.$$

The arrows in Figure 2 indicate the motion of a particle that is at $\sigma(t)$ at "time" t. $\square$

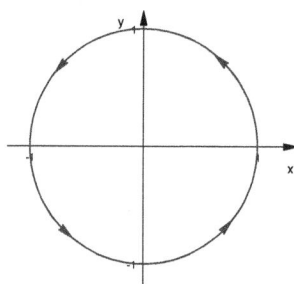

Figure 2: $\sigma(t) = (\cos(t), \sin(t))$

We have to distinguish between a vector-valued function and the curve that is parametrized by that function, as in the following example.

Example 2 Let $\boldsymbol{\sigma_2}(t) = (\cos(2t), \sin(2t))$, where $t \in [0, 2\pi]$.

Since

$$\cos^2(2t) + \sin^2(2t) = 1,$$

the function $\boldsymbol{\sigma_2}$ parametrizes the unit circle, just as the function $\boldsymbol{\sigma}$ of Example 1. But *the function $\boldsymbol{\sigma_2} \neq \boldsymbol{\sigma}$*. For example,

$$\boldsymbol{\sigma_2}(\pi/4) = (\cos(\pi/2), \sin(\pi/2)) = (0, 1),$$

whereas,

$$\boldsymbol{\sigma}(\pi/4) = (\cos(\pi/4), \sin(\pi/4)) = \left(\frac{1}{\sqrt{2}}, \frac{1}{\sqrt{2}}\right).$$

If we imagine that the position of a particle is determined by $\boldsymbol{\sigma_2}(t)$, the particle revolves around the origin twice on the time interval $[0, 2\pi]$, whereas, a particle whose position is determined by $\boldsymbol{\sigma}$ revolves around the origin only once on the "time" interval $[0, 2\pi]$. $\square$

Lines are basic geometric objects. Let $P_0 = (x_0, y_0)$ be a given point and let $\boldsymbol{v} = v_1\mathbf{i} + v_2\mathbf{j}$ be a given vector. With reference to Figure 2, a line in the direction of $\boldsymbol{v}$ that passes through the point P_0 can be parametrized by

$$\boldsymbol{\sigma}(t) = \overrightarrow{OP_0} + t\boldsymbol{v} = x_0\mathbf{i} + y_0\mathbf{j} + t(v_1\mathbf{i} + v_2\mathbf{j})$$
$$= (x_0 + tv_1)\mathbf{i} + (y_0 + tv_2)\mathbf{j}.$$

Thus, the component functions of $\boldsymbol{\sigma}$ are

$$x(t) = x_0 + tv_1 \text{ and } y(t) = y_0 + tv_2$$

Here, the parameter t varies on the entire number line.

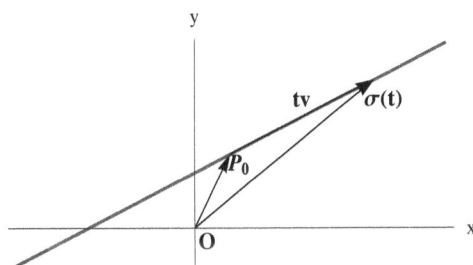

Figure 3: A line through P_0 in the direction of $\boldsymbol{v}$

Example 3 Determine a parametric representation of the line in the direction of the vector $\boldsymbol{v} = \mathbf{i} - 2\mathbf{j}$ that passes through the point $P_0 = (3, 4)$.

Solution

$$\boldsymbol{\sigma}(t) = \overrightarrow{OP_0} + t\boldsymbol{v} = 3\mathbf{i} + 4\mathbf{j} + t(\mathbf{i} - 2\mathbf{j})$$
$$= (3 + t)\mathbf{i} + (4 - 2t)\mathbf{j}.$$

Thus, the component functions of $\boldsymbol{\sigma}$ are

$$x(t) = 3 + t \text{ and } y(t) = 4 - 2t.$$

Figure 4 shows the line.

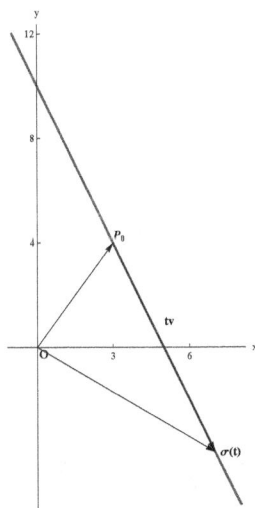

Figure 4

We can eliminate the parameter t and obtain the relationship between x and y:

$$t = x - 3 \Rightarrow y = 4 - 2(x - 3) = -2x + 10.$$

Note that the same line can be represented parametrically by other functions. Indeed, the direction can be specified by any scalar multiple of v, and we can pick an arbitrary point on the line. For example, if we set $t = 1$ in the above representation, $Q_0 = (4, 2)$ is a point on the line, and $w = 2v = 2\mathbf{i} - 4\mathbf{j}$ is a scalar multiple of v. The function

$$\tilde{\sigma}(u) = \overrightarrow{OQ_0} + uw = 4\mathbf{i} + 2\mathbf{j} + u(2\mathbf{i} - 4\mathbf{j})$$
$$= (4 + 2u)\mathbf{i} + (2 - 4u)\mathbf{j}$$

is another parametric representation of the same line. $\square$

Example 4 Imagine that a circle of radius r rolls along a line, and that $P(\theta)$ is a point on the circle, as in Figure 5 ($\theta = 0$ when P is at the origin).

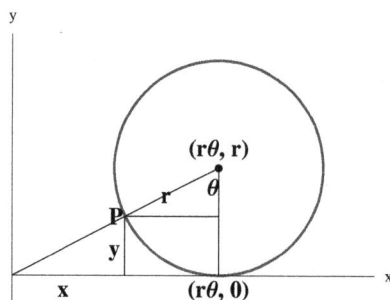

Figure 5

We have

$$x(\theta) = r\theta - r\sin(\theta) = r(\theta - \sin(\theta))$$

and

$$y(\theta) = r - r\cos(\theta) = r(1 - \cos(\theta)),$$

as you can confirm. The parameter θ can be any real number. Let's set

$$\sigma(\theta) = (x(\theta), y(\theta)) = (r(\theta - \sin(\theta)), r(1 - \cos(\theta))), \, -\infty < \theta < +\infty.$$

The curve that is parametrized by the function $\sigma : \mathbb{R} \to \mathbb{R}^2$ is referred to as a **cycloid**. Figure 6 shows the part of the cycloid that is parametrized by

$$\sigma(\theta) = (2\theta - 2\sin(\theta), 2 - 2\cos(\theta)), \text{ where } \theta \in [0, 8\pi].$$

$\square$

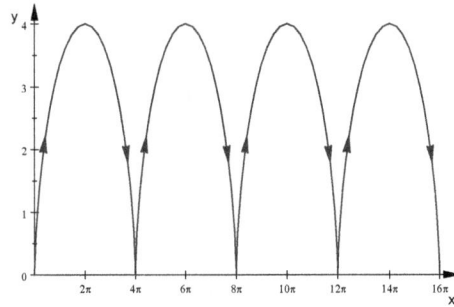

Figure 6

Similarly, we can consider parametrized curves in three dimensions:

Definition 2 Assume that σ is a function from an interval J to the three-dimensional space $\mathbb{R}^3$. If $\sigma(t) = (x(t), y(t), z(t))$, then $x(t)$, $y(t)$ and $z(t)$ are the component functions of σ. The letter t is the parameter. The image of σ, i.e., the set of points $C = \{(x(t), y(t), z(t)) : t \in J\}$ is said to be **the curve that is parametrized by the function σ**.

We will use the notation $\sigma : J \to \mathbb{R}^3$ to indicate that σ is a function from J into $\mathbb{R}^3$. We can identify the point $\sigma(t) = (x(t), y(t), z(t))$ with its position vector, and regard σ as a **vector-valued function** whose values are three-dimensional vectors. We may use the notation

$$\sigma(t) = x(t)\,\mathbf{i} + y(t)\,\mathbf{j} + z(t)\,\mathbf{k}.$$

You can imagine that $\sigma(t)$ is the position at time t of a particle in motion.

Example 5 Let $\sigma(t) = (\cos(t), \sin(t), t)$, where $t \in \mathbb{R}$. The curve that is parametrized by σ is a **helix**.

Since

$$\cos^2(t) + \sin^2(t) = 1,$$

the curve that is parametrized by σ lies on the cylinder $x^2 + y^2 = 1$. Figure 7 shows part of the helix.□

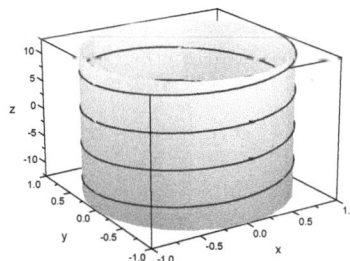

Figure 7

Tangent Vectors

Definition 3 **The limit of $\boldsymbol{\sigma} : J \to \mathbb{R}^2$ at t_0 is the vector $\boldsymbol{w}$ if**

$$\lim_{t \to t_0} ||\boldsymbol{\sigma}(t) - \boldsymbol{w}|| = 0$$

In this case we write

$$\lim_{t \to t_0} \boldsymbol{\sigma}(t) = \boldsymbol{w}.$$

Proposition 1 Assume that $\boldsymbol{\sigma}(t) = x(t)\,\mathbf{i} + y(t)\,\mathbf{j}$. We have

$$\lim_{t \to t_0} \boldsymbol{\sigma}(t) = \boldsymbol{w} = w_1\mathbf{i} + w_2\mathbf{j}$$

if and only if $\lim_{t \to t_0} x(t) = w_1$ and $\lim_{t \to t_0} y(t) = w_2$. Thus,

$$\lim_{t \to t_0} (x(t)\,\mathbf{i} + y(t)\,\mathbf{j}) = \left(\lim_{t \to t_0} x(t)\right)\mathbf{i} + \left(\lim_{t \to t_0} y(t)\right)\mathbf{j}.$$

Proof

Assume that

$$\lim_{t \to t_0} x(t) = w_1 \text{ and } \lim_{t \to t_0} y(t) = w_2.$$

Since

$$||\boldsymbol{\sigma}(t) - \boldsymbol{w}||^2 = (x(t) - w_1)^2 + (y(t) - w_2)^2,$$

we have

$$\lim_{t \to t_0} ||\boldsymbol{\sigma}(t) - \boldsymbol{w}||^2 = \lim_{t \to t_0} (x(t) - w_1)^2 + \lim_{t \to t_0} (y(t) - w_2)^2 = 0.$$

Thus, $\lim_{t \to t_0} \boldsymbol{\sigma}(t) = \boldsymbol{w} = w_1\mathbf{i} + w_2\mathbf{j}$.

Conversely, assume that $\lim_{t \to t_0} \boldsymbol{\sigma}(t) = \boldsymbol{w} = w_1\mathbf{i} + w_2$. Since

$$|x(t) - w_1| \le ||\boldsymbol{\sigma}(t) - \boldsymbol{w}|| \text{ and } |y(t) - w_2| \le ||\boldsymbol{\sigma}(t) - \boldsymbol{w}||,$$

and $\lim_{t \to t_0} ||\boldsymbol{\sigma}(t) - \boldsymbol{w}|| = 0$, we have

$$\lim_{t \to t_0} |x(t) - w_1| = 0 \text{ and } \lim_{t \to t_0} |y(t) - w_2| = 0.$$

Therefore,

$$\lim_{t \to t_0} x(t) = w_1 \text{ and } \lim_{t \to t_0} y(t) = w_2.$$

■

Definition 4 **The derivative of $\boldsymbol{\sigma} : J \to \mathbf{R}^2$ at t is**

$$\lim_{\Delta t \to 0} \frac{\boldsymbol{\sigma}(t + \Delta t) - \boldsymbol{\sigma}(t)}{\Delta t}$$

and will be denoted by

$$\frac{d\boldsymbol{\sigma}}{dt}, \ \frac{d\boldsymbol{\sigma}}{dt}(t), \ \frac{d\boldsymbol{\sigma}(t)}{dt} \text{ or } \boldsymbol{\sigma}'(t).$$

If C is the curve that is parametrized by $\boldsymbol{\sigma}$ and $\boldsymbol{\sigma}'(t)$ is not the zero vector, then $\boldsymbol{\sigma}'(t)$ **is a vector that is tangent to** C **at** $\boldsymbol{\sigma}(t)$.

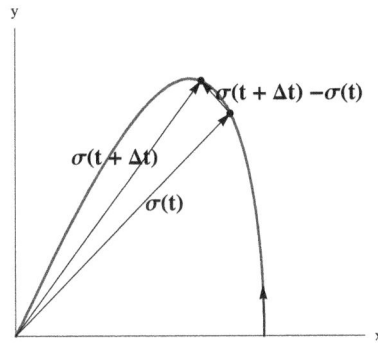

Figure 8

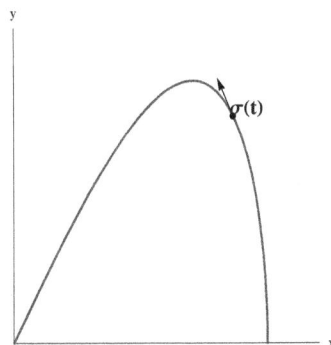

Figure 9

Proposition 2 If $\boldsymbol{\sigma}(t) = x(t)\mathbf{i} + y(t)\mathbf{j}$ and $x'(t)$, $y'(t)$ exist, then

$$\frac{d\boldsymbol{\sigma}}{dt} = \frac{dx}{dt}\mathbf{i} + \frac{dy}{dt}\mathbf{j}.$$

Proof

$$\frac{\boldsymbol{\sigma}(t+\Delta t) - \boldsymbol{\sigma}(t)}{\Delta t} = \frac{1}{\Delta t}\left(x(t+\Delta t)\mathbf{i} + y(t+\Delta t)\mathbf{j} - x(t)\mathbf{i} - y(t)\mathbf{j}\right)$$
$$= \frac{x(t+\Delta t) - x(t)}{\Delta t}\mathbf{i} + \frac{y(t+\Delta t) - y(t)}{\Delta t}\mathbf{j}.$$

Therefore,

$$\frac{d\boldsymbol{\sigma}}{dt} = \lim_{\Delta t \to 0}\frac{\boldsymbol{\sigma}(t+\Delta t) - \boldsymbol{\sigma}(t)}{\Delta t} = \lim_{\Delta t \to 0}\left(\frac{x(t+\Delta t) - x(t)}{\Delta t}\right)\mathbf{i} + \lim_{\Delta t \to 0}\left(\frac{y(t+\Delta t) - y(t)}{\Delta t}\right)\mathbf{j}$$
$$= \frac{dx}{dt}\mathbf{i} + \frac{dy}{dt}\mathbf{j}$$

∎

Example 6 Let $\boldsymbol{\sigma}(t) = \cos(t)\mathbf{i} + \sin(t)j.$ Determine $\boldsymbol{\sigma}'(t).$

Solution

$$\boldsymbol{\sigma}'\left(t\right) = \left(\frac{d}{dt}\cos\left(t\right)\right)\mathbf{i} + \left(\frac{d}{dt}\sin\left(t\right)\right)\mathbf{j} = -\sin\left(t\right)\mathbf{i} + \cos\left(t\right)\mathbf{j}.$$

Note that $\boldsymbol{\sigma}'\left(t\right)$ is orthogonal to $\boldsymbol{\sigma}\left(t\right)$:

$$\boldsymbol{\sigma}'\left(t\right)\cdot\boldsymbol{\sigma}\left(t\right) = -\sin\left(t\right)\cos\left(t\right) + \cos\left(t\right)\sin\left(t\right) = 0.$$

$\square$

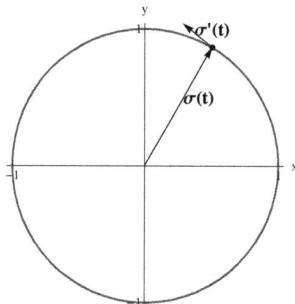

Figure 10

Definition 5 Let C be the curve that is parametrized by the function $\boldsymbol{\sigma} : J \to \mathbb{R}^2$. If $\boldsymbol{\sigma}'\left(t\right) \neq 0$, **the unit tangent to** C at $\boldsymbol{\sigma}\left(t\right)$ is

$$\mathbf{T}\left(t\right) = \frac{\boldsymbol{\sigma}'\left(t\right)}{||\boldsymbol{\sigma}'\left(t\right)||}$$

The curve C is said to be **smooth** at $\boldsymbol{\sigma}\left(t\right)$ if $\boldsymbol{\sigma}'\left(t\right) \neq \mathbf{0}$.

With reference to the curve of Example 6, $\boldsymbol{\sigma}'\left(t\right) = -\sin\left(t\right)\mathbf{i} + \cos\left(t\right)\mathbf{j}$ is of unit length, since

$$||\boldsymbol{\sigma}'\left(t\right)|| = \sqrt{\sin^2\left(t\right) + \cos^2\left(t\right)} = 1.$$

Therefore, $\boldsymbol{T}\left(t\right) = \boldsymbol{\sigma}'\left(t\right) = -\sin\left(t\right)\mathbf{i} + \cos\left(t\right)\mathbf{j}.$

Definition 6 If C is the curve that is parametrized by $\boldsymbol{\sigma} : J \to \mathbb{R}^2$ and $\boldsymbol{\sigma}'\left(t_0\right) \neq \mathbf{0}$, the line that is in the direction of $\boldsymbol{\sigma}'\left(t_0\right)$ and passes through $\boldsymbol{\sigma}\left(t_0\right)$ is **the tangent line to** C at $\boldsymbol{\sigma}\left(t_0\right)$.

Note that a parametric representation of the tangent line is the function $\boldsymbol{L} : \mathbb{R} \to \mathbb{R}^2$, where

$$\boldsymbol{L}\left(u\right) = \boldsymbol{\sigma}\left(t_0\right) + u\boldsymbol{\sigma}'\left(t_0\right).$$

Another parametric representation is

$$\boldsymbol{l}(u) = \boldsymbol{\sigma}\left(t_0\right) + u\mathbf{T}\left(t_0\right).$$

Example 7 Let $\boldsymbol{\sigma}\left(\theta\right) = \left(2\theta - 2\sin\left(\theta\right), 2 - 2\cos\left(\theta\right)\right)$, as in Example 4, and let C be the curve that is parametrized by $\boldsymbol{\sigma}$.

a) Determine $\mathbf{T}(\theta)$. Indicate the values of θ such that $\mathbf{T}(\theta)$ exists.

b) Determine parametric representations of the tangent line to C at $\boldsymbol{\sigma}(\pi/2)$.

Solution

a)

$$\boldsymbol{\sigma}'(\theta) = (2 - 2\cos(\theta))\,\mathbf{i} + 2\sin(\theta)\,\mathbf{j}$$

Therefore,

$$\|\boldsymbol{\sigma}'(\theta)\| = \sqrt{(2 - 2\cos(\theta))^2 + 4\sin^2(\theta)}$$
$$= 2\sqrt{1 - 2\cos(\theta) + \cos^2(\theta) + \sin^2(\theta)}$$
$$= 2\sqrt{2}\sqrt{1 - \cos(\theta)}.$$

Thus, $\boldsymbol{\sigma}'(\theta) = 0$ if $\cos(\theta) = 1$, i.e., if $\theta = 2n\pi$, $n = 0, \pm 1, \pm 2, \pm 3, \ldots$. If θ is not an integer multiple of 2π, the unit tangent to C at $\boldsymbol{\sigma}(\theta)$ is

$$\mathbf{T}(\theta) = \frac{\boldsymbol{\sigma}'(\theta)}{\|\boldsymbol{\sigma}'(\theta)\|} = \frac{1}{2\sqrt{2}\sqrt{1 - \cos(\theta)}}\left((2 - 2\cos(\theta))\,\mathbf{i} + 2\sin(\theta)\,\mathbf{j}\right).$$

Note that

$$\boldsymbol{\sigma}(2n\pi) = (2\theta - 2\sin(\theta), 2 - 2\cos(\theta))|_{\theta = 2n\pi} = (4n\pi, 0).$$

A tangent line to C does not exist at such points. Figure 6 is consistent with this observation. The curve C appears to have a **cusp** at $(4n\pi, 0)$. Let's consider the case $\theta = 2\pi$. If the curve is the graph of the equation $y = y(x)$, by the chain rule,

$$\frac{dy}{dx} = \frac{\dfrac{dy}{d\theta}}{\dfrac{dx}{d\theta}} = \frac{2\sin(\theta)}{2(1 - \cos(\theta))}.$$

Therefore,

$$\lim_{x \to 4\pi-} \frac{dy}{dx} = \lim_{\theta \to 2\pi-} \frac{2\sin(\theta)}{2(1 - \cos(\theta))}.$$

We have

$$\lim_{\theta \to 2\pi} \sin(\theta) = 0 \text{ and } \lim_{\theta \to 2\pi}(1 - \cos(\theta)) = 0.$$

By L'Hôpital's rule,

$$\lim_{\theta \to 2\pi-} \frac{2\sin(\theta)}{2(1 - \cos(\theta))} = \lim_{\theta \to 2\pi} \frac{2\cos(\theta)}{2\sin(\theta)}.$$

Since $\sin(\theta) < 0$ if $\theta < 2\pi$ and θ is close to 2π, and

$$\lim_{\theta \to 2\pi} 2\cos(\theta) = 2 > 0, \quad \lim_{\theta \to 2\pi-} 2\sin(\theta) = 0,$$

we have

$$\lim_{x \to 4\pi-} \frac{dy}{dx} = \lim_{\theta \to 2\pi-} \frac{2\sin(\theta)}{2(1 - \cos(\theta))} = \lim_{\theta \to 2\pi} \frac{2\cos(\theta)}{2\sin(\theta)} = -\infty.$$

Similarly,

$$\lim_{x \to 4\pi+} \frac{dy}{dx} = +\infty.$$

Therefore, the curve has a cusp at $(4\pi, 0)$.

b)

We have

$$\boldsymbol{\sigma}\left(\pi/2\right) = \left(2\theta - 2\sin\left(\theta\right), 2 - 2\cos\left(\theta\right)\right)|_{\theta=\pi/2} = \left(\pi - 2, 2\right),$$

and

$$\mathbf{T}\left(\pi/2\right) = \frac{1}{2\sqrt{2}\sqrt{1-\cos\left(\theta\right)}}\left(\left(2 - 2\cos\left(\theta\right)\right)\mathbf{i} + 2\sin\left(\theta\right)j\right)\Bigg|_{\pi/2}$$

$$= \frac{1}{2\sqrt{2}}\left(2\mathbf{i} + 2\mathbf{j}\right) = \frac{1}{\sqrt{2}}\left(\mathbf{i} + \mathbf{j}\right).$$

A parametric representation of the line that is tangent to C at $\boldsymbol{\sigma}\left(\pi/2\right)$ is

$$\begin{aligned} \boldsymbol{L}\left(u\right) &= \boldsymbol{\sigma}\left(\pi/2\right) + u\boldsymbol{\sigma}'\left(\pi/2\right) \\ &= \left(\pi - 2\right)\mathbf{i} + 2\mathbf{j} + u\left(2\left(\mathbf{i} + \mathbf{j}\right)\right) \\ &= \left(\pi - 2 + 2u\right)i + \left(2 + 2u\right)\mathbf{j} \end{aligned}$$

$\square$

The above definitions extend to curves in the three-dimensional space in the obvious manner: if $\boldsymbol{\sigma} : J \to \mathbb{R}^3$, the limit of $\boldsymbol{\sigma}$ is the vector $\boldsymbol{w}$ if

$$\lim_{t\to t_0}\left\|\boldsymbol{\sigma}\left(t\right) - \boldsymbol{w}\right\| = 0,$$

and we write

$$\lim_{t\to t_0}\boldsymbol{\sigma}\left(t\right) = \boldsymbol{w}.$$

If $\boldsymbol{\sigma}\left(t\right) = x\left(t\right)\mathbf{i} + y\left(t\right)\mathbf{j} + z\left(t\right)\mathbf{k}$, then

$$\lim_{t\to t_0}\boldsymbol{\sigma}\left(t\right) = w_1\mathbf{i} + w_2\mathbf{j} + w_3\mathbf{k}$$

if and only if

$$\lim_{t\to t_0}x\left(t\right) = w_1, \ \lim_{t\to t_0}y\left(t\right) = w_2 \text{ and } \lim_{t\to t_0}z\left(t\right) = w_3.$$

Thus,

$$\lim_{t\to t_0}\left(x\left(t\right)\mathbf{i} + y\left(t\right)\mathbf{j} + z\left(t\right)\mathbf{k}\right) = \left(\lim_{t\to t_0}x\left(t\right)\right)\mathbf{i} + \left(\lim_{t\to t_0}y\left(t\right)\right)\mathbf{j} + \left(\lim_{t\to t_0}z\left(t\right)\right)\mathbf{k}.$$

The derivative of $\boldsymbol{\sigma}$ is

$$\frac{d\boldsymbol{\sigma}}{dt} = \lim_{\Delta t\to 0}\frac{\boldsymbol{\sigma}\left(t + \Delta t\right) - \boldsymbol{\sigma}\left(t\right)}{\Delta t} = \frac{dx}{dt}\mathbf{i} + \frac{dy}{dt}\mathbf{j} + \frac{dz}{dt}\mathbf{k},$$

and the unit tangent is

$$\mathbf{T}\left(t\right) = \frac{1}{\left\|\boldsymbol{\sigma}'\left(t\right)\right\|}\boldsymbol{\sigma}'\left(t\right),$$

provided that $\left\|\boldsymbol{\sigma}'\left(t\right)\right\| \neq 0$.

Example 8 Let $\boldsymbol{\sigma}\left(t\right) = \left(\cos\left(t\right), \sin\left(t\right), t\right)$ as in Example 5, and let C be the curve that is parametrized by $\boldsymbol{\sigma}$.

a) Determine $\mathbf{T}(t)$. Indicate the values of t such that $\mathbf{T}(t)$exists.
b) Determine a parametric representation of the tangent line to C at $\boldsymbol{\sigma}(\pi/3)$.
Solution

a)
$$\boldsymbol{\sigma}'(t) = -\sin(t)\mathbf{i} + \cos(t)\mathbf{j} + \mathbf{k}.$$

Therefore,
$$\|\boldsymbol{\sigma}'(t)\| = \sqrt{\sin^2(t) + \cos^2(t) + 1} = \sqrt{2}.$$

Thus, $\mathbf{T}(t)$ exists for each $t \in R$ and
$$\mathbf{T}(t) = \frac{1}{\|\boldsymbol{\sigma}'(t)\|}\boldsymbol{\sigma}'(t) = \frac{1}{\sqrt{2}}\left(-\sin(t)\mathbf{i} + \cos(t)\mathbf{j} + \mathbf{k}\right).$$

b) We have
$$\boldsymbol{\sigma}(\pi/3) = \cos(\pi/3)\mathbf{i} + \sin(\pi/3)\mathbf{j} + \frac{\pi}{3}\mathbf{k} = \frac{1}{2}\mathbf{i} + \frac{\sqrt{3}}{2}\mathbf{j} + \frac{\pi}{3}\mathbf{k},$$

and
$$\boldsymbol{\sigma}'(\pi/3) = -\sin(\pi/3)\mathbf{i} + \cos(\pi/3)\mathbf{j} + \mathbf{k} = -\frac{\sqrt{3}}{2}\mathbf{i} + \frac{1}{2}\mathbf{j} + \mathbf{k}.$$

A parametric representation of the line that is tangent to C at $\sigma(\pi/6)$ is
$$\mathbf{L}(u) = \boldsymbol{\sigma}(\pi/3) + u\boldsymbol{\sigma}'(\pi/3)$$
$$= \frac{1}{2}\mathbf{i} + \frac{\sqrt{3}}{2}\mathbf{j} + \frac{\pi}{3}\mathbf{k} + u\left(-\frac{\sqrt{3}}{2}\mathbf{i} + \frac{1}{2}\mathbf{j} + \mathbf{k}\right)$$
$$= \left(\frac{1}{2} - \frac{\sqrt{3}}{2}u\right)\mathbf{i} + \left(\frac{\sqrt{3}}{2} + \frac{1}{2}u\right)\mathbf{j} + \left(\frac{\pi}{3} + u\right)\mathbf{k}.$$

$\square$

Velocity

Assume that $\boldsymbol{\sigma}$ is a function from an interval on the number line into the plane or three-dimensional space, and that a particle in motion is at the point $\boldsymbol{\sigma}(t)$ at time t. Let Δt denote a time increment. The **average velocity** of the particle over the time interval determined by t and $t + \Delta t$ is
$$\frac{\text{displacement}}{\text{elapsed time}} = \frac{\boldsymbol{\sigma}(t + \Delta t) - \boldsymbol{\sigma}(t)}{\Delta t}.$$

We define the instantaneous velocity at the instant t as the limit of the average velocity as Δt approaches 0:

Definition 7 If $\boldsymbol{\sigma} : J \to \mathbb{R}^2$ or $\boldsymbol{\sigma} : J \to \mathbb{R}^3$ and a particle is at the point $\boldsymbol{\sigma}(t)$ at time t, **the velocity v(t) of the particle at the instant t is**
$$\mathbf{v}(t) = \frac{d\boldsymbol{\sigma}}{dt} = \lim_{\Delta t \to 0}\frac{\boldsymbol{\sigma}(t + \Delta t) - \boldsymbol{\sigma}(t)}{\Delta t}.$$

Note that $\mathbf{v}(t)$is a vector in the direction of the tangent to C at $\boldsymbol{\sigma}(t)$ if $\mathbf{v}(t) \neq 0$.

Example 9 Let $\boldsymbol{\sigma}(t) = (2t - 2\sin(t))\mathbf{i} + (2 - 2\cos(t))\mathbf{j}$, as in Example 7, and let C be the curve that is parametrized by $\boldsymbol{\sigma}$. Determine $\mathbf{v}(3\pi/4)$, $\mathbf{v}(\pi)$ and $\mathbf{v}(2\pi)$.

Solution
As in Example 7,

$$\mathbf{v}(t) = \boldsymbol{\sigma}'(t) = (2 - 2\cos(t))\mathbf{i} + 2\sin(t)\mathbf{j}.$$

Therefore,

$$\mathbf{v}(3\pi/4) = \left(2 - 2\cos\left(\frac{3\pi}{4}\right)\right)\mathbf{i} + 2\sin\left(\frac{3\pi}{4}\right)\mathbf{j} = \left(2 - 2\left(-\frac{\sqrt{2}}{2}\right)\right)\mathbf{i} + 2\left(\frac{\sqrt{2}}{2}\right)\mathbf{j}$$

$$= \left(2 + \sqrt{2}\right)\mathbf{i} + \sqrt{2}\mathbf{j},$$

$$\mathbf{v}(\pi) = (2 - 2\cos(\pi))\mathbf{i} + 2\sin(\pi)\mathbf{j} = 4\mathbf{i},$$

and

$$\mathbf{v}(2\pi) = (2 - 2\cos(2\pi))\mathbf{i} + 2\sin(2\pi)\mathbf{j} = 0.$$

Note that the direction of the motion is not defined at $t = 2\pi$ since $\mathbf{v}(2\pi) = 0$ (and the curve C is not smooth at $\boldsymbol{\sigma}(2\pi)$). $\square$

Problems

In problems 1-4 assume that a curve C is parametrized by the function $\boldsymbol{\sigma}$.
a) Determine $\boldsymbol{\sigma}'(t)$, $\boldsymbol{\sigma}'(t_0)$ and the unit tangent $\mathbf{T}(t_0)$ to C at $\boldsymbol{\sigma}(t_0)$.
b) Determine the vector-valued function $\boldsymbol{L}$ that parametrizes the tangent line to C at $\boldsymbol{\sigma}(t_0)$ (specify the direction of the line by the vector $\boldsymbol{\sigma}'(t_0)$).

1.
$$\boldsymbol{\sigma}(t) = \left(t, \sin^2(4t)\right), \ -\infty < t < +\infty, \ t_0 = \pi/3.$$

2.
$$\boldsymbol{\sigma}(t) = (2\cos(3t), \sin(t)), \ 0 \le t \le 2\pi, \ t_0 = \pi/4.$$

3.
$$\boldsymbol{\sigma}(t) = \left(\frac{3t}{t^2+1}, \frac{3t^2}{t^2+1}\right), \ -\infty < t < \infty, \ t_0 = 1.$$

4.
$$\boldsymbol{\sigma}(t) = (4\cos(3t), 4\sin(3t), 2t), \ -\infty < t < \infty, \ t_0 = \pi/2.$$

In problems 5 and 6, the position of a particle at time t is $\sigma(t)$. Determine the velocity function $\mathbf{v}(t)$, the velocity at the instant t_0, and the speed at t_0. Express the result in the **ij**-notation.
5
$$\boldsymbol{\sigma}(t) = \left(e^{-t}\sin(t), e^{-t}\cos(t)\right), \ t_0 = \pi$$

6.
$$\sigma(t) = (t, \arctan(t)), \ t_0 = 1$$

12.2   Acceleration and Curvature

Acceleration

Acceleration is the rate of change of velocity:

Definition 1 Let $\mathbf{v}(t)$ be the **velocity** of an object at time t. The **acceleration a(t)** of the object at time t is **the derivative of v** at that instant:

$$\mathbf{a}(t) = \mathbf{v}'(t) = \frac{d\mathbf{v}}{dt}.$$

If $\boldsymbol{\sigma}(t)$ is the position of the object at time t, then $\mathbf{v}(t) = \boldsymbol{\sigma}'(t)$. Therefore, $\mathbf{a}(t)$ **is the second derivative of** $\boldsymbol{\sigma}$:

$$\mathbf{a}(t) = \boldsymbol{\sigma}''(t) = \frac{d^2\boldsymbol{\sigma}}{dt^2}.$$

Example 1 Let $\boldsymbol{\sigma}(t) = (\cos(2t), \sin(2t))$ be the position of an object at time t. Determine the acceleration of the object.

Solution

Note that $\boldsymbol{\sigma}(t)$ is the position vector of a point on unit circle, since

$$||\boldsymbol{\sigma}(t)||^2 = \cos^2(2t) + \sin^2(2t) = 1.$$

We have

$$\mathbf{v}(t) = \frac{d\boldsymbol{\sigma}}{dt} = -2\sin(2t)\mathbf{i} + 2\cos(2t)\mathbf{j},$$

so that

$$\mathbf{a}(t) = \frac{d\mathbf{v}}{dt} = -4\cos(2t)\mathbf{i} - 4\sin(2t)\mathbf{j} = -4\boldsymbol{\sigma}(t).$$

Since $\mathbf{a}(t) = -4\boldsymbol{\sigma}(t)$, and $\boldsymbol{\sigma}(t)$ is in the radial direction, the acceleration vector points towards the origin. In particular, acceleration is not the $\mathbf{0}$ vector, even though the speed of the object is constant:

$$||\mathbf{v}(t)|| = \sqrt{4\sin^2(2t) + 4\cos^2(2t)} = 2.$$

There is no paradox, since acceleration is the rate of change of the velocity vector, and that vector is not constant, even though its magnitude is constant. Figure 1 indicates the directions of the velocity and acceleration vectors (the vectors have been scaled to fit the picture). □

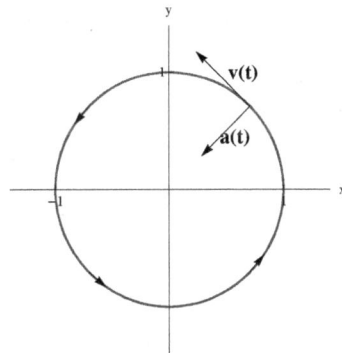

Figure 1

Example 2 Let

$$\boldsymbol{\sigma}(t) = (\sin(2t), \cos(3t))$$

a) Determine $a(t)$.
b) Determine $\mathbf{a}(\pi/4)$ and $\mathbf{a}(3\pi/4)$.

Solution

a) The velocity at the instant t is

$$\mathbf{v}(t) = \frac{d\boldsymbol{\sigma}}{dt} = 2\cos(2t)\mathbf{i} - 3\sin(3t)\mathbf{j},$$

so that acceleration at t is

$$\mathbf{a}\left(t\right) = \frac{d\boldsymbol{v}}{dt} = -4\sin\left(2t\right)\mathbf{i} - 9\cos\left(3t\right)\mathbf{j}.$$

b)

$$\mathbf{a}(\pi/4) = -4\sin(\pi/2)\mathbf{i} - 9\cos(3\pi/4)\mathbf{j} = -4\mathbf{i} + \frac{9}{2}\sqrt{2}\mathbf{j},$$

$$\mathbf{a}\left(3\pi/4\right) = -4\sin\left(3\pi/2\right)\mathbf{i} - 9\cos\left(9\pi/4\right)\mathbf{j} = 4\mathbf{i} - \frac{9}{2}\sqrt{2}\mathbf{j}$$

Figure 2 shows $\mathbf{a}\left(\pi/4\right)$ and $\mathbf{a}\left(3\pi/4\right)$ (the vectors have been scaled to fit the picture). $\square$

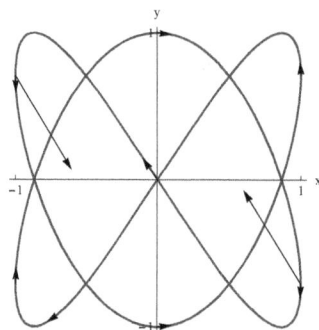

Figure 2

The Moving Frame

We can attach a special rectangular coordinate system to any point on a parametrized curve, with one of the axes along the unit tangent to the curve at that point. It is useful to determine the components of the acceleration vector with respect to this **"moving frame "**.

Let's begin by stating a general fact. If the magnitude of a vector-valued function is a constant, its derivative at any point is orthogonal to the vector at that point:

Proposition 1 If $\|\mathbf{V}\left(t\right)\|$ is constant then $\mathbf{V}'\left(t\right)$ is orthogonal to $\mathbf{V}\left(t\right)$.

Proof

Since $\|\mathbf{V}\left(t\right)\|^2$ is a constant,

$$\frac{d}{dt}\|\mathbf{V}\left(t\right)\|^2 = 0.$$

The product rule is applicable to the dot product:

$$\frac{d}{dt}\left(\mathbf{V}\left(t\right) \cdot \mathbf{W}\left(t\right)\right) = \left(\frac{d}{dt}\mathbf{V}\left(t\right)\right) \cdot \mathbf{W}\left(t\right) + \mathbf{V}\left(t\right) \cdot \left(\frac{d}{dt}\mathbf{W}\left(t\right)\right)$$

(exercise). Therefore,

$$0 = \frac{d}{dt}\|\mathbf{V}\left(t\right)\|^2 = \frac{d}{dt}\left(\mathbf{V}\left(t\right) \cdot \mathbf{V}\left(t\right)\right) = 2\mathbf{V}'\left(t\right) \cdot \mathbf{V}\left(t\right).$$

Thus, $\mathbf{V}'\left(t\right) \perp \mathbf{V}\left(t\right)$, as claimed. $\blacksquare$

Assume that a curve C is parametrized by the function $\boldsymbol{\sigma} : J \to \mathbb{R}^n$, where $n = 2$ or $n = 3$. Let $\mathbf{T}\left(t\right)$ be the unit tangent to the curve C at $\boldsymbol{\sigma}\left(t\right)$. As a corollary to Proposition 1, $\mathbf{T}'\left(t\right)$ **is orthogonal to** $\mathbf{T}\left(t\right)$, since $\|\mathbf{T}\left(t\right)\| = 1$ for each $t \in J$.

Definition 2 The principal unit normal is

$$\mathbf{N}(t) = \frac{\mathbf{T}'(t)}{||\mathbf{T}'(t)||}.$$

Thus, $\mathbf{N}(t)$ is a unit vector that is perpendicular to the unit tangent $\mathbf{T}(t)$.

Example 3 Let $\boldsymbol{\sigma}(t) = (\cos(2t), \sin(2t))$, $t \in [0, 2\pi]$, so that $\boldsymbol{\sigma}$ parametrizes the unit circle. Determine $\mathbf{T}(t)$ and $\mathbf{N}(t)$.

Solution

We have

$$\boldsymbol{\sigma}'(t) = -2\sin(2t)\mathbf{i} + 2\cos(2t)\mathbf{j},$$

so that

$$||\boldsymbol{\sigma}'(t)|| = \sqrt{4\sin^2(2t) + 4\cos^2(2t)} = 2.$$

Therefore,

$$\mathbf{T}(t) = \frac{\boldsymbol{\sigma}'(t)}{||\boldsymbol{\sigma}'(t)||} = \frac{1}{2}(-2\sin(2t)\mathbf{i} + 2\cos(2t)\mathbf{j}) = -\sin(2t)\mathbf{i} + \cos(2t)\mathbf{j}.$$

Thus,

$$\mathbf{T}'(t) = -2\cos(2t)\mathbf{i} - 2\sin(2t)\mathbf{j}.$$

Therefore,

$$||\mathbf{T}'(t)|| = 2.$$

Thus,

$$\mathbf{N}(t) = \frac{\mathbf{T}'(t)}{||\mathbf{T}'(t)||} = \frac{1}{2}(-2\cos(2t)\mathbf{i} - 2\sin(2t)\mathbf{j}) = -\cos(2t)\mathbf{i} + \sin(2t)\mathbf{j} = -\boldsymbol{\sigma}(t).$$

Figure 3 shows $\mathbf{T}(\pi/8)$ and $\mathbf{N}(\pi/8)$ (the vectors have been scaled to fit the picture). bNote that the unit normal points towards the origin. $\square$

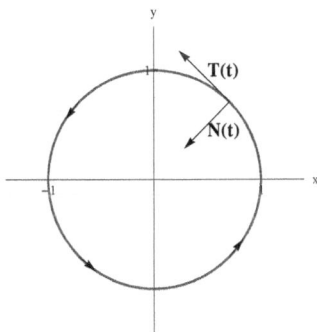

Figure 3

Definition 3 Let $\boldsymbol{\sigma} : J \subset \mathbb{R} \to \mathbb{R}^3$ and let C be the curve that is parametrized by $\boldsymbol{\sigma}$. **The unit binormal to** C **at** $\boldsymbol{\sigma}(t)$ **is**

$$\mathbf{B}(t) = \mathbf{T}(t) \times \mathbf{N}(t)$$

The plane that is determined by $\mathbf{T}(t)$ and $\mathbf{N}(t)$ is referred to as **the osculating plane for the curve** C at $\boldsymbol{\sigma}(t)$.

Thus, **the unit binormal $\mathbf{B}(t)$ is orthogonal to the osculating plane at $\boldsymbol{\sigma}(t)$.**

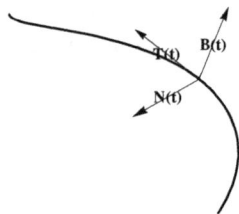

Figure 4: The moving frame

Example 4 Let $\boldsymbol{\sigma}(t) = (\cos(2t), \sin(2t), t)$, where $t \in [0, 2\pi]$, and let C be the curve that is parametrized by $\boldsymbol{\sigma}$. Determine the unit binormal to C at $\boldsymbol{\sigma}(\pi/4)$.

Solution

We have

$$\frac{d\boldsymbol{\sigma}}{dt} = -2\sin(2t)\,\mathbf{i} + 2\cos(2t)\,\mathbf{j} + \mathbf{k},$$

and

$$\left\|\frac{d\boldsymbol{\sigma}}{dt}\right\| = \sqrt{5},$$

so that

$$\mathbf{T}(t) = \frac{1}{\sqrt{5}}\frac{d\boldsymbol{\sigma}}{dt} = -\frac{2}{\sqrt{5}}\sin(2t)\,\mathbf{i} + \frac{2}{\sqrt{5}}\cos(2t)\,\mathbf{j} + \frac{1}{\sqrt{5}}\mathbf{k}.$$

Therefore,

$$\frac{d\mathbf{T}}{dt} = -\frac{4}{\sqrt{5}}\cos(2t)\,\mathbf{i} - \frac{4}{\sqrt{5}}\sin(2t)\,\mathbf{j}.$$

Thus,

$$\left\|\frac{d\mathbf{T}}{dt}\right\| = \frac{4}{\sqrt{5}},$$

so that

$$\mathbf{N}(t) = \frac{\sqrt{5}}{4}\left(-\frac{4}{\sqrt{5}}\cos(2t)\,\mathbf{i} - \frac{4}{\sqrt{5}}\sin(2t)\,\mathbf{j}\right) = -\cos(2t)\,\mathbf{i} - \sin(2t)\,\mathbf{j}.$$

(Note that $\mathbf{N}(t) \perp \mathbf{T}(t)$, as it should be). Therefore,

$$
\mathbf{B}(t) = \mathbf{T}(t) \times \mathbf{N}(t) = \begin{vmatrix} \mathbf{i} & \mathbf{j} & \mathbf{k} \\ -\frac{2}{\sqrt{5}}\sin(2t) & \frac{2}{\sqrt{5}}\cos(2t) & \frac{1}{\sqrt{5}} \\ -\cos(2t) & -\sin(2t) & 0 \end{vmatrix}
$$

$$
= \begin{vmatrix} \frac{2}{\sqrt{5}}\cos(2t) & \frac{1}{\sqrt{5}} \\ -\sin(2t) & 0 \end{vmatrix} \mathbf{i} - \begin{vmatrix} -\frac{2}{\sqrt{5}}\sin(2t) & \frac{1}{\sqrt{5}} \\ -\cos(2t) & 0 \end{vmatrix} \mathbf{j}
$$

$$
+ \begin{vmatrix} -\frac{2}{\sqrt{5}}\sin(2t) & \frac{2}{\sqrt{5}}\cos(2t) \\ -\cos(2t) & -\sin(2t) \end{vmatrix} \mathbf{k}
$$

$$
= \frac{1}{\sqrt{5}}\sin(2t)\,\mathbf{i} - \frac{1}{\sqrt{5}}\cos(2t)\,\mathbf{j} + \left(\frac{2}{\sqrt{5}}\sin^2(2t) + \frac{2}{\sqrt{5}}\cos^2(2t)\right)\mathbf{k}
$$

$$
= \frac{1}{\sqrt{5}}\sin(2t)\,\mathbf{i} - \frac{1}{\sqrt{5}}\cos(2t)\,\mathbf{j} + \frac{2}{\sqrt{5}}\mathbf{k}
$$

(Confirm that $\mathbf{B}(t)$ is orthogonal to $\mathbf{T}(t)$ and $\mathbf{N}(t)$) .□

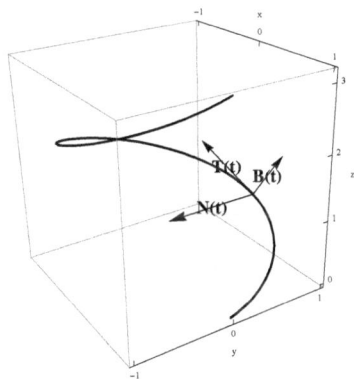

Figure 5

Arc Length

Assume that C is a smooth curve that is parametrized by $\boldsymbol{\sigma} : [a, b] \to \mathbb{R}^2$, where $\boldsymbol{\sigma}(t) = (x(t), y(t))$. As we discussed in Section 10.4, the arc length of C is

$$
\int_a^b \sqrt{\left(\frac{dx}{dt}\right)^2 + \left(\frac{dy}{dt}\right)^2}\, dt,
$$

provided that the curve C is traversed exactly once by $\boldsymbol{\sigma}(t)$ as t varies from a to b. If $\boldsymbol{\sigma}(t)$ is the position of an object at time t, then

$$
\mathbf{v}(t) = \frac{d\boldsymbol{\sigma}}{dt} = \frac{dx}{dt}\mathbf{i} + \frac{dy}{dt}\mathbf{j}
$$

is the **velocity** and

$$
\|\mathbf{v}(t)\| = \|\boldsymbol{\sigma}'(t)\| = \sqrt{\left(\frac{dx}{dt}\right)^2 + \left(\frac{dy}{dt}\right)^2}
$$

is the **speed** of the object at time t. Therefore,

$$
\int_a^b \|\mathbf{v}(t)\|\, dt = \int_a^b \sqrt{\left(\frac{dx}{dt}\right)^2 + \left(\frac{dy}{dt}\right)^2}\, dt
$$

is **the distance traveled by the object** over the time interval $[a, b]$. We can state that **the distance that is traveled by the object during the time interval $[a, b]$ is the integral of the speed of the object over the interval $[a, b]$.** We can identify the **arc length of the curve** C that is parametrized by σ with distance traveled, provided that C is traversed exactly once by $\boldsymbol{\sigma}(t)$ as t varies from a to b.

Similarly, if $\boldsymbol{\sigma}(t) = (x(t), y(t), z(t))$ parametrizes a curve C in $\mathbb{R}^3$ and C is traversed by $\boldsymbol{\sigma}(t)$ exactly once as t varies from a to b, the arc length of C is

$$\int_a^b \left\| \frac{d\boldsymbol{\sigma}}{dt} \right\| dt = \int_a^b \sqrt{\left(\frac{dx}{dt} \right)^2 + \left(\frac{dy}{dt} \right)^2 + \left(\frac{dz}{dt} \right)^2} \, dt.$$

If an object is at the point $\boldsymbol{\sigma}(t)$ at the instant t, the speed of the object at that instant is $\|\mathbf{v}(t)\| = \|\boldsymbol{\sigma}'(t)\|$, so that the distance traveled by the object over the time interval $[a, b]$ is calculated by the same integral.

Example 5 Let C be parametrized by $\boldsymbol{\sigma}(t) = (\cos(2t), \sin(2t), t)$, where $t \in [0, 2\pi]$. Calculate the arc length of C.

Solution

We have
$$\frac{d\boldsymbol{\sigma}}{dt} = -2\sin(2t)\, \boldsymbol{i} + 2\cos(2t)\, \mathbf{j} + \mathbf{k}.$$

Therefore,
$$\|\boldsymbol{\sigma}'(t)\| = \sqrt{4\sin^2(2t) + 4\cos^2(2t) + 1} = \sqrt{5}.$$

Thus, arc length of the path C is
$$\int_0^{2\pi} \|\boldsymbol{\sigma}'(t)\| \, dt = \int_0^{2\pi} \sqrt{5} \, dt = \sqrt{5}\,(2\pi) = 2\sqrt{5}\pi.$$

$\square$

Definition 4 Given a smooth function $\boldsymbol{\sigma}:[a, b] \to \mathbb{R}^n$, where $n = 2$ or $n = 3$, **the arc length function corresponding to σ** is

$$s(t) = \int_a^t \|\boldsymbol{\sigma}'(\tau)\| \, d\tau.$$

Assume that the curve C is traversed exactly once by $\boldsymbol{\sigma}(t)$ as t varies from a to b, and the arc length of C is L. By the Fundamental Theorem of Calculus,

$$\frac{ds}{dt} = \|\boldsymbol{\sigma}'(t)\| > 0,$$

since σ is smooth. Therefore the arc length function is an increasing function from $[a, b]$ to $[0, L]$. and has an inverse. If we denote the inverse of the arc length function by $t(s)$, the function $\boldsymbol{\sigma}(t(s))$, where s varies from 0 to L provides another parametrization of C, and will be referred to as **the parametrization C with respect to arc length.**

Example 6 Let $\boldsymbol{\sigma}(t) = \cos(2t)\, \mathbf{i} + \sin(2t)\, \mathbf{j}, \ 0 \leq t \leq \pi$. and let C be the unit circle that is parametrized by $\boldsymbol{\sigma}$. Determine the parametrization of C with respect to arc length.

Solution

We have

$$s\left(t\right) = \int_0^t \left\|\boldsymbol{\sigma}'\left(\tau\right)\right\| d\tau = \int_0^t \left\|-2\sin\left(2\tau\right)\mathbf{i} + 2\cos\left(2\tau\right)\mathbf{j}\right\| d\tau$$
$$= \int_0^t \sqrt{4\sin^2\left(2\tau\right) + 4\cos^2\left(2\tau\right)} d\tau = \int_0^t 2 d\tau = 2t.$$

Since $0 \le t \le \pi$, we have $0 \le s \le 2\pi$. In this case, the arc length function is simply the linear function $s = 2t$ on the interval $[0, \pi]$. The inverse is $t\left(s\right) = s/2$, where $s \in [0, 2\pi]$. The parametrization of C with respect to arc length is

$$\boldsymbol{\sigma}\left(t\left(s\right)\right) = \boldsymbol{\sigma}\left(s/2\right) = \cos\left(s\right)\mathbf{i} + \sin\left(s\right)\mathbf{j},$$

where $0 \le s \le 2\pi$.$\square$

In many cases, it is not feasible to obtain the arc length function and/or its inverse explicitly. Nevertheless, the existence of the parametrization is a useful concept, as we will see in the following subsections.

Curvature

Intuitively, the **curvature of a curve is a measure of the rate at which it bends** as a point moves along that curve. The parametrization of the curve with respect to arc length provides a way to quantify curvature that is independent of a particular parametrization: Curvature is an "intrinsic" geometric property of a curve.

Definition 5 Assume that C is parametrized by $\boldsymbol{\sigma}$. **The curvature of** C **at** $\boldsymbol{\sigma}\left(t\right)$ **is** $\left\|d\mathbf{T}/ds\right\|$

Thus, **curvature is the magnitude of the rate of change of the unit tangent with respect to arc length.** If we denote the curvature of C at $\boldsymbol{\sigma}\left(t\right)$ by $\kappa\left(t\right)$, we have

$$\kappa\left(t\right) = \left\|\frac{d\mathbf{T}}{ds}\right\|,$$

where $d\mathbf{T}/ds$ is evaluated at $\boldsymbol{\sigma}\left(t\right)$.

Proposition 2 The curvature is given by the expression

$$\kappa\left(t\right) = \frac{1}{\left\|\boldsymbol{\sigma}'\left(t\right)\right\|} \left\|\frac{d\mathbf{T}}{dt}\right\|$$

Proof

Since $s'\left(t\right) > 0$, the inverse $t\left(s\right)$ exists. We have $\mathbf{T} = \mathbf{T}\left(t\left(s\right)\right)$. The chain rule for functions of a single variable leads to the expression

$$\frac{d\mathbf{T}}{ds} = \frac{dt}{ds}\frac{d\mathbf{T}}{dt},$$

as you can confirm as an exercise. Since $t(s)$ denotes the inverse of the function defined by $s(t)$, we have

$$\frac{dt}{ds} = \frac{1}{\dfrac{ds}{dt}}.$$

Therefore,

$$\frac{d\mathbf{T}}{ds} = \frac{dt}{ds}\frac{d\mathbf{T}}{dt} = \frac{1}{\frac{ds}{dt}}\frac{d\mathbf{T}}{dt} = \frac{1}{||\boldsymbol{\sigma}'(t)||}\frac{d\mathbf{T}}{dt}.$$

Since the principal unit normal is

$$\mathbf{N}(t) = \frac{\dfrac{d\mathbf{T}}{dt}}{\left\|\dfrac{d\mathbf{T}}{dt}\right\|},$$

we have

$$\frac{d\mathbf{T}}{dt} = \left\|\frac{d\mathbf{T}}{dt}\right\|\mathbf{N}(t).$$

Therefore,

$$\kappa(t) = \left\|\frac{d\mathbf{T}}{ds}\right\| = \frac{1}{||\boldsymbol{\sigma}'(t)||}\frac{d\mathbf{T}}{dt} = \frac{1}{||\boldsymbol{\sigma}'(t)||}\left\|\frac{d\mathbf{T}}{dt}\right\|\,||\mathbf{N}(t)|| = \frac{1}{||\boldsymbol{\sigma}'(t)||}\left\|\frac{d\mathbf{T}}{dt}\right\|,$$

since $||\mathbf{N}(t)|| = 1$. ■

Remark 1 Assume that C is a parametrized curve in the plane, and the unit tangent is expressed as

$$\mathbf{T}(t) = \cos(\alpha(t))\,\mathbf{i} + \sin(\alpha(t))\,\mathbf{j},$$

where $\alpha(t)$ is the angle of inclination of $\mathbf{T}(t)$ with respect to the positive direction of the x-axis. *Then*

$$\kappa(t) = \left|\frac{d\alpha}{ds}\right|.$$

Indeed,

$$\frac{d\mathbf{T}}{ds} = -\sin(\alpha(t))\frac{d\alpha}{ds}\mathbf{i} + \cos(\alpha(t))\frac{d\alpha}{ds}\mathbf{j} = \frac{d\alpha}{ds}\left(-\sin(\alpha(t))\mathbf{i} + \cos(\alpha(t))\mathbf{j}\right) = \frac{d\alpha}{ds}\mathbf{u}(t),$$

where $||\mathbf{u}(t)|| = 1$. Therefore,

$$\left\|\frac{d\mathbf{T}}{ds}\right\| = \left|\frac{d\alpha}{ds}\right|$$

Thus, curvature is indeed a measure of the rate at which the tangent to the curve turns. ◊

Definition 6 Assume that the curve C is parametrized by the function $\boldsymbol{\sigma}$. **The radius of curvature of** C **at** $\boldsymbol{\sigma}(t)$ **is** $1/\kappa(t)$.

Example 7 Let $\boldsymbol{\sigma}(t) = (a\cos(t), a\sin(t))$, where $a > 0$, and let C be the curve that is parametrized by $\boldsymbol{\sigma}$. Determine the curvature and radius of curvature of C at $\boldsymbol{\sigma}(t)$.

Solution

Note that C is the circle of radius a centered at the origin. We have

$$\boldsymbol{\sigma}'(t) = -a\sin(t)\,\mathbf{i} + a\cos(t)\mathbf{j},$$

so that

$$||\boldsymbol{\sigma}'(t)|| = a,$$

and

$$\mathbf{T}(t) = \frac{\boldsymbol{\sigma}'(t)}{||\boldsymbol{\sigma}'(t)||} = \frac{1}{a}\left(-a\sin(t)\mathbf{i} + a\cos(2t)\mathbf{j}\right) = -\sin(t)\mathbf{i} + \cos(t)\mathbf{j}.$$

Therefore,

$$\mathbf{T}'(t) = -\cos(t)\mathbf{i} - \sin(t)\mathbf{j}.$$

Thus

$$\kappa(t) = \frac{1}{||\boldsymbol{\sigma}'(t)||}\left|\left|\frac{d\mathbf{T}}{dt}\right|\right| = \frac{1}{a}||-\cos(t)\mathbf{i} - \cos(t)|| = \frac{1}{a}.$$

Therefore, the curvature of the circle at any point is the reciprocal of its radius. The radius of curvature at any point is

$$\frac{1}{\kappa(t)} = \frac{1}{\frac{1}{a}} = a,$$

as it should be. $\square$

Example 8 Assume that the curve C is parametrized by $\boldsymbol{\sigma}(t) = (\cos(2t), \sin(2t), t)$. Determine the curvature and the radius of curvature of C at $\boldsymbol{\sigma}(t)$.

Solution

As in Example 4

$$\frac{d\boldsymbol{\sigma}}{dt} = -2\sin(2t)\mathbf{i} + 2\cos(2t)\mathbf{j} + \mathbf{k},$$

and

$$\left|\left|\frac{d\boldsymbol{\sigma}}{dt}\right|\right| = \sqrt{5}$$

so that

$$\mathbf{T}(t) = \frac{1}{\sqrt{5}}\frac{d\boldsymbol{\sigma}}{dt} = -\frac{2}{\sqrt{5}}\sin(2t)\mathbf{i} + \frac{2}{\sqrt{5}}\cos(2t)\mathbf{j} + \frac{1}{\sqrt{5}}\mathbf{k}.$$

Therefore,

$$\left|\left|\frac{d\mathbf{T}}{dt}\right|\right| = \frac{4}{\sqrt{5}},$$

Thus,

$$\kappa(t) = \frac{1}{||\boldsymbol{\sigma}'(t)||}\left|\left|\frac{d\mathbf{T}}{dt}\right|\right| = \frac{1}{\sqrt{5}}\left(\frac{4}{\sqrt{5}}\right) = \frac{4}{5}$$

and the radius of curvature is

$$\frac{1}{\kappa(t)} = \frac{5}{4}$$

at any t. $\square$

Tangential and Normal Components of Acceleration

Proposition 3 We have

$$\mathbf{a}(t) = \left(\frac{d}{dt}||\mathbf{v}(t)||\right)\mathbf{T}(t) + ||\mathbf{v}(t)||\left|\left|\frac{d\mathbf{T}}{dt}\right|\right|\mathbf{N}(t).$$

In particular, acceleration is in the **osculating plane** that is determined by $\mathbf{T}(t)$ and $\mathbf{N}(t)$.

Proof

$$\mathbf{a}\left(t\right) = \frac{d}{dt}\mathbf{v}\left(t\right) = \frac{d}{dt}\left(\left\|\mathbf{v}\left(t\right)\right\| \frac{\mathbf{v}\left(t\right)}{\left\|\mathbf{v}\left(t\right)\right\|}\right)$$

$$= \frac{d}{dt}\left(\left\|\mathbf{v}\left(t\right)\right\| \mathbf{T}\left(t\right)\right)$$

$$= \left(\frac{d}{dt}\left\|\mathbf{v}\left(t\right)\right\|\right)\mathbf{T}\left(t\right) + \left\|\mathbf{v}\left(t\right)\right\|\left(\frac{d}{dt}\mathbf{T}\left(t\right)\right).$$

Since

$$\mathbf{N}\left(t\right) = \frac{\dfrac{d\mathbf{T}}{dt}}{\left\|\dfrac{d\mathbf{T}}{dt}\right\|},$$

we have

$$\frac{d\mathbf{T}}{dt} = \left\|\frac{d\mathbf{T}}{dt}\right\|\mathbf{N}\left(t\right).$$

Therefore,

$$\mathbf{a}\left(t\right) = \left(\frac{d}{dt}\left\|\mathbf{v}\left(t\right)\right\|\right)\mathbf{T}\left(t\right) + \left\|\mathbf{v}\left(t\right)\right\|\left(\frac{d}{dt}\mathbf{T}\left(t\right)\right)$$

$$= \left(\frac{d}{dt}\left\|\mathbf{v}\left(t\right)\right\|\right)\mathbf{T}\left(t\right) + \left\|\mathbf{v}\left(t\right)\right\|\left\|\frac{d\mathbf{T}}{dt}\right\|\mathbf{N}\left(t\right).$$

■

Proposition 4

$$\mathbf{a}\left(t\right) = \left(\frac{d}{dt}\left\|\mathbf{v}\left(t\right)\right\|\right)\mathbf{T}\left(t\right) + \kappa\left(t\right)\left\|\mathbf{v}\left(t\right)\right\|^2\mathbf{N}\left(t\right) = \frac{d^2 s}{dt^2}\mathbf{T}\left(t\right) + \kappa\left(t\right)\left(\frac{ds}{dt}\right)^2\mathbf{N}\left(t\right)$$

and

$$\kappa\left(t\right) = \frac{\left\|\boldsymbol{\sigma}''\left(t\right) \times \boldsymbol{\sigma}'\left(t\right)\right\|}{\left\|\boldsymbol{\sigma}'\left(t\right)\right\|^3} = \frac{\left\|\mathbf{a}\left(t\right) \times \mathbf{v}\left(t\right)\right\|}{\left\|\mathbf{v}\left(t\right)\right\|^3}$$

Remark 2 Note that the tangential component of acceleration is the "scalar acceleration" $s''\left(t\right)$. The normal component is due to curvature, and it is curvature$\times$(speed)^2. The above formula for curvature is convenient for calculations. Also note that

$$\left\|\mathbf{a}\left(t\right)\right\|^2 = \left(\frac{d}{dt}\left\|\mathbf{v}\left(t\right)\right\|\right)^2 + \left(\kappa\left(t\right)\left\|\mathbf{v}\left(t\right)\right\|^2\right)^2,$$

since $\mathbf{T}\left(t\right) \perp \mathbf{N}\left(t\right)$.◊

The Proof of Proposition4

By Proposition 3,

$$\mathbf{a}\left(t\right) = \left(\frac{d}{dt}\left\|\mathbf{v}\left(t\right)\right\|\right)\mathbf{T}\left(t\right) + \left\|\mathbf{v}\left(t\right)\right\|\left\|\frac{d\mathbf{T}}{dt}\right\|\mathbf{N}\left(t\right).$$

Since

$$\kappa\left(t\right) = \frac{1}{\left\|\mathbf{v}\left(t\right)\right\|}\left\|\frac{d\mathbf{T}}{dt}\right\|,$$

we have

$$\left\| \frac{d\mathbf{T}}{dt} \right\| = \kappa(t) \left\| \mathbf{v}(t) \right\|.$$

Therefore,

$$\mathbf{a}(t) = \left(\frac{d}{dt} \left\| \mathbf{v}(t) \right\| \right) \mathbf{T}(t) + \left\| \mathbf{v}(t) \right\| \left\| \frac{d\mathbf{T}}{dt} \right\| \mathbf{N}(t)$$

$$= \left(\frac{d}{dt} \left\| \mathbf{v}(t) \right\| \right) \mathbf{T}(t) + \left\| \mathbf{v}(t) \right\| \kappa(t) \left\| \mathbf{v}(t) \right\| \mathbf{N}(t)$$

$$= \left(\frac{d}{dt} \left\| \mathbf{v}(t) \right\| \right) \mathbf{T}(t) + \kappa(t) \left\| \mathbf{v}(t) \right\|^2 \mathbf{N}(t).$$

Now,

$$\mathbf{a}(t) \times \mathbf{v}(t) = \left(\left(\frac{d}{dt} \left\| \mathbf{v}(t) \right\| \right) \mathbf{T}(t) + \kappa(t) \left\| \mathbf{v}(t) \right\|^2 \mathbf{N}(t) \right) \times \left(\left\| \mathbf{v}(t) \right\| \mathbf{T}(t) \right)$$

$$= \left(\frac{d}{dt} \left\| \mathbf{v}(t) \right\| \right) \left\| \mathbf{v}(t) \right\| \mathbf{T}(t) \times \mathbf{T}(t) + \kappa(t) \left\| \mathbf{v}(t) \right\|^3 \mathbf{N}(t) \times \mathbf{T}(t)$$

$$= \kappa(t) \left\| \mathbf{v}(t) \right\|^3 \mathbf{N}(t) \times \mathbf{T}(t),$$

since $\mathbf{T}(t) \times \mathbf{T}(t) = \mathbf{0}$. Therefore,

$$\left\| \mathbf{a}(t) \times \mathbf{v}(t) \right\| = \kappa(t) \left\| \mathbf{v}(t) \right\|^3 \left\| \mathbf{N}(t) \times \mathbf{T}(t) \right\| = \kappa(t) \left\| \mathbf{v}(t) \right\|^3,$$

since $\left\| \mathbf{N}(t) \times \mathbf{T}(t) \right\| = 1$. Thus,

$$\kappa(t) = \frac{\left\| \mathbf{a}(t) \times \mathbf{v}(t) \right\|}{\left\| \mathbf{v}(t) \right\|^3} = \frac{\left\| \boldsymbol{\sigma}''(t) \times \boldsymbol{\sigma}'(t) \right\|}{\left\| \boldsymbol{\sigma}'(t) \right\|^3},$$

as claimed. ∎

Example 9 Let $\boldsymbol{\sigma}(t) = (\cos(2t), \sin(2t), t)$, as in Example.8. Calculate $\kappa(t)$ by making use of Proposition 4, and determine the tangential and normal components of acceleration.

Solution

We have

$$\boldsymbol{\sigma}'(t) = -2\sin(2t)\,\mathbf{i} + 2\cos(2t)\,\mathbf{j} + \mathbf{k},$$

and

$$\boldsymbol{\sigma}''(t) = -4\cos(2t)\,\mathbf{i} - 4\sin(2t)\,\mathbf{j}.$$

Therefore,

$$\left\| \boldsymbol{\sigma}'(t) \right\| = \sqrt{5}.$$

We also have

$$\boldsymbol{\sigma}''(t) \times \boldsymbol{\sigma}'(t) = \begin{vmatrix} \mathbf{i} & \mathbf{j} & \mathbf{k} \\ -4\cos(2t) & -4\sin(2t) & 0 \\ -2\sin(2t) & 2\cos(2t) & 1 \end{vmatrix}$$

$$= \begin{vmatrix} -4\sin(2t) & 0 \\ 2\cos(2t) & 1 \end{vmatrix} \mathbf{i} - \begin{vmatrix} -4\cos(2t) & 0 \\ -2\sin(2t) & 1 \end{vmatrix} \mathbf{j} + \begin{vmatrix} -4\cos(2t) & -4\sin(2t) \\ -2\sin(2t) & 2\cos(2t) \end{vmatrix} \mathbf{k}$$

$$= -4\sin(2t)\,\mathbf{i} + 4\cos(2t)\,\mathbf{j} - 8\mathbf{k}.$$

Therefore,

$$\|\boldsymbol{\sigma}''(t) \times \boldsymbol{\sigma}'(t)\| = \sqrt{16 + 64} = \sqrt{80}.$$

Thus,

$$\kappa(t) = \frac{\|\boldsymbol{\sigma}''(t) \times \boldsymbol{\sigma}'(t)\|}{\|\boldsymbol{\sigma}'(t)\|^3} = \frac{\sqrt{80}}{5\sqrt{5}} = \frac{4\sqrt{5}}{5\sqrt{5}} = \frac{4}{5}.$$

We have

$$\frac{ds}{dt} = \|\boldsymbol{v}(t)\| = \|\boldsymbol{\sigma}'(t)\| = 5,$$

so that the tangential component of acceleration is 0:

$$\frac{d^2 s}{dt^2} = 0.$$

The normal component of acceleration is

$$\kappa(t) \left(\frac{ds}{dt}\right)^2 = \left(\frac{4}{5}\right)(25) = 20.$$

$\square$

Proposition 5 If $\boldsymbol{\sigma}(t) = x(t)\boldsymbol{i} + y(t)\boldsymbol{j}$ and C is the planar curve that is parametrized by $\boldsymbol{\sigma}$, the curvature of C at $\boldsymbol{\sigma}(t)$ can be expressed as

$$\kappa(t) = \frac{|x'(t)y''(t) - y'(t)x''(t)|}{\left((x'(t))^2 + (y'(t))^2\right)^{3/2}}.$$

If C is the graph of $y = f(x)$,

$$\kappa(x) = \frac{|f'(x)|}{\left(1 + (f'(x))^2\right)^{3/2}}$$

Proof

Since

$$\boldsymbol{\sigma}'(t) = x'(t)\mathbf{i} + y'(t)\mathbf{j},$$
$$\boldsymbol{\sigma}''(t) = x''(t)\mathbf{i} + y''(t)\mathbf{j},$$

we have

$$\|\boldsymbol{\sigma}'(t)\| = \left((x'(t))^2 + (y'(t))^2\right)^{1/2},$$

and

$$\boldsymbol{\sigma}''(t) \times \boldsymbol{\sigma}'(t) = \begin{vmatrix} \mathbf{i} & \mathbf{j} & \mathbf{k} \\ x'(t) & y'(t) & 0 \\ x''(t) & y''(t) & 0 \end{vmatrix}$$

$$= \begin{vmatrix} y'(t) & 0 \\ y''(t) & 0 \end{vmatrix} \mathbf{i} - \begin{vmatrix} x'(t) & 0 \\ x''(t) & 0 \end{vmatrix} \mathbf{i} + \begin{vmatrix} x'(t) & y'(t) \\ x''(t) & y''(t) \end{vmatrix} \mathbf{k}$$

$$= (x'(t)y''(t) - y'(t)x''(t))\mathbf{k}.$$

Therefore,

$$\|\boldsymbol{\sigma}''(t) \times \boldsymbol{\sigma}'(t)\| = |x'(t)\,y''(t) - y'(t)\,x''(t)|.$$

Thus,

$$\kappa(t) = \frac{\|\boldsymbol{\sigma}''(t) \times \boldsymbol{\sigma}'(t)\|}{\|\boldsymbol{\sigma}'(t)\|^3} = \frac{|x'(t)\,y''(t) - y'(t)\,x''(t)|}{\left(\left((x'(t))^2 + (y'(t))^2\right)^{1/2}\right)^3} = \frac{|x'(t)\,y''(t) - y'(t)\,x''(t)|}{\left((x'(t))^2 + (y'(t))^2\right)^{3/2}}.$$

In particular, if $\sigma(x) = (x, f(x))$,

$$\kappa(x) = \frac{|x'(x)\,y''(x) - y'(x)\,x''(x)|}{\left((x'(x))^2 + (y'(x))^2\right)^{3/2}} = \frac{|f''(x)|}{\left(1 + (f'(x))^2\right)^{3/2}}.$$

■

Example 10 Let $\boldsymbol{\sigma}(\theta) = (2\theta - 2\sin(\theta), 2 - 2\cos(\theta))$ and let C be the curve that is parametrized by $\boldsymbol{\sigma}$. Determine the curvature of C at $\boldsymbol{\sigma}(\pi/4)$ and $\boldsymbol{\sigma}(\pi/2)$.

Solution

We have

$$x(\theta) = 2\theta - 2\sin(\theta) \text{ and } y(\theta) = 2 - 2\cos(\theta).$$

Therefore,

$$x'(\theta) = 2 - 2\cos(\theta),\ x''(\theta) = 2\sin(\theta),$$
$$y'(\theta) = 2\sin(\theta),\ y''(\theta) = 2\cos(\theta).$$

Thus,

$$x'(\theta)\,y''(\theta) - y'(\theta)\,x''(\theta) = (2 - 2\cos(\theta))(2\cos(\theta)) - (2\sin(\theta))(2\sin(\theta))$$
$$= 4\cos(\theta) - 4\cos^2(\theta) - 4\sin^2(\theta)$$
$$= 4\cos(\theta) - 4,$$

and

$$(x'(\theta))^2 + (y'(\theta))^2 = (2 - 2\cos(\theta))^2 + (2\sin(\theta))^2$$
$$= 4 - 4\cos(\theta) + 4\cos^2(\theta) + 4\sin^2(\theta)$$
$$= 8 - 4\cos(\theta).$$

Thus,

$$\kappa(\theta) = \frac{4\,|\cos(\theta) - 1|}{8\,(2 - \cos(\theta))^{3/2}} = \frac{4\,(1 - \cos(\theta))}{8\,(2 - \cos(\theta))^{3/2}}$$

Therefore,

$$\kappa(\pi/4) = \frac{4\,(1 - \cos(\pi/4))}{8\,(2 - \cos(\pi/4))^{3/2}} = \frac{1 - \dfrac{\sqrt{2}}{2}}{2\left(2 - \dfrac{\sqrt{2}}{2}\right)^{3/2}} \cong 9.961\,74 \times 10^{-2}$$

and

$$\kappa(\pi/2) = \frac{4\,(1 - \cos(\pi/2))}{8\,(2 - \cos(\pi/2))^{3/2}} = \frac{1}{2\,(2^{3/2})} \cong 0.176\,777.$$

□

Problems

In problems 1-4 the position of a particle at time t is $\boldsymbol{\sigma}(t)$. Determine the acceleration function $\boldsymbol{a}(t)$ and the acceleration at the instant t_0.

1.
$$\boldsymbol{\sigma}(t) = (6 + 3\cos(4t), 6 + 3\sin(4t)), \ t_0 = \pi/12$$

2.
$$\boldsymbol{\sigma}(t) = (6t - 6\sin(2t), 6t - 6\cos(2t)), \ t_0 = 0$$

3.
$$\boldsymbol{\sigma}(t) = (t, \arctan(t)), \ t_0 = 1$$

4.
$$\boldsymbol{\sigma}(t) = \left(te^t, e^t\right), \ t_0 = 2$$

In problems In problems 5 and 6, assume that a curve C is parametrized by the function $\boldsymbol{\sigma}$. Determine the unit tangent $\mathbf{T}(t_0)$, the principal unit normal $\mathbf{N}(t_0)$ and the unit binormal $\mathbf{B}(t_0)$, if applicable:

5.
$$\boldsymbol{\sigma}(t) = (4 + 2\cos(t), 3 + 2\sin(t)), \ t_0 = \pi/6$$

6.
$$\boldsymbol{\sigma}(t) = (\cos(3t), \sin(3t), 4t), \ t_0 = \pi/2.$$

In problems 7 and 8, assume that a curve C is parametrized by the function $\boldsymbol{\sigma}$.
a) Express the arc length function $s(t)$ in terms of an integral,
b) [C] Calculate $s(t_0)$ approximately with the help of the numerical integrator of your computational utility:

7.
$$\boldsymbol{\sigma}(t) = (2\cos(t), 3\sin(t)), \ t \geq 0, \ t_0 = \pi$$

8.
$$\boldsymbol{\sigma}(t) = \left(te^t, e^t\right), \ t \geq 0, \ t_0 = 1.$$

In problems 9 and 10 assume that a curve C is parametrized by the function $\boldsymbol{\sigma}$. Calculate the curvature at t_0.
Hint: It is practical to use the formula

$$\kappa(t) = \frac{||\boldsymbol{\sigma}''(t) \times \boldsymbol{\sigma}'(t)||}{||\boldsymbol{\sigma}'(t)||^3}$$

9.
$$\boldsymbol{\sigma}(t) = (3\cos(t), 2\sin(t), 0), \ t \geq 0, \ t_0 = \pi/2$$

10.
$$\boldsymbol{\sigma}(t) = \left(t, e^t, 0\right), \ t \geq 0, \ t_0 = 2.$$

In problems 11 and 12, $\boldsymbol{\sigma}(t)$ is the position of an object at time t. Determine the tangential component

$$\frac{d}{dt}||\mathbf{v}(t)||\bigg|_{t=t_0}$$

and the normal component

$$\kappa(t_0)||\mathbf{v}(t_0)||^2$$

of the acceleration of the object at t_0.

Hint: It is practical to use the formula

$$\kappa\left(t\right) = \frac{\left\|\boldsymbol{\sigma}''\left(t\right) \times \boldsymbol{\sigma}'\left(t\right)\right\|}{\left\|\boldsymbol{\sigma}'\left(t\right)\right\|^{3}}$$

11.

$$\boldsymbol{\sigma}\left(t\right) = \left(\cos\left(2t\right), \sin\left(2t\right), 0\right), \ t \geq 0, \ t_0 = \pi/6$$

12.

$$\boldsymbol{\sigma}\left(t\right) = \left(t, t^2, 0\right), \ t \geq 0, \ t_0 = 1.$$

12.3 Real-Valued Functions of Several Variables

In this section we will introduce scalar-valued (i.e., real-valued) functions of several independent variables. The emphasis will be on functions of two variables. We will be able to visualize the behavior of such functions since their graphs are surfaces in the three-dimensional space.

Real-Valued Functions of Two Variables

Definition 1 A **real-valued (or scalar-valued) function** f two independent variables x and y is a rule that assigns a unique real number $f\left(x, y\right)$ to each point $\left(x, y\right)$ in a subset D of $\mathbb{R}^2$. The set D is the **domain** of f and the number $f\left(x, y\right)$ is the value of f at $\left(x, y\right)$. The **range** of f is the set of all possible values of f. The **graph** of f is the surface in $\mathbb{R}^3$ that consists of points of the form $\left(x, y, f\left(x, y\right)\right)$. If we set $z = f\left(x, y\right)$, then z is the dependent variable of f and the graph of f is a subset of the Cartesian coordinate plane where the axes are labeled as x, y and z.

We may refer to a real-valued function f with domain D by the notation $f : D \to \mathbb{R}$. As in the case of a function of a single variable, we may refer to "the function $f\left(x, y\right)$" if the rule that defines the function is defined by the single expression $f\left(x, y\right)$. In such a case, it should be understood that the domain of f is its natural domain, i.e., the set of all $\left(x, y\right) \in \mathbb{R}^2$ such that $f\left(x, y\right)$ is defined.

Example 1 Let $f\left(x, y\right) = x^2 + y^2$ for each $\left(x, y\right) \in \mathbb{R}^2$. The domain of f is the entire plane $\mathbb{R}^2$. The range of f is the set of all nonnegative numbers $[0, +\infty)$. The graph of f in the xyz-space is the surface that consists of points $\left(x, y, z\right)$ where $\left(x, y\right)$ is an arbitrary point in $\mathbb{R}^2$ and $x = x^2 + y^2$. Figure 1 shows the part of the graph of f in the viewing window $[-3, 3] \times [-3, 3] \times [0, 18]$. Even though such a picture depends on the viewing window, we may simply refer to "the graph of f". $\square$

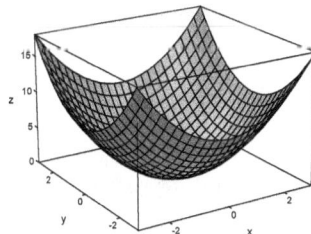

Figure 1

Example 2 Let $f(x,y) = x^2 - y^2$ for each $(x,y) \in \mathbb{R}^2$. Figure 2 shows the graph of f. $\square$

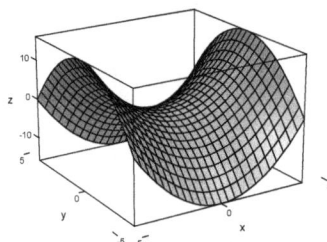

Figure 2

Assume that $f : D \subset \mathbb{R}^2 \to \mathbb{R}$. A **horizontal slice** of the graph of f is the intersection of the graph of f with a plane that is horizontal to the xy-plane. The projection of such a curve onto the xy-plane is a **level curve of** f. Thus, a level curve of f is a curve in the xy-plane that is the graph of an equation $f(x,y) = c$, where c is a constant. We may also consider vertical slices of the graph of f in order to have a better idea about the behavior of the function. Vertical slices corresponding to $x = c$ or $y = c$, where c is a constant, are especially useful for visualization.

Example 3 Let $z = f(x,y) = x^2 + y^2$, as in 1. The level curve of f corresponding to $z = c > 0$ is the graph of the equation

$$x^2 + y^2 = c.$$

This is a circle of radius $\sqrt{c}$ centered at the origin. As c increases, the radius if the circle increases. Figure 3 displays some level curves of f. The darker shading indicates smaller values of f.

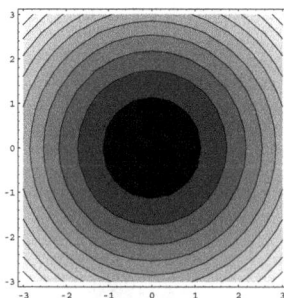

Figure 3

A vertical slice of the graph of f corresponding to the plane $x = c$ projects onto the yz-plane as the graph of the equation

$$z = c^2 + y^2.$$

This is a parabola. Figure 4 displays some of these parabolas.

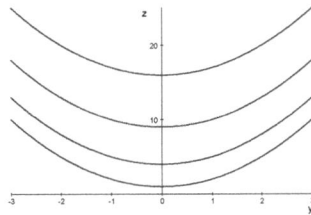

Figure 4

Similarly, a vertical slice of the graph of f corresponding to the plane $y = c$ projects onto the xz-plane as the parabola that is the graph of the equation

$$z = x^2 + c^2.$$

□

Example 4 Let $z = f(x, y) = x^2 - y^2$, as in Example 2. The level curve of f corresponding to $z = c$ is the graph of the equation

$$x^2 - y^2 = c.$$

This is a hyperbola. Figure 5 displays some of these hyperbolas.

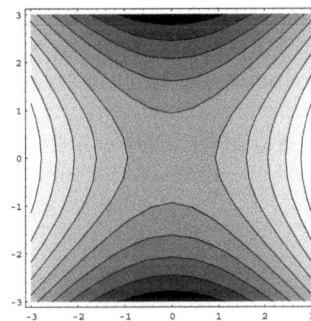

Figure 5

As in the previous example, the darker shading indicates that the function is decreasing, and the lighter color indicates that the function is increasing. This is consistent with the saddle shape of the graph as displayed in Figure 2

A vertical slice of the graph of f corresponding to the plane $x = c$ projects onto the yz-plane as the graph of the equation

$$z = c^2 - y^2.$$

This is a parabola. Figure 6 displays some of these parabolas.

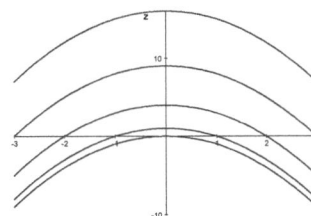

Figure 6

Similarly, a vertical slice of the graph of f corresponding to the plane $y = c$ projects onto the xz-plane as the parabola that is the graph of the equation

$$z = x^2 - c^2.$$

Figure 7 displays some of these parabolas.

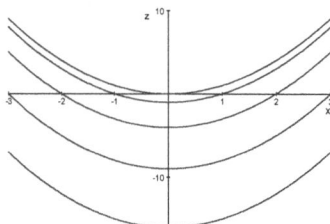

Figure 7

□

Real-Valued Functions of Three or More Variables

As in the case of two independent variables, a function f of three independent variables x, y and z is a rule that assigns a real number $f(x, y, z)$ to each (x, y, z) in a subset D of $\mathbb{R}^3$, referred to as the domain of f. If we set $w = f(x, y, z)$, then the graph of f is the subset of the four-dimensional space $\mathbb{R}^4$ that consists of points of the form $(x, y, z, f(x, y, z))$, where (x, y, z) ranges over D. Since we cannot visualize the four-dimensional space, we cannot display the graph of f as in the case of two independent variables. Nevertheless, we can consider the level surfaces of f, and that is helpful in the understanding of the function: A **level surface** of $f(x, y, z)$ consists of the points in $\mathbb{R}^3$ such that

$$f(x, y, z) = c,$$

where c is some constant.

Example 5 Let

$$f(x, y, z) = x^2 - y^2 - z^2.$$

Figure 8 and Figure 9 display the level surfaces of f such that $f(x, y, z) = 1$ and $f(x, y, z) = -1$, respectively. □

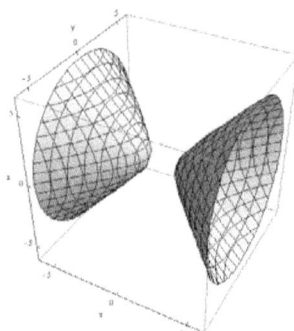

Figure 8

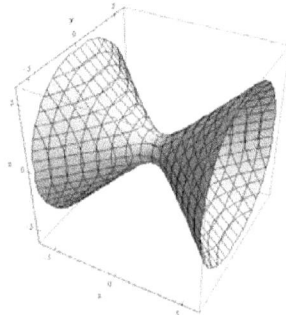

Figure 9

In many applications, functions of more than three variables are encountered as well. Even though we will discuss the ideas of differential and integral calculus mainly within the contexts of functions of two or three variables, we will indicate possible extensions to more than three variables.

Problems

[C] In problems 1-6, you will need to use your plotting device:
a) Plot the graph of the function f as the graph of the equation $z = f(x, y)$ in an appropriate viewing window,
b) Plot some of the level curves of f.

1.
$$f(x, y) = (x - 2)^2 + 2(y - 1)^2$$

2.
$$f(x, y) = 3(x + 1)^2 + (y - 2)^2$$

3.
$$f(x, y) = 4x^2 - y^2$$

4.
$$f(x, y) = (y - 2)^2 - (x + 1)^2$$

5.
$$f(x, y) = (x^2 + y^2) e^{-\sqrt{x^2 + y^2}}$$

6.
$$f(x, y) = \sin(x)\cos(y)$$

[C] In problems 7-10 you will need to use your plotting device. Plot some of the level surfaces of $f(x, y, z)$.

7.
$$f(x, y, z) = x^2 + y^2 + z^2$$

8.
$$f(x, y, z) = 4x^2 + 9y^2 + 16z^2$$

9.
$$f(x, y, z) = x^2 - y^2 + z^2$$

10.
$$f(x, y, z) = 4x^2 - 9y^2 - 16z^2$$

12.4 Partial Derivatives

In this section we will discuss the rate of change of a function of several variables when one the variables varies. This leads to **the partial derivatives** of the functions with respect to its independent variables. We begin with the notion of the limit of a such a function at a point.

Limits and Continuity

As in the case of a real-valued function of a single variable, the limit of $f(x, y)$ as (x, y) approaches the point (x_0, y_0) is the number L if $f(x, y)$ is as close to L as desired provided that $(x, y) \neq (x_0, y_0)$ and (x, y) is sufficiently close to (x_0, y_0). In this case we write

$$\lim_{(x,y) \to (x_0,y_0)} f(x, y) = L.$$

Here is the precise definition:

Definition 1 Assume that f is a real-valued function of two variables and $f(x, y)$ is defined at (x, y) if the distance of (x, y) from (x_0, y_0) is small enough, with the possible exception of (x_0, y_0) itself. **The limit of** f at (x_0, y_0) is the number L if, given any $\varepsilon > 0$ there exists $\delta > 0$ such that

$$(x, y) \neq (x_0, y_0) \text{ and } ||(x - x_0, y - y_0)|| < \delta$$

$\Rightarrow$

$$|f(x, y) - L| < \varepsilon.$$

The definition has the following geometric interpretation: Given any $\varepsilon > 0$ there exists a disk D_δ of radius δ centered at (x_0, y_0) such that for any $(x, y) \in D_\delta$ we have $|f(x, y) - f(x_0, y_0)| < \varepsilon$.

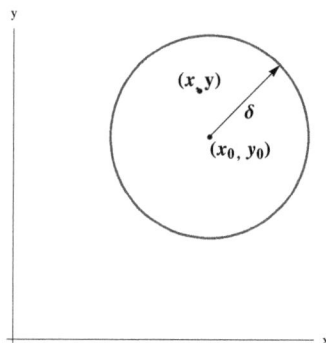

Figure 1

As in the case of a function of a single variable, the real-valued function $f(x, y)$ is said to be **continuous** at (x_0, y_0) if

$$\lim_{(x,y) \to (x_0,y_0)} f(x, y) = f(x_0, y_0).$$

The definitions of the limits and continuity of real-valued functions of more than two variables are similar. For example,

$$\lim_{(x,y,z) \to (x_0,y_0,z_0)} f(x, y, z) = L$$

if, given any $\varepsilon > 0$ there exists $\delta > 0$ such that

$$(x, y, z) \neq (x_0, y_0, z_0) \text{ and } ||(x - x_0.y - y_0, z - z_0)|| < \delta$$

$\Rightarrow$

$$|f(x, y, z) - L| < \varepsilon.$$

As a rule of the thumb, **the functions that we will deal with will be continuous at almost each point of their domain.** At the end of this section you can find an example that illustrates the rigorous proof of the continuity of a function of two variables. Such proofs are best left to a course in advanced Calculus.

The Definition and Evaluation of Partial Derivatives

Let's begin with functions of two variables.

Definition 2 The partial derivative of f with respect to x at (x, y) is

$$\frac{\partial f}{\partial x}(x, y) = \lim_{\Delta x \to 0} \frac{f(x + \Delta x, y) - f(x, y)}{\Delta x},$$

and the partial derivative of f with respect to y at (x, y) is

$$\frac{\partial f}{\partial y}(x, y) = \lim_{\Delta y \to 0} \frac{f(x, y + \Delta y) - f(x, y)}{\Delta y}.$$

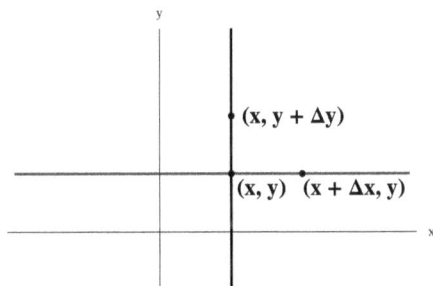

Figure 2

We will also use the notations

$$f_x, \ \partial_x f, \ f_y \text{ and } \partial_y f$$

Since

$$\frac{f(x + \Delta x, y) - f(x, y)}{\Delta x}$$

is **the average rate of change of** f as the first independent variable varies from x to $x + \Delta x$, and

$$\frac{\partial f}{\partial x}(x, y) = \lim_{\Delta x \to 0} \frac{f(x + \Delta x, y) - f(x, y)}{\Delta x},$$

we can interpret $f_x(x, y)$ as **the rate of change of** f **with respect to** x at (x, y). Similarly, we can interpret $f_y(x, y)$ as **the rate of change of** f **with respect to** y at (x, y).

Since y is kept fixed in the evaluation of $f_x(x, y)$, **we can simply apply the rules for the differentiation of a function of a single variable to evaluate** $f_x(x, y)$ **by treating** y **as a constant.** Similarly, we treat x as a constant and apply the familiar rules of differentiation in order to evaluate $f_y(x, y)$.

Example 1 Let $f(x, y) = 36 - 4x^2 - y^2$. Evaluate the partial derivatives $\partial_x f$ and $\partial_y f$.

Solution

In order to evaluate f_x, we treat y as a constant. Thus,

$$\frac{\partial f}{\partial x} = \frac{\partial}{\partial x}\left(36 - 4x^2 - y^2\right) = -4\frac{\partial}{\partial x}\left(x^2\right) - \frac{\partial}{\partial x}\left(y^2\right) = -8x.$$

Similarly, we treat x as a constant to evaluate f_y:

$$\frac{\partial f}{\partial y} = \frac{\partial}{\partial y}\left(36 - 4x^2 - y^2\right) = -2y.$$

□

Example 2 Let $f(x,y) = \cos(x-y)$. Evaluate the partial derivatives $\partial_x f$ and $\partial_y f$.

Solution

By the chain rule,

$$\frac{\partial f}{\partial x}(x,y) = \frac{\partial}{\partial x}\cos(x-y) = \left(\frac{d}{du}\cos(u)\Big|_{u=x-y}\right)\left(\frac{\partial}{\partial x}(x-y)\right)$$
$$= -\sin(x-y)(1) = -\sin(x-y),$$

and

$$\frac{\partial f}{\partial y}(x,y) = \frac{\partial}{\partial y}\cos(x-y) = \left(\frac{d}{du}\cos(u)\Big|_{u=x-y}\right)\left(\frac{\partial}{\partial y}(x-y)\right)$$
$$= -\sin(x-y)(-1) = \sin(x-y).$$

$\square$

The idea of partial differentiation extends to functions of more than two variables in a straightforward manner. In order to evaluate the partial derivative with respect to a specific variable, we treat all the other variables as constants, and apply the rules for the differentiation of functions of a single variable.

Example 3 Let
$$f(x,y,z) = \sqrt{x^2 + y^2 + z^2},$$
so that $f(x,y,z)$ is the distance of the point (x,y,z) from the origin. Determine $\partial_x f$, $\partial_y f$ and $\partial_z f$.

Solution

We have

$$\frac{\partial f}{\partial x}(x,y,z) = \frac{\partial}{\partial x}\left(x^2 + y^2 + z^2\right)^{1/2}$$
$$= \frac{1}{2}\left(x^2 + y^2 + z^2\right)^{-1/2}\left(\frac{\partial}{\partial x}\left(x^2 + y^2 + z^2\right)\right)$$
$$= \frac{1}{2}\left(x^2 + y^2 + z^2\right)^{-1/2}(2x)$$
$$= \frac{x}{\sqrt{x^2 + y^2 + z^2}}.$$

Similarly,
$$\frac{\partial f}{\partial y}(x,y,z) = \frac{y}{\sqrt{x^2 + y^2 + z^2}} \text{ and } \frac{\partial f}{\partial z}(x,y,z) = \frac{z}{\sqrt{x^2 + y^2 + z^2}}.$$

$\square$

The Geometric Interpretation of Partial Derivatives

We can visualize the graph of $z = f(x,y)$ as a surface in the three-dimensional space with Cartesian coordinates x, y and z. Let's pick a point (x_0, y_0) in the xy-plane and consider the plane $y = y_0$. This is a plane that is parallel to the xz-plane and intersects the graph of f along a curve, as illustrated in Figure 3. We will refer to the curve as **the x-coordinate curve through the point** $(x_0, y_0, f(x_0, y_0))$.

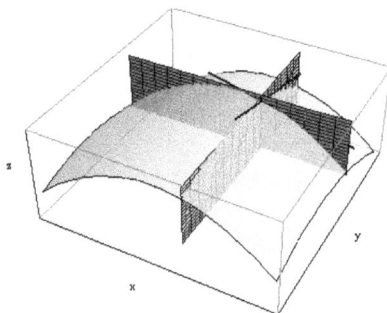

Figure 3

Since y is kept fixed at y_0 and $z = f(x, y_0)$ on the x-coordinate curve through the point $(x_0, y_0, f(x_0, y_0))$, this curve can be parametrized by the function

$$\boldsymbol{\sigma}_1(x) = (x, y_0, f(x, y_0)).$$

The assignment $x \to f(x, y_0)$ defines a function of the single variable x whose derivative at x_0 is $\partial_x f(x_0, y_0)$, by the definition of the partial derivative. Therefore, we can determine a **tangent vector to the x-coordinate curve through the point at** $(x_0, y_0, f(x_0, y_0))$ as

$$\frac{d\boldsymbol{\sigma}_1}{dx}(x_0) = \left(1, 0, \frac{\partial f}{\partial x}(x_0, y_0)\right) = \mathbf{i} + \frac{\partial f}{\partial x}(x_0, y_0)\,\mathbf{k}.$$

Similarly, the plane $x = x_0$ is parallel to the yz-plane and intersects the graph of f along a curve that can be parametrized by the function

$$\boldsymbol{\sigma}_2(y) = (x_0, y, f(x_0, y)).$$

We will refer to this curve as **the y-coordinate curve** that passes through $(x_0, y_0, f(x_0, y_0))$. A **tangent vector to the y-coordinate curve through the point at** $(x_0, y_0, f(x_0, y_0))$ can be determined as

$$\frac{d\boldsymbol{\sigma}_2}{dy}(y_0) = \left(0, 1, \frac{\partial f}{\partial y}(x_0, y_0)\right) = \mathbf{j} + \frac{\partial f}{\partial y}(x_0, y_0)\,\mathbf{k}.$$

Example 4 Let $f(x, y) = 36 - 4x^2 - y^2$, as in Example 1. Determine the lines that are tangent at $(1, 3, f(1, 3))$ to the coordinate curves passing through that point.

Solution

As in Example 1

$$\frac{\partial f}{\partial x} = -8x \text{ and } \frac{\partial f}{\partial y} = -2y.$$

Thus,

$$\frac{\partial f}{\partial x}(1, 3) = -8x\big|_{x=1} = -8,$$

and

$$\frac{\partial f}{\partial y}(1, 3) = -2y\big|_{y=3} = -6.$$

Therefore, a tangent vector to the x-coordinate curve passing through $(1, 3, f(1, 3)) = (1, 3, 23)$ can be determined as

$$\left(1, 0, \frac{\partial f}{\partial x}(1, 3)\right) = (1, 0, -8) = \mathbf{i} - 8\mathbf{k}.$$

The line that is tangent to this curve at $(1, 3, f(1, 3))$ can be parametrized by

$$
\begin{aligned}
L_1(u) &= (1, 3, f(1, 3)) + u(1, 0, -8) \\
&= (1, 3, 23) + u(1, 0, -8) = (1 + u, 3, 23 - 8u),
\end{aligned}
$$

where $u \in \mathbb{R}$.

Similarly, the vector

$$
\left(0, 1, \frac{\partial f}{\partial y}(1, 3)\right) = (0, 1, -6) = \mathbf{j} - 6\mathbf{k}
$$

is tangent to the y-coordinate curve passing through $(1, 3, f(1, 3)) = (1, 3, 23)$. The line that is tangent to this curve at $(1, 3, f(1, 3))$ can be parametrized by

$$
\begin{aligned}
L_2(u) &= (1, 3, f(1, 3)) + u(0, 1, -6) \\
&= (1, 3, 23) + u(0, 1, -6) = (1, 3 + u, 23 - 6u).
\end{aligned}
$$

Figure 4 illustrates the relevant coordinate curves and tangent lines. $\square$

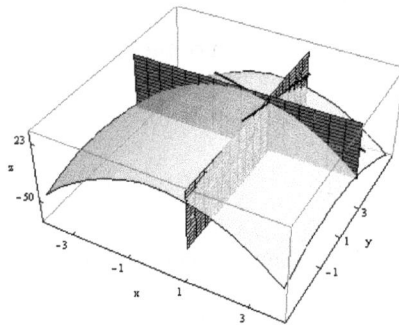

Figure 4

Example 5 Let $f(x, y) = \cos(x - y)$ as in Example 2. Determine the lines that are tangent at $(\pi, 2\pi/3, f(\pi, 2\pi/3))$ to the coordinate curves passing through that point.

Solution

As in Example 2,

$$
\frac{\partial f}{\partial x}(x, y) = -\sin(x - y),
$$

and

$$
\frac{\partial f}{\partial y}(x, y) = \sin(x - y).
$$

Thus,

$$
\frac{\partial f}{\partial x}(\pi, 2\pi/3) = -\sin(x - y)|_{x=\pi,\ y=2\pi/3} = -\sin\left(\pi - \frac{2\pi}{3}\right) = -\sin\left(\frac{\pi}{3}\right) = -\frac{\sqrt{3}}{2},
$$

and

$$
\frac{\partial f}{\partial y}(\pi, 2\pi/3) = \sin(x - y)|_{x=\pi,\ y=2\pi/3} = \sin\left(\pi - \frac{2\pi}{3}\right) = \sin\left(\frac{\pi}{3}\right) = \frac{\sqrt{3}}{2}.
$$

Therefore, the vector

$$
\left(1, 0, \frac{\partial f}{\partial x}(\pi, 2\pi/3)\right) = \left(1, 0, -\frac{\sqrt{3}}{2}\right)
$$

is tangent to the x-coordinate curve at $(\pi, 2\pi/3, f(\pi, 2\pi/3))$.
We have

$$f(\pi, 2\pi/3) = \cos\left(\pi - \frac{2\pi}{3}\right) = \cos\left(\frac{\pi}{3}\right) = \frac{1}{2}.$$

Thus,the tangent line to the x-coordinate curve at $(\pi, 2\pi/3, f(\pi, 2\pi/3))$.can be parametrized by

$$\mathbf{L_1}(u) = \left(\pi, \frac{2\pi}{3}, f(\pi, 2\pi/3)\right) + u\left(1, 0, -\frac{\sqrt{3}}{2}\right) = \left(\pi, \frac{2\pi}{3}, \frac{1}{2}\right) + u\left(1, 0, -\frac{\sqrt{3}}{2}\right)$$

$$= \left(\pi + u, \frac{2\pi}{3}, \frac{1}{2} - \frac{\sqrt{3}}{2}u\right).$$

The vector

$$\left(0, 1, \frac{\partial f}{\partial y}(\pi, 2\pi/3)\right) = \left(0, 1, \frac{\sqrt{3}}{2}\right)$$

is tangent to the y-coordinate curve at $(\pi, 2\pi/3, 1/2)$.. The tangent line can be parameterized by

$$\mathbf{L_2}(u) = \left(\pi, \frac{2\pi}{3}, \frac{1}{2}\right) + u\left(0, 1, \frac{\sqrt{3}}{2}\right) = \left(\pi, \frac{2\pi}{3} + u, \frac{1}{2} + \frac{\sqrt{3}}{2}u\right).$$

Figure 5 illustrates the relevant coordinate curves and tangent lines. $\square$

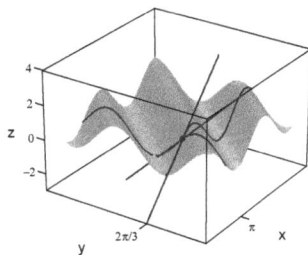

Figure 5

Higher-Order Partial Derivatives

If f is a function of x and y, f_x and f_y are also functions of x and y. Therefore, we can differentiate them with respect to x and y. Thus, we can compute

$$(f_x)_x = \frac{\partial}{\partial x}\left(\frac{\partial f}{\partial x}\right).$$

This is a **second-order partial derivative of** f. We set

$$f_{xx}(x, y) = \frac{\partial^2 f}{\partial x^2}(x, y) = \partial_x \partial_x f(x, y) = \frac{\partial}{\partial x}\left(\frac{\partial f}{\partial x}\right)$$

and refer to this second-order partial derivative as **the second partial derivative of** f **with respect to** x**.**

Similarly, **the second partial derivative of** f **with respect to** y is

$$f_{yy}(x,y) = \frac{\partial^2 f}{\partial y^2}(x,y) = \partial_y \partial_y f(x,y) = \frac{\partial}{\partial y}\left(\frac{\partial f}{\partial y}\right).$$

The **mixed second-order partial derivatives** are

$$f_{xy}(x,y) = \frac{\partial^2 f}{\partial y \partial x}(x,y) = \partial_y \partial_x f(x,y) = \frac{\partial}{\partial y}\left(\frac{\partial f}{\partial x}\right),$$

and

$$f_{yx}(x,y) = \frac{\partial^2 f}{\partial x \partial y}(x,y) = \partial_x \partial_y f(x,y) = \frac{\partial}{\partial x}\left(\frac{\partial f}{\partial y}\right).$$

Example 6 Let $f(x,y) = e^{-x^2 - 2y^2}$. Determine the second-order partial derivatives of f.

Solution

We have

$$\frac{\partial f}{\partial x}(x,y) = -2xe^{-x^2-2y^2} \text{ and } \frac{\partial f}{\partial y}(x,y) = -4ye^{-x^2-2y^2}.$$

Therefore,

$$\begin{aligned}
\frac{\partial^2 f}{\partial x^2}(x,y) = \frac{\partial}{\partial x}\left(-2xe^{-x^2-2y^2}\right) &= -2e^{-x^2-2y^2} - 2x\left(-2xe^{-x^2-2y^2}\right)\\
&= -2e^{-x^2-2y^2} + 4x^2 e^{-x^2-2y^2},
\end{aligned}$$

$$\begin{aligned}
\frac{\partial^2 f}{\partial y^2}(x,y) = \frac{\partial}{\partial y}\left(-4ye^{-x^2-2y^2}\right) &= -4e^{-x^2-2y^2} - 4y\left(-4ye^{-x^2-4y^2}\right)\\
&= -4e^{-x^2-2y^2} + 16y^2 e^{-x^2-2y^2},
\end{aligned}$$

$$\frac{\partial^2 f}{\partial y \partial x}(x,y) = \frac{\partial}{\partial y}\left(-2xe^{-x^2-2y^2}\right) = -2x\left(-4ye^{-x^2-2y^2}\right) = 8xye^{-x^2-2y^2},$$

and

$$\frac{\partial^2 f}{\partial x \partial y}(x,y) = \frac{\partial}{\partial x}\left(-4ye^{-x^2-2y^2}\right) = -4y\left(-2xe^{-x^2-2y^2}\right) = 8xye^{-x^2-2y^2}.$$

□

Note that

$$\frac{\partial^2 f}{\partial x \partial y}(x,y) = \frac{\partial^2 f}{\partial y \partial x}(x,y)$$

in the above example. That is not an accident:

Theorem 1 If $\partial_x \partial_y f$ and $\partial_y \partial_x f$ **are continuous in some open set** D**, then they are equal on** D**.**

The proof of this theorem is left to a course in advanced calculus.

The generalization of second-order partial derivatives to functions of more than two variables is straightforward. Partial derivatives of order higher than two are also defined and calculated in the obvious manner. For example,

$$\frac{\partial^3 f}{\partial x^2 \partial y}(x,y) = \frac{\partial^2}{\partial x^2}\left(\frac{\partial f}{\partial y}\right)(x,y).$$

A Rigorous Proof of Continuity (Optional)

Example 7 If $f(x, y) = x^2 - y^2$ then f is continuous at $(3, 2)$.

Solution

Let's set $x = 3 + h$ and $y = 2 + k$. Then

$$
\begin{aligned}
|f(3+h, 2+k) - f(3, 2)| &= \left|(3+h)^2 - (2+k)^2 - 5\right| \\
&= \left|h^2 + 6h - k^2 - 4k\right| \\
&\leq |h|^2 + 6|h| + |k|^2 + 4|k| \\
&= |h|(|h| + 6) + |k|(|k| + 4).
\end{aligned}
$$

Let's impose the restriction that $dist((x, y), (3, 2)) < 1$ so that $|h| < 1$ and $|k| < 1$. Then

$$
|f(3+h, 2+k) - f(3, 2)| < 7|h| + 5|k| < 7(|h| + |k|).
$$

Note that

$$
(|h| + |k|)^2 \leq (2 \max(|h|, |k|))^2 = 4 \max\left(|h|^2, |k|^2\right) \leq 4\left(|h|^2 + |k|^2\right)
$$

so that

$$
(|h| + |k|) \leq 2\sqrt{|h|^2 + |k|^2} = 2dist((x, y), (3, 2)).
$$

Therefore,

$$
|f(3+h, 2+k) - f(3, 2)| < 7(|h| + |k|) \leq 14 dist((x, y), (3, 2)).
$$

Thus, if

$$
dist((x, y), (3, 2)) < \delta = \min\left(1, \frac{\varepsilon}{14}\right)
$$

then

$$
|f(x, y) - f(3, 2)| < \varepsilon.
$$

Therefore f is continuous at $(3, 2)$. $\square$

Problems

In problems 1-14 compute the partial derivatives of the given function with respect to all of the independent variables:

1.
$$f(x, y) = 4x^2 + 9y^2$$

2.
$$f(x, y) = 6x^2 - 5y^2$$

3.
$$f(x, y) = \sqrt{2x^2 + y^2}$$

4.
$$f(r, \theta) = r^2 \cos(\theta)$$

5.
$$f(x, y) = e^{-x^2 - y^2}$$

6.
$$f(x, y) = \ln\left(x^2 + y^2\right)$$

7.
$$f(x, y) = \sin\left(\sqrt{x^2 - y^2}\right)$$

8.
$$f(r, \theta) = e^{r^2} \tan(\theta)$$

9.
$$f(x, y) = \arctan\left(\frac{y}{\sqrt{x^2 + y^2}}\right)$$

10.
$$f(x, y) = \arccos\left(\frac{x}{\sqrt{x^2 + y^2}}\right)$$

11.

$$f\left(x,y,z\right)=\frac{1}{\sqrt{x^2+y^2+z^2}}$$

13.

$$f\left(x,y,z\right)=\arcsin\left(\frac{1}{\sqrt{x^2+y^2+z^2}}\right)$$

12.

$$f\left(x,y,z\right)=xyze^{x-y+2z}$$

14.

$$f\left(\rho,\varphi,\theta\right)=\rho\cos\left(\varphi\right)\sin\left(\theta\right)$$

In problems 15 - 18, let C_1 be the x-coordinate curve and let C_2 be the y-coordinate curve that passes through $(x_0,y_0,f(x_0,y_0))$.
a) Determine tangent vectors to C_1 and C_2 at $(x_0,y_0,f\left(x_0,y_0\right))$,
b) Determine parametric representations of the lines that are tangent to C_1 and C_2 at $(x_0,y_0,f\left(x_0,y_0\right))$.

15.

$$f\left(x,y\right)=\sqrt{x^2+y^2},\ \left(x_0,y_0\right)=\left(3,1\right).$$

17.

$$f\left(x,y\right)=e^{x^2+y^2},\ \left(x_0,y_0\right)=\left(1,0\right).$$

16.

$$f\left(x,y\right)=10-x^2-y^2,\ \left(x_0,y_0\right)=\left(2,1\right).$$

18.

$$f\left(x,y\right)=x^2-y^2,\ \left(x_0,y_0\right)=\left(3,2\right).$$

In problems 19-24, compute the indicated higher-order partial derivatives:

19.

$$f\left(x,y\right)=4x^2+9y^2,\ \frac{\partial^2 f}{\partial x^2},\ \frac{\partial^2 f}{\partial x\partial y}$$

22.

$$f\left(x,y\right)=\sin\left(\sqrt{x^2-y^2}\right),\ \frac{\partial^2 f}{\partial y^2},\ \frac{\partial^2 f}{\partial x\partial y}$$

20.

$$f\left(x,y\right)=\frac{1}{\sqrt{x^2+y^2}},\ \frac{\partial^2 f}{\partial y^2},\ \frac{\partial^2 f}{\partial y\partial x}$$

23.

$$f\left(x,y,z\right)=xyze^{x-y+2z},\ \frac{\partial^2 f}{\partial x\partial z},\ \frac{\partial^2 f}{\partial y\partial z}$$

21.

$$f\left(x,y\right)=e^{-x^2+y^2},\ \frac{\partial^2 f}{\partial x^2},\ \frac{\partial^2 f}{\partial y\partial x}$$

24.

$$f\left(\rho,\varphi,\theta\right)=\rho\cos\left(\varphi\right)\sin\left(\theta\right),\ \frac{\partial^2 f}{\partial\varphi\partial\theta},\ \frac{\partial^2 f}{\partial\rho\partial\theta}$$

In problems 25 and 26, evaluate f_{xy} and f_{yx} and confirm that they are equal:

25.

$$f\left(x,y\right)=\ln\left(x^2+4y^2\right)$$

26.

$$f\left(x,y\right)=x^2ye^{x^2-2y^2}$$

12.5   Linear Approximations and the Differential

Local Linear Approximations

Assume that f is a function of a single variable that is differentiable at x_0. **The linear approximation to f based at x_0** is the linear function

$$L_{x_0}\left(x\right)=f\left(x_0\right)+f'\left(x_0\right)\left(x-x_0\right).$$

The graph of L_{x_0} is **the tangent line to the graph of f at $(x_0,f\left(x_0\right))$**. We saw that $L_{x_0}\left(x\right)$ approximates $f\left(x\right)$ very well if x is close to the basepoint x_0. Indeed,

$$f\left(x_0+\Delta x\right)=L_{x_0}\left(x_0+\Delta x\right)+\Delta xQ\left(\Delta x\right),$$

where
$$\lim_{\Delta x \to 0} Q(\Delta x) = 0.$$

Thus, the magnitude of the error in the approximation of $f(x_0 + \Delta x)$ by $L_{x_0}(x_0 + \Delta x)$ is much smaller than $|\Delta x|$ if $|\Delta x|$ is small.

What is the appropriate generalization of the idea of local linear approximation to scalar functions of more than one variable? Let's begin with functions of two variables. Assume that $f(x,y)$ is defined for each (x,y) in some disk containing the point (x_0, y_0), and that the partial derivatives $\partial_x f(x_0, y_0)$ and $\partial_y f(x_0, y_0)$ exist. We would like to determine a linear function L that approximates f well near (x_0, y_0). It is natural to require that $L(x_0, y_0) = f(x_0, y_0)$. Thus, we can express L as

$$L(x,y) = f(x_0, y_0) + A(x - x_0) + B(x - x_0),$$

where A and B are constants. The graph of L is a plane that passes through the point $(x_0, y_0, f(x_0, y_0))$. It is also natural to require that

$$\frac{\partial L}{\partial x}(x_0, y_0) = \frac{\partial f}{\partial x}(x_0, y_0) \text{ and } \frac{\partial L}{\partial y}(x_0, y_0) = \frac{\partial f}{\partial y}(x_0, y_0).$$

We have

$$\frac{\partial L}{\partial x}(x,y) = A \text{ and } \frac{\partial L}{\partial y}(x,y) = B$$

for each $(x,y) \in \mathbb{R}^2$. Therefore, we will set

$$A = \frac{\partial f}{\partial x}(x_0, y_0) \text{ and } B = \frac{\partial f}{\partial y}(x_0, y_0).$$

Thus,

$$L(x,y) = f(x_0, y_0) + \frac{\partial f}{\partial x}(x_0, y_0)(x - x_0) + \frac{\partial f}{\partial y}(x_0, y_0)(y - y_0).$$

Definition 1 The linear approximation to f based at (x_0, y_0) is the linear function

$$L_{(x_0, y_0)}(x,y) = f(x_0, y_0) + \frac{\partial f}{\partial x}(x_0, y_0)(x - x_0) + \frac{\partial f}{\partial y}(x_0, y_0)(y - y_0).$$

The graph of $L_{(x_0, y_0)}$ is **the tangent plane to the graph of f at $(x_0, y_0, f(x_0, y_0))$.**

Example 1 Let $f(x,y) = 36 - 4x^2 - y^2$. Determine the linear approximation to f based at $(1,3)$.

Solution

We have

$$\frac{\partial f}{\partial x}(x,y) = -8x \text{ and } \frac{\partial f}{\partial y}(x,y) = -2y.$$

Therefore,

$$\frac{\partial f}{\partial x}(1,3) = -8 \text{ and } \frac{\partial f}{\partial y}(1,3) = -6,$$

and $f(1,3) = 23$. Therefore, the linear approximation to f based at $(1,3)$ is

$$L_{(1,3)}(x,y) = f(1,3) + \frac{\partial f}{\partial x}(1,3)(x - 1) + \frac{\partial f}{\partial y}(1,3)(y - 3)$$
$$= 23 - 8(x - 1) - 6(y - 3).$$

Figure 1 illustrates the graphs of f and $L_{(1,3)}$. Note that the graph of $L_{(1,3)}$ is consistent with the intuitive idea of a tangent plane to the surface that is the graph of f. $\square$

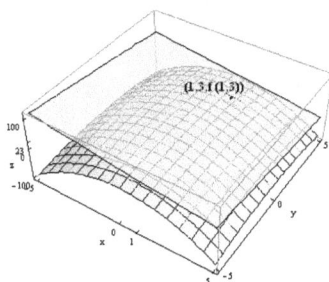

Figure 1

Remark 1 If we set

$$z = L_{(x_0,y_0)}(x,y) = f(x_0,y_0) + \frac{\partial f}{\partial x}(x_0,y_0)(x-x_0) + \frac{\partial f}{\partial y}(x_0,y_0)(y-y_0),$$

we can express the equation of the tangent plane to the graph of f at $(x_0, y_0, f(x_0, y_0))$ as

$$-\frac{\partial f}{\partial x}(x_0,y_0)(x-x_0) - \frac{\partial f}{\partial y}(x_0,y_0)(y-y_0) + (z - f(x_0,y_0)) = 0.$$

Thus, the vector

$$\mathbf{N}_{(x_0,y_0)} = -\frac{\partial f}{\partial x}(x_0,y_0)\mathbf{i} - \frac{\partial f}{\partial y}(x_0,y_0)\mathbf{j} + \mathbf{k}$$

is orthogonal to the tangent plane. With reference to Example 1,

$$\mathbf{N}_{(1.3)} = -\frac{\partial f}{\partial x}(1,3)\mathbf{i} - \frac{\partial f}{\partial y}(1.3)\mathbf{j} + \mathbf{k} = -(-8)\mathbf{i} - (-6)\mathbf{j} + \mathbf{k} = 8\mathbf{i} + 6\mathbf{j} + \mathbf{k}.$$

In Section 12.4 we saw that the vector

$$\mathbf{i} + \frac{\partial f}{\partial x}(x_0,y_0)\mathbf{k}$$

is tangential to the x-coordinate curve that passes through $(x_0, y_0, f(x_0, y_0))$ at that point. Similarly, the vector

$$\mathbf{j} + \frac{\partial f}{\partial y}(x_0,y_0)\mathbf{k}$$

is tangential to the y-coordinate curve that passes through $(x_0, y_0, f(x_0, y_0))$ at that point. It is reasonable to declare that the plane that is tangent to the graph of f at $(x_0, y_0, f(x_0, y_0))$ is spanned by these vectors. A normal vector to that plane can be determined as the cross product of these tangent vectors. Indeed,

$$\mathbf{N}_{(x_0,y_0)} = \begin{vmatrix} \mathbf{i} & \mathbf{j} & \mathbf{k} \\ 1 & 0 & \partial_x f(x_0,y_0) \\ 0 & 1 & \partial_y f(x_0,y_0) \end{vmatrix} = -\partial_x f(x_0,y_0)\mathbf{i} - \partial_y f(x_0,y_0)\mathbf{j} + \mathbf{k}$$

Thus, we are led to the same idea of a tangent plane geometrically and via the idea of a local linear approximation. $\lozenge$

We say that a function f of a single variable is differentiable at x_0 if $f'(x_0)$ exists. Shall we say that a function of two variables is differentiable at a point if its partial derivatives exist at that point? In the case of a function of a single variable, we know that differentiability implies continuity. It is natural to require that a function of two variables should be continuous at a point if it is declared to be differentiable at that point. The following example shows that it is not suitable to define differentiability in terms of the existence of partial derivatives:

Example 2 Let

$$f(x,y) = \begin{cases} \dfrac{x^2 y}{x^4 + y^2} & \text{if} \quad (x,y) \neq (0,0), \\ 0 & \text{if} \quad (x,y) = (0,0) \end{cases}$$

Both partial derivatives of f exist at $(0,0)$:

$$\frac{\partial f}{\partial x}(0,0) = \lim_{h \to 0} \frac{f(h,0) - f(0,0)}{h} = \lim_{h \to 0} (0) = 0,$$
$$\frac{\partial f}{\partial y}(0,0) = \lim_{h \to 0} \frac{f(0,h) - f(0,0)}{h} = \lim_{h \to 0} (0) = 0.$$

But the function is not continuous at $(0,0)$. Indeed,

$$\lim_{y \to 0} f(0,y) = 0, \ \lim_{x \to 0} f(x,0) = 0$$

but

$$\lim_{x \to 0} f(x,x^2) = \lim_{x \to 0} \frac{x^4}{2x^4} = \lim_{x \to 0} \frac{1}{2} = \frac{1}{2} \neq 0.$$

Figure 2 shows the graph of f. The picture gives an indication of the erratic behavior of the function near the origin. $\square$

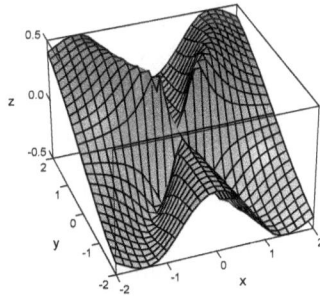

Figure 2

We will define the differentiability of a function of two variables by generalizing the approximation property of local linear approximations for a function of a single variable. Note that the value of $L_{(x_0,y_0)}(x,y)$ at $(x_0 + \Delta x, y_0 + \Delta y)$ is

$$L_{(x_0,y_0)}(x_0 + \Delta x, y_0 + \Delta y) = f(x_0,y_0) + \frac{\partial f}{\partial x}(x_0,y_0)(x_0 + \Delta x - x_0) + \frac{\partial f}{\partial y}(x_0,y_0)(y_0 + \Delta y - y_0)$$

$$= f(x_0,y_0) + \frac{\partial f}{\partial x}(x_0,y_0)\Delta x + \frac{\partial f}{\partial y}(x_0,y_0)\Delta y.$$

Definition 2 Assume that $f(x, y)$ is defined for each (x, y) in some open disk that contains (x_0, y_0) and that the partial derivatives of f at (x_0, y_0) exist. We say that f **is differentiable at** (x_0, y_0) if

$$f(x_0 + \Delta x, y_0 + \Delta x) = L_{(x_0, y_0)}(x_0 + \Delta x, y_0 + \Delta y) + \|(\Delta x, \Delta y)\| Q(\Delta x, \Delta y)$$

$$= f(x_0, y_0) + \frac{\partial f}{\partial x}(x_0, y_0)\Delta x + \frac{\partial f}{\partial y}(x_0, y_0)\Delta y + \|(\Delta x, \Delta y)\| Q(\Delta x, \Delta y),$$

where

$$\lim_{\Delta x \to 0, \Delta y \to 0} Q(\Delta x, \Delta y) = 0.$$

Intuitively, **we require that the magnitude of the error in the approximation of** $f(x_0 + \Delta x, y_0 + \Delta x)$ **by** $L_{(x_0, y_0)}(x_0 + \Delta x, y_0 + \Delta y)$ **to be much smaller than the distance of** $(x_0 + \Delta x, y_0 + \Delta y)$ **from the basepoint** (x_0, y_0) **near that point.**

Example 3 Let $f(x, y) = 36 - 4x^2 - y^2$, as in Example 1. Show that f is differentiable at the origin.

Solution

We showed that

$$L_{(1,3)}(x, y) = 23 - 8(x - 1) - 6(y - 3).$$

Therefore

$$L(1 + \Delta x, 3 + \Delta y) = 23 - 8(1 + \Delta x - 1) - 6(3 + \Delta y - 3) = 23 - 8\Delta x - 6\Delta y.$$

Thus

$$f(1 + \Delta x, 3 + \Delta y) - L(1 + \Delta x, 3 + \Delta y)$$
$$= 36 - 4(1 + \Delta x)^2 - (3 + \Delta y)^2 - (23 - 8\Delta x - 6\Delta y)$$
$$= 36 - 4 - 8\Delta x - 4(\Delta x)^2 - 9 - 6\Delta y - (\Delta y)^2 - 23 + 8\Delta x + 6\Delta y$$
$$= -4(\Delta x)^2 - (\Delta y)^2.$$

With reference to Definition 2, the error in the approximation is .

$$-4(\Delta x)^2 - (\Delta y)^2 = \|(\Delta x, \Delta y)\| Q(\Delta x, \Delta y),$$

so that

$$Q(\Delta x, \Delta y) = \frac{4(\Delta x)^2 + (\Delta y)^2}{\|(\Delta x, \Delta y)\|}$$

We need to show that $\lim_{\Delta x \to 0, \Delta y \to 0} Q(\Delta x, \Delta y) = 0$. Indeed,

$$Q(\Delta x, \Delta y) = \frac{4(\Delta x)^2 + (\Delta y)^2}{\|(\Delta x, \Delta y)\|} \leq \frac{4\left((\Delta x)^2 + (\Delta y)^2\right)}{\|(\Delta x, \Delta y)\|} = 4\frac{\|(\Delta x, \Delta y)\|^2}{\|(\Delta x, \Delta y)\|} = 4\|(\Delta x, \Delta y)\|.$$

Therefore,

$$\lim_{\Delta x \to 0, \Delta y \to 0} Q(\Delta x, \Delta y) = 0.$$

Thus, f is differentiable at $(0, 0)$. $\square$

Proposition 1 If f is differentiable at (x_0, y_0) then f is continuous at (x_0, y_0).

Proof

Since f is differentiable at (x_0, y_0),

$$f(x_0 + \Delta x, y_0 + \Delta x) = f(x_0, y_0) + \frac{\partial f}{\partial x}(x_0, y_0)\Delta x + \frac{\partial f}{\partial y}(x_0, y_0)\Delta y + \|(\Delta x, \Delta y)\| Q(\Delta x, \Delta y),$$

where $\lim_{\Delta x \to 0, \Delta y \to 0} Q(\Delta x, \Delta y) = 0$. Therefore,

$$\lim_{(\Delta x, \Delta y) \to (0,0)} f(x_0 + \Delta x, y_0 + \Delta x) = f(x_0, y_0) + \frac{\partial f}{\partial x}(x_0, y_0)\left(\lim_{(\Delta x, \Delta y) \to (0,0)} \Delta x\right)$$

$$+ \frac{\partial f}{\partial y}(x_0, y_0)\left(\lim_{(\Delta x, \Delta y) \to (0,0)} \Delta y\right) + \lim_{\Delta x \to 0, \Delta y \to 0} \|(\Delta x, \Delta y)\| Q(\Delta x, \Delta y)$$

$$= f(x_0, y_0),$$

Thus, f is continuous at (x_0, y_0). ∎

Proposition 1 shows that the function of Example 2 is not differentiable at $(0,0)$ since it is not continuous at $(0,0)$.

In general, it is not easy to show that a function is differentiable at a point in accordance with Definition 2. On the other hand, the continuity of the partial derivatives ensures differentiability:

Theorem 1 Assume that the partial derivatives of f exist in some open disk that contains (x_0, y_0), and that they are continuous at (x_0, y_0). Then f is differentiable at (x_0, y_0).

The proof of Theorem 1 is left to a course in advanced calculus.

Example 4 Let

$$f(x, y) = \sqrt{x^2 + y^2}.$$

a) Show that f is differentiable at $(3, 4)$ and determine the linear approximation to f based at $(3, 4)$.
b) Make use of the result of part a) in order to approximate $f(3.1, 3.9)$. Compare the absolute error with the distance from the basepoint.

Solution

a) We have

$$\frac{\partial f}{\partial x}(x, y) = \frac{\partial}{\partial x}(x^2 + y^2)^{1/2} = \frac{1}{2}(x^2 + y^2)^{-1/2}(2x) = \frac{x}{\sqrt{x^2 + y^2}}.$$

Similarly,

$$\frac{\partial f}{\partial y}(x, y) = \frac{y}{\sqrt{x^2 + y^2}}.$$

Thus, the partial derivatives of f are continuous at any point other than the origin. By Theorem 1 f is differentiable at any point other than the origin. In particular, f is differentiable at $(3, 4)$. We have

$$\frac{\partial f}{\partial x}(3, 4) = \frac{3}{5}, \quad \frac{\partial f}{\partial y}(3, 4) = \frac{4}{5} \text{ and } f(3, 4) = 5.$$

Thus, the linear approximation to f based at $(3, 4)$ is

$$L(x, y) = f(3, 4) + \frac{\partial f}{\partial x}(3, 4)(x - 3) + \frac{\partial f}{\partial y}(3, 4)(y - 4)$$

$$= 5 + \frac{3}{5}(x - 3) + \frac{4}{5}(y - 4).$$

The graph of L is the tangent plane to the graph of f at $(3, 4, f(3, 4)) = (3, 4, 5)$, as shown in Figure 3.

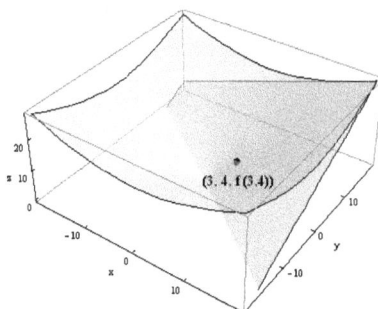

Figure 3

b) We have

$$f(3.1, 3.9) = \sqrt{(3.1)^2 + (3.9)^2} \cong 4.981\,97,$$

rounded to 6 significant digits. The corresponding value of the linear approximation is

$$L(3.1, 3.9) = 5 + \frac{3}{5}(0.1) + \frac{4}{5}(-0.1) = 4.98.$$

Therefore, the absolute error is approximately 1.97×10^{-3}. This number is much smaller than the distance of $(3.1, 3.9)$ from the basepoint (3.4):

$$\sqrt{(0.1)^2 + (-0.1)^2} = \sqrt{2 \times 10^{-2}} = \sqrt{2} \times 10^{-1} \cong 0.14$$

□

The Differential

Recall that the differential of a function of a single variable is a convenient way of keeping track of local linear approximations as the basepoint varies. If f is differentiable at x,

$$f(x + \Delta x) - f(x) \cong df(x, \Delta x) = \frac{df}{dx}(x)\,\Delta x,$$

and $df(x, \Delta x)$ is the change along the tangent line $(x, f(x))$ corresponding to the increment Δx.

Assume that f is a function of two variables that is differentiable at (x, y). For example, the partial derivatives of f are continuous in some disk centered at (x, y). Let L be the linear approximation to f based at (x, y), so that

$$L(u, v) = f(x, y) + \frac{\partial f}{\partial x}(x, y)(u - x) + \frac{\partial f}{\partial y}(x, y)(v - y).$$

Therefore,

$$L(x + \Delta x, y + \Delta y) = f(x, y) + \frac{\partial f}{\partial x}(x, y)\,\Delta x + \frac{\partial f}{\partial y}(x, y)\,\Delta y.$$

Thus,

$$f(x + \Delta x, y + \Delta y) = f(x, y) + \frac{\partial f}{\partial x}(x, y)\,\Delta x + \frac{\partial f}{\partial y}(x, y)\,\Delta y + ||(\Delta x, \Delta y)||\,Q(\Delta x, \Delta y),$$

where

$$\lim_{\Delta x \to 0, \Delta y \to 0} Q(\Delta x, \Delta y) = 0.$$

Therefore,

$$f(x + \Delta x, y + \Delta y) - f(x, y) \cong \frac{\partial f}{\partial x}(x, y) \Delta x + \frac{\partial f}{\partial y}(x, y) \Delta y,$$

and the magnitude of the error is much smaller than $||(\Delta x, \Delta y)||$ if $||(\Delta x, \Delta y)||$ is small. The right-hand side defines the differential of f:

Definition 3 Assume that f is differentiable at (x, y). The **differential of** f is defined by the expression

$$df(x, y, \Delta x, \Delta y) = \frac{\partial f}{\partial x}(x, y) \Delta x + \frac{\partial f}{\partial y}(x, y) \Delta y.$$

Thus, the differential of a function of two variables is a function of four variables. For a given (x, y), the function

$$(\Delta x, \Delta y) \to \frac{\partial f}{\partial x}(x, y) \Delta x + \frac{\partial f}{\partial y}(x, y) \Delta y$$

is a linear function. We have

$$f(x + \Delta x, y + \Delta y) - f(x, y) \cong df(x, y, \Delta x, \Delta y)$$

if $||(\Delta x, \Delta y)||$ is small.

It is traditional to denote Δx and Δy as by dx and dy within the context of differentials. Thus,

$$df(x, y, dx, dy) = \frac{\partial f}{\partial x}(x, y) \, dx + \frac{\partial f}{\partial y}(x, y) \, dy.$$

It is also traditional to be cryptic and simply write

$$df = \frac{\partial f}{\partial x}(x, y) \, dx + \frac{\partial f}{\partial y}(x, y) \, dy$$

as the differential of the function f.

Example 5 Let $f(x, y) = \sqrt{x^2 + y^2}$, as in Example 4.

a) Express the differential of f.
b) Make use of the differential in order to approximate $f(1.9, 5.2)$.

Solution

a) As in Example 4,

$$\frac{\partial f}{\partial x}(x, y) = \frac{x}{\sqrt{x^2 + y^2}} \text{ and } \frac{\partial f}{\partial y}(x, y) = \frac{y}{\sqrt{x^2 + y^2}}.$$

Therefore,

$$df(x, y, \Delta x, \Delta y) = \frac{\partial f}{\partial x}(x, y) \Delta x + \frac{\partial f}{\partial y}(x, y) \Delta y = \frac{x}{\sqrt{x^2 + y^2}} \Delta x + \frac{y}{\sqrt{x^2 + y^2}} \Delta y.$$

In the traditional notation.

$$df = \frac{x}{\sqrt{x^2 + y^2}} dx + \frac{y}{\sqrt{x^2 + y^2}} dy$$

b) It is reasonable to set $x = 2$ and $y = 5$ in order to approximate $f(1.9, 5.2)$. Thus,

$$f(1.9, 5.2) - f(2, 5) \cong df(2, 5, -0.1, 0.2) = \frac{\partial f}{\partial x}(2, 5)(-0.1) + \frac{\partial f}{\partial y}(2, 5)(0.2)$$

$$= \frac{2}{\sqrt{29}}(-0.1) + \frac{5}{\sqrt{29}}(0.2) \cong 0.148\,556.$$

Therefore,

$$f(1.9, 5.2) \cong f(2, 5) + 0.148\,556. = \sqrt{29} + 0.148\,556 \cong 5.533\,72.$$

We have

$$f(1.9, 5.2) = \sqrt{(1.9)^2 + (5.2)^2} \cong 5.536\,24,$$

rounded to 6 significant digits. Thus, the absolute error in the approximation of $f(1.9, 5.2)$ via the differential is approximately 2.5×10^{-3}. This is much smaller than the distance of $(1.9, 5.2)$ from the basepoint $(2, 5)$ (which is approximately 0.2236). □

Functions of more than two variables

The generalization of the idea of local linear approximations and the related idea of the differential to functions of more than two variables is straightforward. For example, if f is a function of three variables, the linear approximation to f based at (x_0, y_0, z_0) is

$$L(x, y, z) = f(x_0, y_0, z_0) + \frac{\partial f}{\partial x}(x_0, y_0, z_0)(x - x_0)$$

$$+ \frac{\partial f}{\partial y}(x_0, y_0, z_0)(y - y_0) + \frac{\partial f}{\partial z}(x_0, y_0, z_0)(z - z_0).$$

We say that f is differentiable at (x_0, y_0, z_0) if

$$f(x_0 + \Delta x, y_0 + \Delta x, z_0 + \Delta z) = f(x_0, y_0, z_0) + \frac{\partial f}{\partial x}(x_0, y_0, z_0)\Delta x + \frac{\partial f}{\partial y}(x_0, y_0, z_0)\Delta y + \frac{\partial f}{\partial z}(x_0, y_0, z_0)\Delta z$$

$$+ \|(\Delta x, \Delta y, \Delta z)\| Q(\Delta x, \Delta y, \Delta z),$$

where

$$\lim_{\Delta x \to 0, \Delta y \to 0, \Delta z \to 0} Q(\Delta x, \Delta y, \Delta z) = 0$$

Differentiability implies continuity and a sufficient condition for differentiability is the continuity of the partial derivatives.
The differential of f is

$$df(x, y, z, \Delta x, \Delta y, \Delta z) = \frac{df}{\partial x}(x, y, z)\Delta x + \frac{\partial f}{\partial y}(x, y, z)\Delta y + \frac{\partial f}{\partial z}(x, y, z)\Delta z.$$

We have

$$f(x + \Delta x, y + \Delta x, z + \Delta z) - f(x, y, z) = df(x, y, z, \Delta x, \Delta y, \Delta z) + \|(\Delta x, \Delta y, \Delta z)\| Q(\Delta x, \Delta y, \Delta z),$$

where

$$\lim_{\Delta x \to 0, \Delta y \to 0, \Delta z \to 0} Q(\Delta x, \Delta y, \Delta z) = 0$$

Cryptically,

$$df = \frac{df}{\partial x}dx + \frac{\partial f}{\partial y}dy + \frac{\partial f}{\partial z}dz.$$

As in the case of two variables,

$$f(x + dx, y + dy, z + dz) - f(x, y, z) \cong df$$

if $\|(dx, dy, dz)\|$ is small.

Example 6 Let $f(x, y, z) = \sqrt{x^2 + y^2 - z^2}$.

a) Determine the linear approximation to f based at $(2, 2, 1)$. Make use of the result to approximate $(2.1, 1.8, 0.9)$

b) Determine the differential of f. Make use of the result to approximate $f(-3.1, 1.1, 1.9)$.

Solution

a) We have

$$\frac{\partial f}{\partial x}(x, y, z) = \frac{x}{\sqrt{x^2 + y^2 - z^2}}, \ \frac{\partial f}{\partial y}(x, y, z) = \frac{y}{\sqrt{x^2 + y^2 - z^2}},$$

$$\frac{\partial f}{\partial z}(x, y, z) = -\frac{z}{\sqrt{x^2 + y^2 - z^2}}.$$

Therefore,

$$\frac{\partial f}{\partial x}(2, 2, 1) = \frac{2}{\sqrt{7}}, \ \frac{\partial f}{\partial y}(2, 2, 1) = \frac{2}{\sqrt{7}} \text{ and } \frac{\partial f}{\partial z}(2, 2, 1) = -\frac{1}{\sqrt{7}}.$$

Thus, the linear approximation to f based at $(2, 2, 1)$ is

$$L(x, y, z) = f(2, 2, 1) + \frac{\partial f}{\partial x}(2, 2, 1)(x - 2) + \frac{\partial f}{\partial y}(2, 2, 1)(y - 2) + \frac{\partial f}{\partial z}(2, 2, 1)(z - 1)$$

$$= \sqrt{7} + \frac{2}{\sqrt{7}}(x - 2) + \frac{2}{\sqrt{7}}(y - 2) - \frac{1}{\sqrt{7}}(z - 1).$$

Therefore,

$$f(2.1, 1.8, 0.9) \cong L(2.1, 1.8, 0.9)$$

$$= \sqrt{7} + \frac{2}{\sqrt{7}}(0.1) + \frac{2}{\sqrt{7}}(-0.2) - \frac{1}{\sqrt{7}}(-0.1) \cong 2.60795$$

Note that

$$f(2.1, 1.8, 0.9) = \sqrt{(2.1)^2 + (1.8)^2 - (0.9)^2} \cong 2.61534.$$

Thus, the absolute error in the approximation is approximately 6.8×10^{-2}. The distance of $(2.1, 1.8, 0.9)$ from the basepoint $(2, 2, 1)$ is

$$\sqrt{(0.1)^2 + (0.2)^2 + (0.1)^2} \cong 0.245$$

We see that the absolute error in the approximation is much smaller than the distance from the basepoint.

b) In the traditional notation, the differential of f is

$$df = \frac{df}{\partial x}dx + \frac{\partial f}{\partial y}dy + \frac{\partial f}{\partial z}dz$$

$$= \frac{x}{\sqrt{x^2 + y^2 - z^2}}dx + \frac{y}{\sqrt{x^2 + y^2 - z^2}}dy - \frac{z}{\sqrt{x^2 + y^2 - z^2}}dz.$$

It is reasonable to choose $(-3, 1, 2)$ as the basepoint in order to approximate $f(-3.1, 1.1, 1.9)$, Thus,

$$f(-3.1, 1.1, 1.9) - f(-3, 1, 1) \cong df(-3, 1, 2, -0.1, 0.1, -0.1)$$

$$= -\frac{3}{\sqrt{6}}(-0.1) + \frac{1}{\sqrt{6}}(0.1) - \frac{2}{\sqrt{6}}(-0.1)$$

$$= \frac{0.6}{\sqrt{6}}.$$

Therefore,

$$f(-3.1, 1.1, 1.9) \cong f(-3, 1, 2) + \frac{0.6}{\sqrt{6}} = \sqrt{6} + \frac{0.6}{\sqrt{6}} \cong 2.694\,44.$$

We have

$$f(-3.1, 1.1, 1.9) = \sqrt{(-3.1)^2 + (1.1)^2 - (1.9)^2} \cong 2.685\,14$$

Thus, the absolute error in the approximation of $f(-3.1, 1.1, 1.9)$ via the differential is approximately 9.3×10^{-3}. This is much smaller than the distance of $(-3.1, 1.1, 1.9)$ from the basepoint $(-3, 1, 2)$ (approximately 0.173). $\square$

Problems

In problems 1-5,
a) Determine the linear approximation to f based at (x_0, y_0) or (x_0, y_0, z_0),
b) Make use of the result of part a) in order to approximate $f(x_1, y_1)$ or $f(x_1, y_1, z_1)$.
c) [C] Calculate the absolute error in the approximation by assuming that your calculator calculates the exact value.

1.

$$f(x, y) = \left(x^2 + y^2\right)^{3/2}, \ (x_0, y_0) = (3, 4), \ (x_1, y_1) = (3.1, 3.9)$$

2.

$$f(x, y) = e^{-x} \sin(y), \ (x_0, y_0) = (0, \pi/2), \ (x_1, y_1) = (0.2, \pi/2 + 0.1)$$

3.

$$f(x, y) = \arctan\left(\frac{y}{x}\right), \ (x_0, y_0) = \left(\sqrt{3}, 1\right), \ (x_1, y_1) = (1.8, 0.8)$$

4.

$$f(x, y) = e^{x^2 + y^2}, \ (x_0, y_0) = (1, 0), \ (x_1, y_1) = (1.1, -0.2)$$

5.

$$f(x, y, z) = \sqrt{x^2 + y^2 + z^2}, \ (x_0, y_0, z_0) = (1, 2, 3), \ (x_1, y_1, z_1) = (0.9, 2.2, 2.9)$$

In problems 6-8,
a) Determine the differential of f,
b) Make use of the differential of f in order to approximate the indicated value of f:
c) [C] Calculate the absolute error in the approximation by assuming that your calculator calculates the exact value.

6.

$$f(x, y) = \sqrt{x^2 + y^2}, \ f(12.1, 4.9)$$

7.

$$f(x, y, z) = \sqrt{xyz}, \ f(0.9, 2.9, 3.1)$$

8.

$$f(x, y) = \arctan\left(x^2 + y^2\right), \ f(0.1, -1.2)$$

12.6 The Chain Rule

In this section we will discuss various versions of the chain rule for functions of several variables. Even though these rules are not as useful for the evaluation of derivatives as the chain rule for functions of a single variable, they can be interpreted in ways that lead to useful general results, as we will se;e within several contexts.

Let's begin with the following version of the chain rule:

Proposition 1 Assume that f is a function of two variables, x and y are functions of a single variable. If x and y are differentiable at t and f is differentiable at $(x(t), y(t))$ then

$$\frac{d}{dt} f(x(t), y(t)) = \frac{\partial f}{\partial x}(x(t), y(t)) \frac{dx}{dt} + \frac{\partial f}{\partial y}(x(t), y(t)) \frac{dy}{dt}$$

A Plausibility Argument:

Since f is differentiable at $(x(t), y(t))$, if $|\Delta t|$ is small,

$$f(x(t+\Delta t), y(t+\Delta t)) - f(x(t), y(t)) \cong \frac{\partial f}{\partial x}(x(t), y(t))(x(t+\Delta t) - x(t))$$

$$+ \frac{\partial f}{\partial y}(x(t), y(t))(y(t+\Delta t) - y(t))$$

$$\cong \frac{\partial f}{\partial x}(x(t), y(t)) \frac{dx}{dt}(t) \Delta t + + \frac{\partial f}{\partial y}(x(t), y(t)) \frac{dy}{dt}(t) \Delta t.$$

Therefore,

$$\frac{f(x(t+\Delta t), y(t+\Delta t)) - f(x(t), y(t))}{\Delta t} \cong \frac{\partial f}{\partial x}(x(t), y(t)) \frac{dx}{dt}(t) + + \frac{\partial f}{\partial y}(x(t), y(t)) \frac{dy}{dt}(t)$$

if $|\Delta t|$ is small. Thus, it is plausible that

$$\frac{d}{dt} f(\sigma(t)) = \lim_{\Delta t \to 0} \frac{f(x(t+\Delta t), y(t+\Delta t)) - f(x(t), y(t))}{\Delta t}$$

$$= \frac{\partial f}{\partial x}(x(t), y(t)) \frac{dx}{dt}(t) + + \frac{\partial f}{\partial y}(x(t), y(t)) \frac{dy}{dt}(t)$$

■

We can express the above version of the chain rule in more practical (albeit imprecise) ways

$$\frac{df}{dt} = \frac{\partial f}{\partial x} \frac{dx}{dt} + \frac{\partial f}{\partial y} \frac{dy}{dt}$$

If $z = f(x, y)$,

$$\frac{dz}{dt} = \frac{\partial z}{\partial x} \frac{dx}{dt} + \frac{\partial z}{\partial y} \frac{dy}{dt}$$

Example 1 Let $f(x, y) = x^2 - y^2$ and let $x(t) = \cos(t)$, $y(t) = \sin(t)$.

a) Determine

$$\frac{d}{dt} f(x(t), y(t))$$

by applying the chain rule and directly.

b) Determine

$$\frac{d}{dt} f\left(x\left(t\right), y\left(t\right)\right)\Big|_{t=\pi/3}.$$

Solution

a) We have $x\left(t\right) = \cos\left(t\right)$ and $y\left(t\right) = \sin\left(t\right)$. Thus.

$$\frac{\partial f}{\partial x} = 2x,\ \frac{\partial f}{\partial y} = -2y,\ \frac{dx}{dt} = -\sin\left(t\right),\ \frac{dy}{dt} = \cos\left(t\right).$$

Therefore,

$$\frac{df}{dt} = \frac{\partial f}{\partial x}\frac{dx}{dt} + \frac{\partial f}{\partial y}\frac{dy}{dt} = -2x\sin\left(t\right) - 2y\cos\left(t\right) = -2\cos\left(t\right)\sin\left(t\right) - 2\sin\left(t\right)\cos\left(t\right)$$

$$= -4\cos\left(t\right)\sin\left(t\right).$$

As for direct evaluation, we have $f\left(\cos\left(t\right), \sin\left(t\right)\right) = \cos^2\left(t\right) - \sin^2\left(t\right)$. Therefore,

$$\frac{df}{dt} = -2\cos\left(t\right)\sin\left(t\right) - 2\sin\left(t\right)\cos\left(t\right) = -4\cos\left(t\right)\sin\left(t\right).$$

b) By a),

$$\frac{df}{dt}\left(\pi/3\right) = -4\cos\left(\pi/3\right)\sin\left(\pi/3\right) = -4\left(\frac{1}{2}\right)\left(\frac{\sqrt{3}}{2}\right) = -4\sqrt{3}.$$

$\square$

Here is another version of the chain rule:

Proposition 2 Assume that f is a function of the single variable and x is a function of two variables. If x is differentiable at (u, v) and f is differentiable at $x\left(u, v\right)$ then

$$\frac{\partial}{\partial u} f\left(x\left(u, v\right)\right) = \frac{df}{dx}\left(x\left(u, v\right)\right)\frac{\partial x}{\partial u}\left(u, v\right)\ \text{ and }\ \frac{\partial}{\partial v} f\left(x\left(u, v\right)\right) = \frac{df}{dx}\left(x\left(u, v\right)\right)\frac{\partial x}{\partial v}\left(u, v\right)$$

Practical notation: If $z = f\left(x\right)$,

$$\frac{\partial z}{\partial u} = \frac{dz}{dx}\frac{\partial x}{\partial u}\ \text{ and }\ \frac{\partial z}{\partial v} = \frac{dz}{dx}\frac{\partial x}{\partial v}$$

The Plausibility of Proposition 2

If $|\Delta u|$ is small,

$$f\left(x\left(u + \Delta u, v\right)\right) - f\left(x\left(u, v\right)\right) \cong f\left(x\left(u, v\right) + \frac{\partial x}{\partial u}\left(u, v\right)\Delta u\right) - f\left(x\left(u, v\right)\right)$$

$$\cong f\left(x\left(u, v\right)\right) + \frac{df}{dx}\left(x\left(u, v\right)\right)\frac{\partial x}{\partial u}\left(u, v\right)\Delta u - f\left(x\left(u, v\right)\right)$$

$$= \frac{df}{dx}\left(x\left(u, v\right)\right)\frac{\partial x}{\partial u}\left(u, v\right)\Delta u$$

Therefore,

$$\frac{f\left(x\left(u + \Delta u, v\right)\right) - f\left(x\left(u, v\right)\right)}{\Delta u} \cong \frac{df}{dx}\left(x\left(u, v\right)\right)\frac{\partial x}{\partial u}\left(u, v\right)$$

if $|\Delta u|$ is small. Thus, we should have

$$\frac{\partial}{\partial u} f\left(x\left(u, v\right)\right) = \lim_{\Delta u \to 0}\frac{f\left(x\left(u + \Delta u, v\right)\right) - f\left(x\left(u, v\right)\right)}{\Delta u} = \frac{df}{dx}\left(x\left(u, v\right)\right)\frac{\partial x}{\partial u}\left(u, v\right)$$

■

Example 2 Let $u = f(x - at)$, where f is a differentiable function of a single variable and a is a constant. Show that u is a solution of the wave equation

$$\frac{\partial^2 u}{\partial t^2} = a^2 \frac{\partial u}{\partial x^2}.$$

Solution

Set $w(x, t) = x - at$, so that $u = f(w(x, t))$. By the chain rule,

$$\frac{\partial u}{\partial t} = \frac{df}{dw}(w(x, t)) \frac{\partial w}{\partial t} = \frac{df}{dw}(w(x, t))(-a) = -a \frac{df}{dw}(w(x, t)).$$

Therefore,

$$\frac{\partial^2 u}{\partial t^2} = -a \frac{\partial}{\partial t}\left(\frac{df}{dw}(w(x, t))\right) = -a\left(\frac{d^2 f}{dw^2}(w(x, t))\frac{\partial w}{\partial t}\right)$$

$$= -a \frac{d^2 f}{dw^2}(w(x, t))(-a) = a^2 \frac{d^2 f}{dw^2}(x - at)$$

Similarly,

$$\frac{\partial u}{\partial x} = \frac{df}{dw}(w(x, t)) \frac{\partial w}{\partial x} = \frac{df}{dw}(w(x, t))(1) = \frac{df}{dw}(w(x, t))$$

and

$$\frac{\partial^2 u}{\partial x^2} = \frac{\partial}{\partial x}\left(\frac{df}{dw}(w(x, t))\right) = \frac{d^2 f}{dw^2}(w(x, t))\frac{\partial w}{\partial x}$$

$$= \frac{d^2 f}{dw^2}(w(x, t))(1) = \frac{d^2 f}{dw^2}(x - at)$$

Therefore,

$$\frac{\partial^2 u}{\partial t^2} = a^2 \frac{d^2 f}{dw^2}(x - at) = a^2 \frac{\partial^2 u}{\partial x^2}$$

$\square$

Here is still another version of the chain rule:

Proposition 3 Assume that $f = f(x, y)$, $x = x(u, v)$ and $y = y(u, v)$. are differentiable functions. Then,

$$\frac{\partial f}{\partial u}(x(u, v), y(u, v)) = \frac{\partial f}{\partial x}(x(u, v), y(u, v))\frac{\partial x}{\partial u}(u, v) + \frac{\partial f}{\partial y}(x(u, v), y(u, v))\frac{\partial y}{\partial u}(u, v),$$

$$\frac{\partial f}{\partial v}(x(u, v), y(u, v)) = \frac{\partial f}{\partial x}(x(u, v), y(u, v))\frac{\partial x}{\partial v}(u, v) + \frac{\partial f}{\partial y}(x(u, v), y(u, v))\frac{\partial y}{\partial v}(u, v).$$

Practical notation:

$$\frac{\partial f}{\partial u} = \frac{\partial f}{\partial x}\frac{\partial x}{\partial u} + \frac{\partial f}{\partial y}\frac{\partial y}{\partial u},$$

$$\frac{\partial f}{\partial v} = \frac{\partial f}{\partial x}\frac{\partial x}{\partial v} + \frac{\partial f}{\partial y}\frac{\partial y}{\partial v}.$$

Practical hint to remember the rule: The differential of f is

$$df = \frac{\partial f}{\partial x}dx + \frac{\partial f}{\partial y}dy.$$

In order to express $\partial_u f$, replace dx by $\partial x/\partial u$ and dy by $\partial y/\partial u$.

Example 3 Let r and θ be polar coordinates so that $x = r\cos(\theta)$ and $y = r\sin(\theta)$. Show that

$$\frac{\partial f}{\partial r} = \cos(\theta)\frac{\partial f}{\partial x} + \sin(\theta)\frac{\partial f}{\partial y}$$

$$\frac{\partial f}{\partial \theta} = -r\sin(\theta)\frac{\partial f}{\partial x} + r\cos(\theta)\frac{\partial f}{\partial y}$$

Solution

By the chain rule, as in Proposition 3,

$$\frac{\partial f}{\partial r} = \frac{\partial f}{\partial x}\frac{\partial x}{\partial r} + \frac{\partial f}{\partial y}\frac{\partial y}{\partial r} = \frac{\partial f}{\partial x}\left(\frac{\partial}{\partial r}\left(r\cos(\theta)\right)\right) + \frac{\partial f}{\partial y}\left(\frac{\partial}{\partial r}\left(r\sin(\theta)\right)\right)$$

$$= \frac{\partial f}{\partial x}\cos(\theta) + \frac{\partial f}{\partial y}\sin(\theta)$$

$$= \cos(\theta)\frac{\partial f}{\partial x} + \sin(\theta)\frac{\partial f}{\partial y},$$

and

$$\frac{\partial f}{\partial \theta} = \frac{\partial f}{\partial x}\frac{\partial x}{\partial \theta} + \frac{\partial f}{\partial y}\frac{\partial y}{\partial \theta} = \frac{\partial f}{\partial x}\left(\frac{\partial}{\partial \theta}\left(r\cos(\theta)\right)\right) + \frac{\partial f}{\partial y}\left(\frac{\partial}{\partial \theta}\left(r\sin(\theta)\right)\right)$$

$$= \frac{\partial f}{\partial x}\left(-r\sin(\theta)\right) + \frac{\partial f}{\partial y}\left(r\cos(\theta)\right)$$

$$= -r\sin(\theta)\frac{\partial f}{\partial x} + r\cos(\theta)\frac{\partial f}{\partial y}.$$

$\square$

Example 4 Show that

$$\frac{\partial^2 f}{\partial x^2} + \frac{\partial^2 f}{\partial y^2} = \frac{\partial^2 f}{\partial r^2} + \frac{1}{r}\frac{\partial f}{\partial r} + \frac{1}{r^2}\frac{\partial^2 f}{\partial \theta^2}$$

Solution

$$\frac{\partial^2 f}{\partial r^2} = \frac{\partial}{\partial r}\left(\frac{\partial f}{\partial r}\right) = \frac{\partial}{\partial r}\left(\cos(\theta)\frac{\partial f}{\partial x} + \sin(\theta)\frac{\partial f}{\partial y}\right)$$

$$= \cos(\theta)\frac{\partial}{\partial r}\left(\frac{\partial f}{\partial x}\right) + \sin(\theta)\frac{\partial}{\partial r}\left(\frac{\partial f}{\partial y}\right)$$

$$= \cos(\theta)\left(\cos(\theta)\frac{\partial}{\partial x}\left(\frac{\partial f}{\partial x}\right) + \sin(\theta)\frac{\partial}{\partial y}\left(\frac{\partial f}{\partial x}\right)\right)$$

$$+ \sin(\theta)\left(\cos(\theta)\frac{\partial}{\partial x}\left(\frac{\partial f}{\partial y}\right) + \sin(\theta)\frac{\partial}{\partial y}\left(\frac{\partial f}{\partial y}\right)\right)$$

$$= \cos^2(\theta)\frac{\partial^2 f}{\partial x^2} + 2\sin(\theta)\cos(\theta)\frac{\partial^2 f}{\partial y\partial x} + \sin^2(\theta)\frac{\partial^2 f}{\partial y^2}.$$

$$\frac{\partial^2 f}{\partial \theta^2} = \frac{\partial}{\partial \theta}\left(\frac{\partial f}{\partial \theta}\right) = \frac{\partial}{\partial \theta}\left(-r\sin\left(\theta\right)\frac{\partial f}{\partial x} + r\cos\left(\theta\right)\frac{\partial f}{\partial y}\right)$$

$$= -r\cos\left(\theta\right)\frac{\partial f}{\partial x} - r\sin\left(\theta\right)\frac{\partial}{\partial \theta}\left(\frac{\partial f}{\partial x}\right) - r\sin\left(\theta\right)\frac{\partial f}{\partial y} + r\cos\left(\theta\right)\frac{\partial}{\partial \theta}\left(\frac{\partial f}{\partial y}\right)$$

$$= -r\cos\left(\theta\right)\frac{\partial f}{\partial x} - r\sin\left(\theta\right)\left(-r\sin\left(\theta\right)\frac{\partial}{\partial x}\left(\frac{\partial f}{\partial x}\right) + r\cos\left(\theta\right)\frac{\partial}{\partial y}\left(\frac{\partial f}{\partial x}\right)\right)$$

$$- r\sin\left(\theta\right)\frac{\partial f}{\partial y} + r\cos\left(\theta\right)\left(-r\sin\left(\theta\right)\frac{\partial}{\partial x}\left(\frac{\partial f}{\partial y}\right) + r\cos\left(\theta\right)\frac{\partial}{\partial y}\left(\frac{\partial f}{\partial y}\right)\right)$$

$$= -r\cos\left(\theta\right)\frac{\partial f}{\partial x} - r\sin\left(\theta\right)\frac{\partial f}{\partial y} + r^2\sin^2\left(\theta\right)\frac{\partial^2 f}{\partial x^2} + r^2\cos^2\left(\theta\right)\frac{\partial^2 f}{\partial y^2}$$

$$- 2r^2\sin\left(\theta\right)\cos\left(\theta\right)\frac{\partial^2 f}{\partial y \partial x}$$

Therefore,

$$\frac{\partial^2 f}{\partial r^2} + \frac{1}{r}\frac{\partial f}{\partial r} + \frac{1}{r^2}\frac{\partial^2 f}{\partial \theta^2}$$

$$= \cos^2\left(\theta\right)\frac{\partial^2 f}{\partial x^2} + 2\sin\left(\theta\right)\cos\left(\theta\right)\frac{\partial^2 f}{\partial y \partial x} + \sin^2\left(\theta\right)\frac{\partial^2 f}{\partial y^2}$$

$$+ \frac{1}{r}\left(\cos\left(\theta\right)\frac{\partial f}{\partial x} + \sin\left(\theta\right)\frac{\partial f}{\partial y}\right)$$

$$+ \frac{1}{r^2}\left(-r\cos\left(\theta\right)\frac{\partial f}{\partial x} - r\sin\left(\theta\right)\frac{\partial f}{\partial y} + r^2\sin^2\left(\theta\right)\frac{\partial^2 f}{\partial x^2} + r^2\cos^2\left(\theta\right)\frac{\partial^2 f}{\partial y^2} - 2r^2\sin\left(\theta\right)\cos\left(\theta\right)\frac{\partial^2 f}{\partial y \partial x}\right)$$

$$= \cos^2\left(\theta\right)\frac{\partial^2 f}{\partial x^2} + 2\sin\left(\theta\right)\cos\left(\theta\right)\frac{\partial^2 f}{\partial y \partial x} + \sin^2\left(\theta\right)\frac{\partial^2 f}{\partial y^2}$$

$$+ \frac{1}{r}\cos\left(\theta\right)\frac{\partial f}{\partial x} + \frac{1}{r}\sin\left(\theta\right)\frac{\partial f}{\partial y}$$

$$- \frac{1}{r}\cos\left(\theta\right)\frac{\partial f}{\partial x} - \frac{1}{r}\sin\left(\theta\right)\frac{\partial f}{\partial y} + \sin^2\left(\theta\right)\frac{\partial^2 f}{\partial x^2} + \cos^2\left(\theta\right)\frac{\partial^2 f}{\partial y^2} - 2\sin\left(\theta\right)\cos\left(\theta\right)\frac{\partial^2 f}{\partial y \partial x}$$

$$= \frac{\partial^2 f}{\partial x^2} + \frac{\partial^2 f}{\partial y^2}$$

□

In general if f is a function of $x, y, z, \ldots$ and

$$x = x\left(u, v, \ldots\right),$$
$$y = y\left(u, v, \ldots\right),$$
$$z = z\left(u, v, \ldots\right),$$

etc., we have

$$\frac{\partial f}{\partial u} = \frac{\partial f}{\partial x}\frac{\partial x}{\partial u} + \frac{\partial f}{\partial y}\frac{\partial y}{\partial u} + \cdots,$$
$$\frac{\partial f}{\partial v} = \frac{\partial f}{\partial x}\frac{\partial x}{\partial v} + \frac{\partial f}{\partial y}\frac{\partial y}{\partial v} + \cdots,$$

etc..

Implicit Differentiation

Assume that y is defined **implicitly as a function of** x by the equation $F(x,y) = 0$. This means that $F(x, y(x)) = 0$ for each x in an interval J. In Chapter 2 we discussed **implicit differentiation** for the computation of dy/dx. The chain rule for a function of two variables enables us to describe the procedure for implicit differentiation in general terms. Indeed,

$$\frac{d}{dx} F(x, y(x)) = 0$$

for each $x \in J$. By the chain rule,

$$0 = \frac{d}{dx} F(x, y(x)) = \frac{\partial F(x,y)}{\partial x} + \frac{\partial F(x,y)}{\partial y} \frac{dy}{dx}$$

Therefore,

$$\frac{dy}{dx} = -\frac{\dfrac{\partial F(x,y)}{\partial x}}{\dfrac{\partial F(x,y)}{\partial y}}$$

The expression makes sense if

$$\frac{\partial F(x,y)}{\partial y} \neq 0.$$

It can be shown that y can be expressed as a function of x in some open interval J that contains the point x_0 such that $y(x_0) = y_0$ if $F(x_0, y_0) = 0$, $F(x,y)$ has continuous partial derivatives in a open disk containing (x_0, y_0) and we have

$$\left. \frac{\partial F(x,y)}{\partial y} \right|_{(x,y)=(x_0,y_0)} \neq 0$$

Under these conditions, we do have

$$\frac{dy}{dx} = -\frac{\dfrac{\partial F(x,y)}{\partial x}}{\dfrac{\partial F(x,y)}{\partial y}}$$

for each x in an open interval that contains x_0. This is a version of **the implicit function theorem** that is proven in a course on advanced calculus.

Example 5 Assume that y is defined as a function of x implicitly by the equation $x^3 + y^3 = 9xy$. Figure 1 shows the graph of the equation.

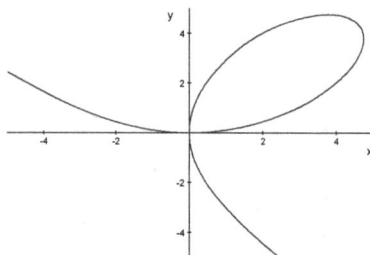

Figure 1

If we set $F(x,y) = x^3 + y^3 - 9xy$, we have $F(x, y(x)) = 0$ for each x in some interval. As in the previous discussion,

$$\frac{dy}{dx} = -\frac{\dfrac{\partial F(x,y)}{\partial x}}{\dfrac{\partial F(x,y)}{\partial y}} = -\frac{3x^2 - 9y}{3y^2 - 9x} = \frac{9y - 3x^2}{3y^2 - 9x}.$$

We have $F(2,4) = 0$, and

$$\frac{\partial F(x,y)}{\partial y}\bigg|_{(x,y)=(2,4)} = 3y^2 - 9x\big|_{(x,y)=(2,4)} = 30 \neq 0.$$

Therefore, the equation $F(x,y) = 0$ defines y as a function of x such that $y(2) = 4$. We have

$$\frac{dy}{dx}\bigg|_{x=2} = \frac{9y - 3x^2}{3y^2 - 9x}\bigg|_{x=2, y=4} = \frac{4}{5}.$$

□

Under certain conditions, an equation of the form $F(x,y,z) = 0$ defines z implicitly as a function of x and y. If $z(x,y)$ is such a function that is differentiable, we can apply the chain rule in order to express its partial derivatives. Indeed,

$$0 = \frac{\partial}{\partial x} F(x, y, z(x,y)) = \frac{\partial F}{\partial x} + \frac{\partial F}{\partial z} \frac{\partial z}{\partial x}$$

and

$$0 = \frac{\partial}{\partial y} F(x, y, z(x,y)) = \frac{\partial F}{\partial y} + \frac{\partial F}{\partial z} \frac{\partial z}{\partial y},$$

where F_x, F_y and F_z are evaluated at $(x, y, z(x,y))$. Therefore,

$$\frac{\partial z}{\partial x} = -\frac{\dfrac{\partial F}{\partial x}}{\dfrac{\partial F}{\partial z}} \quad \text{and} \quad \frac{\partial z}{\partial y} = -\frac{\dfrac{\partial F}{\partial y}}{\dfrac{\partial F}{\partial z}},$$

provided that $F_z \neq 0$ at the relevant point $(x, y, z(x,y))$. A version of **the implicit function theorem** says that the above expressions make sense for each (x,y) in some open disk centered at (x_0, y_0) if $F(x_0, y_0, z_0) = 0$, $F_z(x_0, y_0, z_0) \neq 0$ and $z(x_0, y_0) = z_0$.

Example 6 Assume that $z(x,y)$ is defined implicitly by the equation

$$z^2 - x^2 - 2y^2 - 3 = 0,$$

and $z(2,1) = 3$. Figure 2 shows the graph of the above equation.

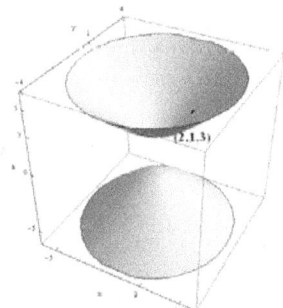

Figure 2

a) Determine $z_x(x, y)$ and $z_y(x, y)$.

b) Evaluate $z_x(2, 1)$ and $z_y(2, 1)$.

Solution

a) If we set $F(x, y, z) = z^2 - x^2 - 2y^2 - 3$, we have $F(x, y, z(x, y)) = 0$. As in the previous discussion,

$$\frac{\partial z}{\partial x} = -\frac{\dfrac{\partial F}{\partial x}}{\dfrac{\partial F}{\partial z}} = \frac{2x}{2z} = -\frac{x}{z}$$

and

$$\frac{\partial z}{\partial y} = -\frac{\dfrac{\partial F}{\partial y}}{\dfrac{\partial F}{\partial z}} = \frac{4y}{2z} = \frac{2y}{z}.$$

(Exercise: Obtain these expressions directly).

b)

$$\frac{\partial z}{\partial x}(2, 1) = -\frac{2}{3} \text{ and } \frac{\partial z}{\partial y}(2, 1) = \frac{2}{3}.$$

Note that

$$z = \sqrt{x^2 + 2y^2 + 3},$$

so that

$$\frac{\partial z}{\partial x} = \frac{1}{2\sqrt{x^2 + 2y^2 + 3}}(2x) = \frac{x}{\sqrt{x^2 + 2y^2 + 3}} = \frac{x}{z}$$

and

$$\frac{\partial z}{\partial y} = \frac{1}{2\sqrt{x^2 + 2y^2 + 3}}(4y) = \frac{2y}{z}$$

as before. $\square$

Problems

In problems 1 and 2, compute

$$\frac{d}{dt}f(x(t), y(t))$$

a) By making use of the chain rule,

b) Directly, from the expression for $f(x(t), y(t))$.

1.

$$f(x, y) = \sqrt{x^2 + y^2}, \ x(t) = \sin(t), \ y(t) = 2\cos(t).$$

2.

$$f(x, y) = e^{-x^2 + y^2}, \ x(t) = 2t - 1, \ y(t) = t + 1$$

In problems 3 and 4, compute

$$\frac{\partial}{\partial u}f(x(u, v)) \text{ and } \frac{\partial}{\partial v}f(x(u, v))$$

a) By making use of the chain rule,

b) Directly, from the expression for $f(x(u, v))$:

3.

$$f(x) = \ln(x), \ x = \frac{1}{\sqrt{u^2 + v^2}}$$

4.

$$f(x) = \sin(x^2), \ x = u - 4v$$

In problems 5 and 6, compute

$$\frac{\partial}{\partial u} f(x(u,v), y(u,v)) \ \text{and} \ \frac{\partial}{\partial v} f(x(u,v), y(u,v))$$

a) By making use of the chain rule,
b) Directly, from the expression for $f(x(u,v), y(u,v))$:

5.

$$f(x,y) = \arcsin\left(\frac{y}{\sqrt{x^2 + y^2}}\right), \ x = u\cos(v), \ y = u\sin(v)$$

(assume that $-\pi/2 < v < \pi/2$).

6.

$$f(x,y) = \arctan\left(\frac{y}{x}\right), \ x = u + 2v, \ y = u - 2v$$

7. Let $z = f(x,y)$ and let r and θ be polar coordinates so that $x = r\cos(\theta)$ and $y = r\sin(\theta)$. Show that

$$\left(\frac{\partial z}{\partial x}\right)^2 + \left(\frac{\partial z}{\partial y}\right)^2 = \left(\frac{\partial z}{\partial r}\right)^2 + \frac{1}{r^2}\left(\frac{\partial z}{\partial \theta}\right)^2$$

8. Let $z = f(x,y)$, $x = e^t\cos(\theta)$ and $y = e^t\sin(\theta)$. Show that

$$\left(\frac{\partial z}{\partial x}\right)^2 + \left(\frac{\partial z}{\partial y}\right)^2 = e^{-2t}\left[\left(\frac{\partial z}{\partial t}\right)^2 + \left(\frac{\partial z}{\partial \theta}\right)^2\right]$$

9.
a) Let $u(x,t) = f(x + at)$, where f is a differentiable function of a single variable and a is a constant. Show that u is a solution of the wave equation

$$\frac{\partial^2 u}{\partial t^2} = a^2 \frac{\partial^2 u}{\partial x^2}.$$

b) Determine $u(x,0)$, $u(x,2)$, $u(4)$ if $f(w) = \sin(w)$ and $a = \pi/4$.
c) [C] if $f(w) = \sin(w)$ plot $u(x,0)$, $u(x,2)$, $u(x,4)$, where $-4\pi \le x \le 4\pi$.

10. Assume that $z(x,y)$ is defined implicitly by the equation

$$z^2 - x^2 - y^2 - 4 = 0,$$

and $z(1,2) = -3$.

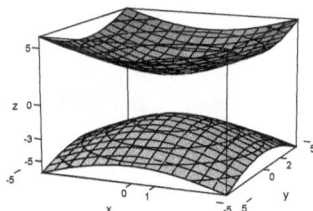

a) Determine $z_x(x, y)$ and $z_y(x, y)$ directly (i.e., without relying on a formula from the text).
b) Evaluate $z_x(1, 2)$ and $z_y(1, 2)$ and determine an equation for the plane that is tangent to the graph of the given equation at $(1, 2, -3)$.

11. Assume that $z(x, y)$ is defined implicitly by the equation

$$z^2 + x^2 - y^2 - 9 = 0,$$

and $z(3, 6) = 6$.

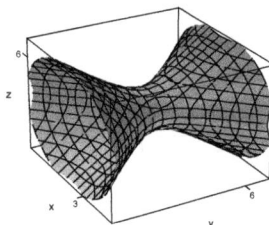

a) Determine $z_x(x, y)$ and $z_y(x, y)$ directly (i.e., without relying on a formula from the text).
b) Evaluate $z_x(3, 6)$ and $z_y(3, 6)$. and determine an equation for the plane that is tangent to the graph of the given equation at $(3, 6, 6)$.

12.7   Directional Derivatives and the Gradient

Directional Derivatives

We will begin by considering functions of two variables. Assume that f has partial derivatives in some open disk centered at (x, y). The partial derivative $\partial_x f(x, y)$ can be interpreted as the rate of change of f in the direction of the standard basis vector $\mathbf{i}$ and the partial derivative $\partial_y f(x, y)$ can be interpreted as the rate of change of f in the direction of the standard basis vector $\mathbf{j}$. We would like to arrive at a reasonable definition of the rate of f at (x, y) in the direction of an arbitrary unit vector $\mathbf{u} = u_1 \mathbf{i} + u_2 \mathbf{j}$.

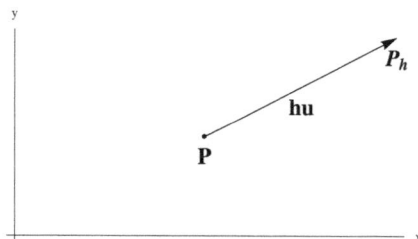

Figure 1

Let $h \neq 0$. Let's assume that $\overrightarrow{OP_h} = \overrightarrow{OP} + h\mathbf{u}$ so that

$$P_h = (x + hu_1, y + hu_2).$$

We define **the average rate of change of** f corresponding to the displacement $\overrightarrow{PP_h}$ as

$$\frac{f(P_h) - f(P)}{h} = \frac{f(x + hu_1, y + hu_2) - f(x, y)}{h}$$

It is reasonable to define **the rate of change of** f **at** P **in the direction of** u as

$$\lim_{h \to 0} \frac{f(P_h) - f(P)}{h}.$$

Another name for this limit is "directional derivative":

Definition 1 The directional derivative of f **at** (x, y) **in the direction of the unit vector u is**

$$D_{\mathbf{u}}f(x, y) = \lim_{h \to 0} \frac{f(x + hu_1, y + hu_2) - f(x, y)}{h}.$$

Note that $D_{\mathbf{i}}f(x, y)$ is $\partial_x f(x, y)$ and $D_{\mathbf{j}}f(x, y)$ is $\partial_y f(x, y)$. We can calculate the directional derivative in an arbitrary direction in terms of $\partial_x f(x, y)$ and $\partial_y f(x, y)$:

Proposition 1 Assume that f **is differentiable at** (x, y) **and that** $u = u_1\mathbf{i} + u_2\mathbf{j}$ **is a unit vector. Then**

$$D_{\mathbf{u}}f(x, y) = \frac{\partial f}{\partial x}(x, y) u_1 + \frac{\partial f}{\partial y}(x, y) u_2.$$

Proof

Set

$$g(t) = f(x + tu_1, y + tu_2), \ t \in \mathbb{R}.$$

Then,

$$g'(0) = \lim_{h \to 0} \frac{g(h) - g(0)}{h} = \lim_{h \to 0} \frac{f(x + hu_1, y + hu_2) - f(x, y)}{h} = D_{\mathbf{u}}f(x, y)$$

Let's set

$$v(x, t) = x + tu_1 \text{ and } w(x, t) = y + tu_2,$$

so that $g(t) = f(v(x, t), w(x, t))$. By the chain rule,

$$\frac{d}{dt}g(t) = \frac{d}{dt}f(x + tu_1, y + tu_2)$$

$$= \frac{\partial f}{\partial v}(v, w)\frac{dv}{dt} + \frac{\partial f}{\partial w}(v, w)\frac{dw}{dt}$$

$$= \frac{\partial f}{\partial v}(x + tu_1, y + tu_2)\frac{d}{dt}(x + tu_1) + \frac{\partial f}{\partial w}(x + tu_1, y + tu_2)\frac{d}{dt}(y + tu_2)$$

$$= \frac{\partial f}{\partial v}(x + tu_1, y + tu_2)u_1 + \frac{\partial f}{\partial w}(x + tu_1, y + tu_2)u_2.$$

Therefore,

$$D_{\mathbf{u}}f(x, y) = g'(0) = \frac{\partial f}{\partial v}(x + tu_1, y + tu_2)u_1 + \frac{\partial f}{\partial w}(x + tu_1, y + tu_2)u_2 \Big|_{t=0}$$

$$= \frac{\partial f}{\partial v}(x, y)u_1 + \frac{\partial f}{\partial w}(x, y)u_2$$

$$= \frac{\partial f}{\partial x}(x, y)u_1 + \frac{\partial f}{\partial y}(x, y)u_2,$$

as claimed. ∎

Example 1 Let $f(x, y) = 36 - 4x^2 - y^2$ and let **u** be the unit vector in the direction of $\mathbf{v} = -\mathbf{i} + 4\mathbf{j}$. Determine $D_{\mathbf{u}}f(1, 3)$.

Solution

We have

$$\frac{\partial f}{\partial x}(x,y) = \frac{\partial}{\partial x}\left(36 - 4x^2 - y^2\right) = -8x \text{ and } \frac{\partial f}{\partial y}(x,y) = \frac{\partial}{\partial y}\left(36 - 4x^2 - y^2\right) = -2y.$$

Therefore,

$$\frac{\partial f}{\partial x}(1,3) = -8 \text{ and } \frac{\partial f}{\partial y}(1,3) = -6.$$

We have

$$\|\mathbf{v}\| = \sqrt{1 + 4^2} = \sqrt{17}.$$

Therefore,

$$\mathbf{u} = \frac{\mathbf{v}}{\|\mathbf{v}\|} = \frac{1}{\sqrt{17}}(-\mathbf{i} + 4\mathbf{j}) = -\frac{1}{\sqrt{17}}\mathbf{i} + \frac{4}{\sqrt{17}}\mathbf{j}.$$

Thus,

$$D_{\mathbf{u}}f(1,3) = \frac{\partial f}{\partial x}(1,3)\,u_1 + \frac{\partial f}{\partial y}(1,3)\,u_2$$

$$= (-8)\left(-\frac{1}{\sqrt{17}}\right) + (-6)\left(\frac{4}{\sqrt{17}}\right) = -\frac{16}{\sqrt{17}}.$$

$\square$

Remark 1 All directional derivatives may exist, but f may not be differentiable, as in the following example:

Let

$$f(x,y) = \begin{cases} \dfrac{xy^2}{x^2 + y^4} & \text{if} \quad (x,y) \neq (0,0), \\ 0 & \text{if} \quad (x,y) = (0,0) \end{cases}$$

Figure 2 shows the graph of f.

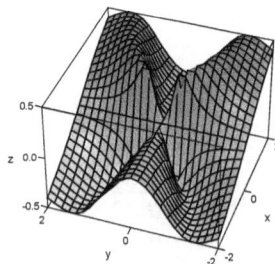

Figure 2

In Section 12.5 we saw that f is not continuous at $(0,0)$ so that it is not differentiable. Nevertheless, the directional derivatives of f exist in all directions at the origin. Indeed, let $\mathbf{u}$ be an arbitrary unit vector, so that $\mathbf{u} = (\cos(\theta), \sin(\theta))$ for some θ. Let's calculate $D_{\mathbf{u}}f(0,0)$, directly from the definition. We have

$$\frac{f(h\cos(\theta), h\sin(\theta))}{h} = \frac{\dfrac{h^3\cos(\theta)\sin^2(\theta)}{h^2\cos^2(\theta) + h^4\sin^4(\theta)}}{h} = \frac{h^3\cos(\theta)\sin^2(\theta)}{h^3\cos^2(\theta) + h^5\sin^4(\theta)}$$

$$= \frac{\cos(\theta)\sin^2(\theta)}{\cos^2(\theta) + h^2\sin^4(\theta)}.$$

Therefore,

$$D_{\mathbf{u}}f(0,0) = \lim_{h\to 0}\frac{f(h\cos(\theta), h\sin(\theta))}{h} = \frac{\cos(\theta)\sin^2(\theta)}{\cos^2(\theta)} = \frac{\sin^2(\theta)}{\cos(\theta)}$$

if $\cos(\theta) \neq 0$. The case $\cos(\theta) = 0$ corresponds to $f_y(0,0)$ or $-f_y(0,0)$. We have

$$f_y(0,0) = \lim_{h\to 0}\frac{f(0,h) - f(0,0)}{h} = \lim_{h\to 0}\frac{0}{h} = 0.$$

Thus, the directional derivatives of f exist in all directions. $\Diamond$

The gradient of a function is a useful notion within the context of directional derivatives, and in many other contexts:

Definition 2 The gradient of f at (x,y) is the vector-valued function that assigns the vector

$$\boldsymbol{\nabla} f(x,y) = \frac{\partial f}{\partial x}(x,y)\,\mathbf{i} + \frac{\partial f}{\partial y}(x,y)\,\mathbf{j}$$

to each (x,y) where the partial derivatives of f exist (read $\boldsymbol{\nabla} f$ as "**del** f").

We can express a directional derivative in terms of the gradient:

Proposition 2 The directional derivative of f at (x,y) in the direction of the unit vector u can be expressed as

$$D_{\mathbf{u}}f(x,y) = \boldsymbol{\nabla} f(x,y) \cdot \mathbf{u}$$

Proof

By Proposition 1,

$$D_{\mathbf{u}}f(x,y) = \frac{\partial f}{\partial x}(x,y)\,u_1 + \frac{\partial f}{\partial y}(x,y)\,u_2$$

By the definition of $\boldsymbol{\nabla} f(x,y)$, the right-hand side can be expressed as a dot product. Therefore,

$$D_{\mathbf{u}}f(x,y) = \left(\frac{\partial f}{\partial x}(x,y)\,\mathbf{i} + \frac{\partial f}{\partial y}(x,y)\,\mathbf{j}\right) \cdot (u_1\mathbf{i} + u_2\mathbf{j})$$

$$= \boldsymbol{\nabla} f(x,y) \cdot \mathbf{u}$$

∎

Thus, **the directional derivative of f at (x,y) in the direction of the unit vector u is the component of $\boldsymbol{\nabla} f(x,y)$ along u.**

Example 2 Let $f(x,y) = x^2 + 4xy + y^2$.

a) Determine the gradient of f.

b) Calculate the directional derivative of f at $(2,1)$ in the directions of the vectors $\mathbf{i} + \mathbf{j}$ and $\mathbf{i} - \mathbf{j}$.

Solution

a) Figure 3 displays the graph of f.

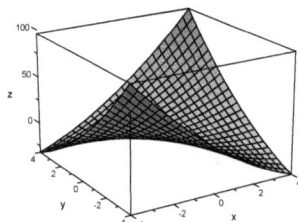

Figure 3

We have

$$\frac{\partial f}{\partial x}(x,y) = \frac{\partial}{\partial x}\left(x^2 + 4xy + y^2\right) = 2x + 4y$$

and

$$\frac{\partial f}{\partial y}(x,y) = \frac{\partial}{\partial y}\left(x^2 + 4xy + y^2\right) = 4x + 2y.$$

Therefore,

$$\boldsymbol{\nabla} f(x,y) = \frac{\partial f}{\partial x}(x,y)\,\mathbf{i} + \frac{\partial f}{\partial y}(x,y)\,\mathbf{j} = (2x + 4y)\,\mathbf{i} + (4x + 2y)\,\mathbf{j}.$$

Figure 4 displays $\boldsymbol{\nabla} f(x,y)$ at certain points (x,y) as arrows located at these points.

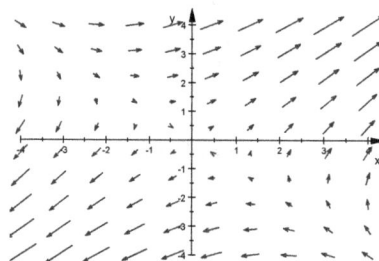

Figure 4

b) A unit vector in the direction of $\mathbf{i} + \mathbf{j}$ is

$$u_1 = \frac{1}{\sqrt{2}}\mathbf{i} + \frac{1}{\sqrt{2}}\mathbf{j},$$

and a unit vector in the direction of $\mathbf{i} - \mathbf{j}$ is

$$u_2 = \frac{1}{\sqrt{2}}\mathbf{i} - \frac{1}{\sqrt{2}}\mathbf{j}.$$

We have

$$\boldsymbol{\nabla} f(2,1) = (2x + 4y)\,\mathbf{i} + (4x + 2y)\,\mathbf{j}\big|_{x=2, y=1} = 8\mathbf{i} + 10\mathbf{j}.$$

Therefore,

$$D_{\mathbf{u}_1} f(2,1) = \nabla f(2,1) \cdot \mathbf{u}_1$$
$$= (8\mathbf{i} + 10\mathbf{j}) \cdot \left(\frac{1}{\sqrt{2}}\mathbf{i} + \frac{1}{\sqrt{2}}\mathbf{j}\right)$$
$$= \frac{8}{\sqrt{2}} + \frac{10}{\sqrt{2}} = \frac{18}{\sqrt{2}} = 9\sqrt{2},$$

and

$$D_{\mathbf{u}_2} f(2,1) = \nabla f(2,1) \cdot \mathbf{u}_2$$
$$= (8\mathbf{i} + 10\mathbf{j}) \cdot \left(\frac{1}{\sqrt{2}}\mathbf{i} - \frac{1}{\sqrt{2}}\mathbf{j}\right)$$
$$= \frac{8}{\sqrt{2}} - \frac{10}{\sqrt{2}} = -\frac{2}{\sqrt{2}} = -\sqrt{2}.$$

Thus, at $(2,1)$ the function increases in the direction of $\mathbf{u}_1$ and decreases in the direction of $\mathbf{u}_2$. Figure 5 shows some level curves of f, the vectors $\mathbf{u}_1$ and $\mathbf{u}_2$, and a unit vector along $\nabla f(2,1)$. $\square$

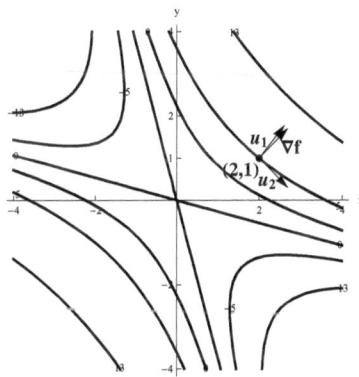

Figure 5

Proposition 3 The maximum value of the directional derivatives of a function f at a point (x,y) is $||\nabla f(x,y)||$, the length of the gradient of f at (x,y), and it is attained in the direction of $\nabla f(x,y)$. The minimum value of the directional derivatives of f at (x,y) is $-||\nabla f(x,y)||$ and it is attained the direction of $-\nabla f(x,y)$.

Proof

Since $D_{\mathbf{u}} f(x,y) = \nabla f(x,y) \cdot \mathbf{u}$, and $||\mathbf{u}|| = 1$, we have

$$D_{\mathbf{u}} f(x,y) = ||\nabla f(x,y)|| \, ||\mathbf{u}|| \cos(\theta) = ||\nabla f(x,y)|| \cos(\theta),$$

where θ is the angle between $\nabla f(x,y)$ and the unit vector u. Since $\cos(\theta) \leq 1$,

$$D_{\mathbf{u}} f(x,y) \leq ||\nabla f(x,y)||.$$

The maximum value is attained when $\cos(\theta) = 1$, i.e., $\theta = 0$. This means that $\mathbf{u}$ is the unit vector in the direction of $\nabla f(x,y)$.

Similarly,

$$D_{\mathbf{u}}f(x,y) = \|\boldsymbol{\nabla}f(x,y)\|\cos(\theta)$$

attains its minimum value when $\cos(\theta) = -1$, i.e., $\theta = \pi$. This means that $u = -\boldsymbol{\nabla}f(x,y)$. The corresponding directional derivative is $-\|\boldsymbol{\nabla}f(x,y)\|$. $\blacksquare$

Example 3 Let $f(x,y) = x^2 + 4xy + y^2$, as in Example 2. Determine the unit vector along which the rate of increase of f at $(2,1)$ is maximized and and the unit vector along which the rate of decrease of f at $(2,1)$ is maximized. Evaluate the corresponding rates of change of f.

Solution

In Example 2 we showed that $\boldsymbol{\nabla}f(2,1) = 8\mathbf{i} + 10\mathbf{j}$. Therefore,

$$\|\boldsymbol{\nabla}f(2,1)\| = \|(8\mathbf{i}+10\mathbf{j})\| = \sqrt{64+100} = \sqrt{164} = 2\sqrt{41}.$$

Thus, the value of f increases at the maximum rate of $2\sqrt{41}$ in the direction of the unit vector

$$\frac{1}{\|\boldsymbol{\nabla}f(2,1)\|}\boldsymbol{\nabla}f(2,1) = \frac{1}{2\sqrt{41}}(8\mathbf{i}+10\mathbf{j}) = \frac{4}{\sqrt{41}}\mathbf{i} + \frac{5}{41}\mathbf{j}.$$

The value of the function decreases at the maximum rate of $2\sqrt{41}$ (i.e., the rate of change is $-2\sqrt{41}$) in the direction of the unit vector

$$-\frac{1}{\|\boldsymbol{\nabla}f(2,1)\|}\boldsymbol{\nabla}f(2,1) = -\frac{4}{\sqrt{41}}\mathbf{i} - \frac{5}{41}\mathbf{j}.$$

$\square$

The Chain Rule and the Gradient

A special case of the chain rule says that

$$\frac{d}{dt}f(x(t),y(t)) = \frac{\partial f}{\partial x}(x(t),y(t))\frac{dx}{dt} + \frac{\partial f}{\partial y}(x(t),y(t))\frac{dy}{dt},$$

as In Proposition 1 of Section 12.6. The notion of the gradient enables us to provide a useful geometric interpretation of the above expression. Let's begin by expressing the chain rule as follows:

Proposition 4 Assume that the curve C in the plane is parametrized by the function $\boldsymbol{\sigma}$ where

$$\boldsymbol{\sigma}(t) = (x(t),y(t)).$$

Then

$$\frac{d}{dt}f(\boldsymbol{\sigma}(t)) = \boldsymbol{\nabla}f(\boldsymbol{\sigma}(t)) \cdot \frac{d\boldsymbol{\sigma}}{dt}$$

Proof

We have

$$\frac{d}{dt}f(\boldsymbol{\sigma}(t)) = \frac{d}{dt}f(x(t),y(t)),$$

$$\boldsymbol{\nabla}f(\boldsymbol{\sigma}(t)) = \frac{\partial f}{\partial x}(x(t),y(t))\mathbf{i} + \frac{\partial f}{\partial y}(x(t),y(t)),$$

and

$$\frac{d\boldsymbol{\sigma}}{dt} = \frac{dx}{dt}\mathbf{i} + \frac{dy}{dt}\mathbf{j}.$$

Therefore,

$$\boldsymbol{\nabla} f\left(\boldsymbol{\sigma}\left(t\right)\right) \cdot \frac{d\boldsymbol{\sigma}}{dt} = \left(\frac{\partial f}{\partial x}\left(x\left(t\right), y\left(t\right)\right)\mathbf{i} + \frac{\partial f}{\partial y}\left(x\left(t\right), y\left(t\right)\right)\right) \cdot \left(\frac{dx}{dt}\mathbf{i} + \frac{dy}{dt}\mathbf{j}\right)$$
$$= \frac{\partial f}{\partial x}\left(x\left(t\right), y\left(t\right)\right)\frac{dx}{dt} + \frac{\partial f}{\partial y}\left(x\left(t\right), y\left(t\right)\right)\frac{dy}{dt}.$$

By the chain rule,

$$\frac{\partial f}{\partial x}\left(x\left(t\right), y\left(t\right)\right)\frac{dx}{dt} + \frac{\partial f}{\partial y}\left(x\left(t\right), y\left(t\right)\right)\frac{dy}{dt} = \frac{d}{dt}f\left(x\left(t\right), y\left(t\right)\right) = \frac{d}{dt}f\left(\boldsymbol{\sigma}\left(t\right)\right).$$

Thus,

$$\frac{d}{dt}f\left(\boldsymbol{\sigma}\left(t\right)\right) = \boldsymbol{\nabla} f\left(\boldsymbol{\sigma}\left(t\right)\right) \cdot \frac{d\boldsymbol{\sigma}}{dt},$$

as claimed. ■

Remark 2 With the notation of Proposition 4, assuming that $\boldsymbol{\sigma}'\left(t\right) \neq \mathbf{0}$,

$$\frac{d}{dt}f\left(\boldsymbol{\sigma}\left(t\right)\right) = \boldsymbol{\nabla} f\left(\boldsymbol{\sigma}\left(t\right)\right) \cdot \frac{d\boldsymbol{\sigma}}{dt} = \boldsymbol{\nabla} f\left(\boldsymbol{\sigma}\left(t\right)\right) \cdot \frac{\frac{d\boldsymbol{\sigma}}{dt}\left(t\right)}{\left\|\frac{d\boldsymbol{\sigma}}{dt}\left(t\right)\right\|} \left\|\frac{d\boldsymbol{\sigma}}{dt}\left(t\right)\right\|$$
$$= \left\|\frac{d\boldsymbol{\sigma}}{dt}\left(t\right)\right\| \boldsymbol{\nabla} f\left(\boldsymbol{\sigma}\left(t\right)\right) \cdot \mathbf{T}\left(t\right),$$

where $\mathbf{T}\left(t\right)$ is the unit tangent to C at $\boldsymbol{\sigma}\left(t\right)$. The quantity $\boldsymbol{\nabla} f\left(\boldsymbol{\sigma}\left(t\right)\right) \cdot \mathbf{T}\left(t\right)$ is $D_{\mathbf{T}(t)}f\left(\boldsymbol{\sigma}\left(t\right)\right)$, the directional derivative of f at $\boldsymbol{\sigma}\left(t\right)$ in the direction of the unit tangent $\mathbf{T}\left(t\right)$, i.e., the component of $\boldsymbol{\nabla} f\left(\boldsymbol{\sigma}\left(t\right)\right)$ along $\mathbf{T}\left(t\right)$. Thus,

$$\frac{d}{dt}f\left(\boldsymbol{\sigma}\left(t\right)\right) = \left\|\frac{d\boldsymbol{\sigma}}{dt}\left(t\right)\right\| D_{\mathbf{T}(t)}f\left(\boldsymbol{\sigma}\left(t\right)\right).$$

If $\boldsymbol{\sigma}\left(t\right)$ is the position of a particle at time t, $\left\|\boldsymbol{\sigma}'\left(t\right)\right\|$ is its speed at t. In this case,

$$\frac{d}{dt}f\left(\boldsymbol{\sigma}\left(t\right)\right) = \text{the speed of the object at } t \times \text{component of } \boldsymbol{\nabla} f\left(\boldsymbol{\sigma}\left(t\right)\right) \text{ along } \mathbf{T}\left(t\right) .$$

$\Diamond$

Proposition 5 Let C be the level curve of f that passes through the point (x_0, y_0). The gradient of f at (x_0, y_0) is perpendicular to C at (x_0, y_0), in the sense that $\boldsymbol{\nabla} f$ is orthogonal to the tangent line to C at (x_0, y_0).

A Plausibility Argument for Proposition 5

We will assume that a segment of C near (x_0, y_0) can be parametrized by the function $\boldsymbol{\sigma} : J \to \mathbb{R}^2$ such that $\boldsymbol{\sigma}\left(t_0\right) = (x_0, y_0)$ and $\boldsymbol{\sigma}'\left(t_0\right) \neq \mathbf{0}$. By Proposition 4,

$$\frac{d}{dt}f\left(\boldsymbol{\sigma}\left(t\right)\right) = \boldsymbol{\nabla} f\left(\boldsymbol{\sigma}\left(t\right)\right) \cdot \frac{d\boldsymbol{\sigma}}{dt}\left(t\right).$$

Since C is a level curve of f, the value of f on C is a constant. Therefore,

$$\frac{d}{dt} f\left(\boldsymbol{\sigma}\left(t\right)\right) = 0$$

for each $t \in J$. In particular,

$$0 = \frac{d}{dt} f\left(\boldsymbol{\sigma}\left(t_0\right)\right) = \boldsymbol{\nabla} f\left(\boldsymbol{\sigma}\left(t_0\right)\right) \cdot \frac{d\boldsymbol{\sigma}}{dt}\left(t_0\right) = \boldsymbol{\nabla} f\left(x_0, y_0\right) \cdot \frac{d\boldsymbol{\sigma}}{dt}\left(t_0\right).$$

This implies that $\boldsymbol{\nabla} f\left(x_0, y_0\right)$ is orthogonal to the vector $\boldsymbol{\sigma}'\left(t_0\right)$ that is tangent to C at $\boldsymbol{\sigma}\left(t_0\right) = \left(x_0, y_0\right)$. ■

Example 4

a) The point $(1, 3)$ is on the curve C that is the graph of the equation $x^2 + 4xy + y^2 = 22$. Determine a vector that is orthogonal to C at $(1, 3)$.
b) Determine the line that is tangent to the C at $(1, 3)$.

Solution

a(We set $f(x, y) = x^2 + 4xy + y^2$, as in Example 2, so that C is a level curve of f. We have

$$\boldsymbol{\nabla} f\left(1, 3\right) = \boldsymbol{\nabla} f\left(1, 3\right) = \left(2x + 4y\right)\mathbf{i} + \left(4x + 2y\right)\mathbf{j}\big|_{x=1, y=3} = 14\mathbf{i} + 10\mathbf{j}.$$

Therefore, $14\mathbf{i} + 10\mathbf{j}$. is orthogonal to C at $(1, 3)$. Figure 6 shows C and a vector along $\nabla f\left(1, 3\right)$. The picture is consistent with our claim.

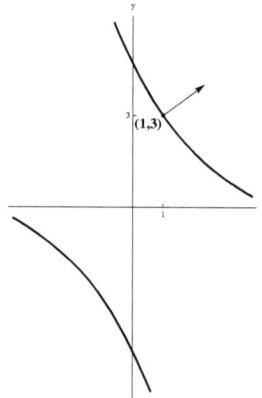

Figure 6

b) If $P_0 = (1, 3)$ and $P = (x, y)$ is an arbitrary point on the line that is tangent to C at P_0, we have

$$\boldsymbol{\nabla} f\left(1, 3\right) \cdot \overrightarrow{P_0 P} = 0.$$

Therefore,

$$\left(14\mathbf{i} + 10\mathbf{j}\right) \cdot \left(\left(x - 1\right)\mathbf{i} + \left(y - 3\right)\right) = 0$$

$\Leftrightarrow$

$$14\left(x - 1\right) + 10\left(y - 3\right) = 0.$$

Figure 7 shows the tangent line.

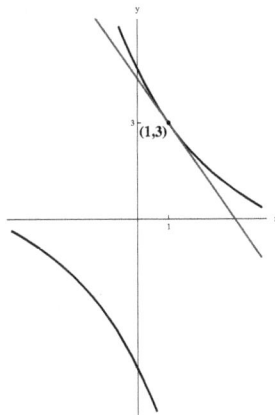

Figure 7

Functions of three or more variables

Our discussion of directional derivatives and the gradient extend to functions of more than two variables in a straightforward fashion. Let's consider functions of three variables to be specific. Thus, assume that f is a scalar function of the variables x, y and z. If $u = u_1 \mathbf{i} + \mathbf{u_2 j} + u_3 \mathbf{k}$ is a three-dimensional unit vector, **the directional derivative of f at (x_0, y_0, z_0) in the direction of u** is

$$D_{\mathbf{u}} f(x_0, y_0, z_0) = \lim_{h \to 0} \frac{f(x_0 + hu_1, y_0 + hu_2, z_0 + hu_3) - f(x_0, y_0, z_0)}{h}$$

The gradient of f at (x, y, z) is

$$\boldsymbol{\nabla} f(x, y, z) = \frac{\partial f}{\partial x}(x, y, z)\,\mathbf{i} + \frac{\partial f}{\partial y}(x, y, z)\,\mathbf{j} + \frac{\partial f}{\partial z}(x, y, z)\,\mathbf{k}.$$

If f is differentiable at (x_0, y_0, z_0),

$$D_{\mathbf{u}} f(x_0, y_0, z_0) = \frac{\partial f}{\partial x}(x_0, y_0, z_0)\,u_1 + \frac{\partial f}{\partial y}(x_0, y_0, z_0)\,u_2 + \frac{\partial f}{\partial z}(x_0, y_0, z_0)\,u_3$$

$$= \boldsymbol{\nabla} f(x_0, y_0; z_0) \cdot \mathbf{u}$$

We have

$$-||\boldsymbol{\nabla} f(x_0, y_0, z_0)|| \leq D_{\mathbf{u}} f(x_0, y_0, z_0) \leq ||\boldsymbol{\nabla} f(x_0, y_0, z_0)||,$$

Thus, the rate of change of f at (x_0, y_0, z_0) has the maximum value $||\boldsymbol{\nabla} f(x_0, y_0, z_0)||$ in the direction of $\boldsymbol{\nabla} f(x_0, y_0, z_0)$. The function decreases at the maximum rate of $||\boldsymbol{\nabla} f(x_0, y_0, z_0)||$ (i.e., the rate of change is $-||\boldsymbol{\nabla} f(x_0, y_0, z_0)||$) at the point (x_0, y_0, z_0) in the direction of $-\boldsymbol{\nabla} f(x_0, y_0, z_0)$. If $\boldsymbol{\sigma} : J \to \mathbb{R}^3$, and $\boldsymbol{\sigma}(t) = (x(t), y(t), z(t))$, by the chain rule,

$$\frac{d}{dt} f(x(t), y(t), z(t))$$
$$= \frac{\partial f}{\partial x}(x(t), y(t), z(t))\frac{dx}{dt}(t) + \frac{\partial f}{\partial y}(x(t), y(t), z(t))\frac{dy}{dt}(t) + \frac{\partial f}{\partial z}(x(t), y(t), z(t))\frac{dz}{dt}(t).$$

so that.

$$\frac{d}{dt} f(\boldsymbol{\sigma}(t)) = \boldsymbol{\nabla} f(\boldsymbol{\sigma}(t)) \cdot \frac{d\boldsymbol{\sigma}}{dt}.$$

If S is **a level surface** of $f = f(x, y, z)$ so that f has a constant value on S, and (x_0, y_0, z_0) lies on S, then $\nabla f(x_0, y_0, z_0)$ is orthogonal to S at (x_0, y_0, z_0), in the sense that $\nabla f(x_0, y_0, z_0)$ is orthogonal at (x_0, y_0, z_0) to any curve on S that passes through (x_0, y_0, z_0). Indeed, if such a curve C is parametrized by $\boldsymbol{\sigma} : J \to \mathbb{R}^3$ and $\boldsymbol{\sigma}(t_0) = (x_0, y_0, z_0)$ we have

$$\frac{d}{dt} f(\boldsymbol{\sigma}(t)) = 0 \text{ for } t \in J,$$

since $f(\boldsymbol{\sigma}(t))$ has a constant value. Therefore,

$$\nabla f(\boldsymbol{\sigma}(t)) \cdot \frac{d\boldsymbol{\sigma}}{dt}(t) = \frac{d}{dt} f(\boldsymbol{\sigma}(t)) = 0.$$

In particular,

$$\nabla f(x_0, y_0, z_0) \cdot \frac{d\boldsymbol{\sigma}}{dt}(t_0) = \nabla f(\boldsymbol{\sigma}(t_0)) \cdot \frac{d\boldsymbol{\sigma}}{dt}(t_0) = 0,$$

so that $\nabla f(x_0, y_0, z_0)$ is orthogonal to the vector $\boldsymbol{\sigma}'(t_0)$ that is tangential to the curve C. It is reasonable to declare that the plane that is spanned by all such tangent vectors and passes through (x_0, y_0, z_0) is **the tangent plane to the level surface S of** f. Since $\nabla f(x_0, y_0 z_0)$ is normal to that plane, the equation of the tangent plane is

$$\nabla f(x_0, y_0 z_0) \cdot ((x - x_0)\mathbf{i} + (y - y_0)\mathbf{j} + (z - z_0)\mathbf{k}) = 0,$$

i.e.,

$$\frac{\partial f}{\partial x}(x_0, y_0, z_0)(x - x_0) + \frac{\partial f}{\partial y}(x_0, y_0, z_0)(y - y_0) + \frac{\partial f}{\partial z}(x_0, y_0, z_0)(z - z_0) = 0.$$

Example 5 Let f be the function such that

$$f(x, y, z) = \frac{x^2}{9} + \frac{y^2}{4} + z^2$$

a) Determine $\nabla f(x, y, z)$ and the directional derivative of f at $(2, 1, 1)$ in the direction of $(1, 1, 1)$.

b) *Determine the tangent plane to the surface*

$$\frac{x^2}{9} + \frac{y^2}{4} + z^2 = \frac{61}{36}$$

at $(2, 1, 1)$.

Solution
a) We have

$$\nabla f(x, y, z) = \frac{\partial f}{\partial x}(x, y, z)\mathbf{i} + \frac{\partial f}{\partial y}(x, y, z)\mathbf{j} + \frac{\partial f}{\partial z}(x, y, z) = \frac{2x}{9}\mathbf{i} + \frac{y}{2}\mathbf{j} + 2z\mathbf{k}.$$

Therefore,

$$\nabla f(2, 1, 1) = \frac{4}{9}\mathbf{i} + \frac{1}{2}\mathbf{j} + 2\mathbf{k}.$$

The unit vector $\mathbf{u}$ along $(1, 1, 1)$ is

$$\frac{1}{\sqrt{3}}\mathbf{i} + \frac{1}{\sqrt{3}}\mathbf{j} + \frac{1}{\sqrt{3}}\mathbf{k}.$$

Therefore,

$$D_{\mathbf{u}} f(2,1,1) = \nabla f(2,1,1) \cdot \mathbf{u}$$

$$= \left(\frac{4}{9}\mathbf{i} + \frac{1}{2}\mathbf{j} + 2\mathbf{k}\right) \cdot \left(\frac{1}{\sqrt{3}}\mathbf{i} + \frac{1}{\sqrt{3}}\mathbf{j} + \frac{1}{\sqrt{3}}\mathbf{k}\right)$$

$$= \frac{4}{9\sqrt{3}} + \frac{1}{2\sqrt{3}} + \frac{2}{\sqrt{3}} = \frac{53}{54}\sqrt{3}.$$

b) We can express the equation of the required plane as

$$\nabla f(2,1,1) \cdot ((x-2)\mathbf{i} + (y-1)\mathbf{j} + (z-1)\mathbf{k}) = 0,$$

i.e.,

$$\left(\frac{4}{9}\mathbf{i} + \frac{1}{2}\mathbf{j} + 2\mathbf{k}\right) \cdot ((x-2)\mathbf{i} + (y-1)\mathbf{j} + (z-1)\mathbf{k}) = 0.$$

Thus,

$$\frac{4}{9}(x-2) + \frac{1}{2}(y-1) + 2(z-1) = 0.$$

Figure 8 illustrates the surface and the above tangent plane. $\square$

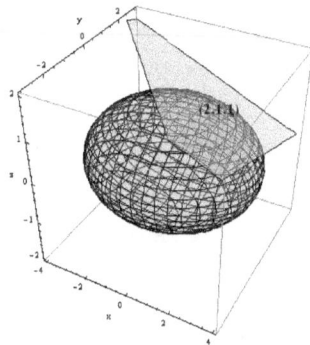

Figure 8

Problems

In problems 1-6,
a) Compute the gradient of f
b) Compute the directional derivative of f at P in the direction of the vector $\mathbf{v}$.

1.
$$f(x,y) = 4x^2 + 9y^2, \ P = (3,4), \ \mathbf{v} = (-2,1)$$

2.
$$f(x,y) = x^2 - 4x - 3y^2 + 6y + 1, \ P = (0,0), \ \mathbf{v} = (1,-1)$$

3.
$$f(x,y) = e^{x^2 - y^2}, \ P = (2,1), \ \mathbf{v} = (-1,3)$$

4.
$$f(x,y) = \sin(x)\cos(y), \ P = (\pi/3, \pi/4), \ v = (2,-3)$$

5.

$$f(x,y,z) = x^2 - y^2 + 2z^2, \ P = (1,-1,2), \ \mathbf{v} = (1,-1,1)$$

6.

$$f(x,y,z) = \frac{1}{\sqrt{x^2 + y^2 + z^2}}, \ P = (2,-2,1), \ \mathbf{v} = (3,4,1)$$

In problems 7 and 8,
a) Determine a vector $\mathbf{v}$ such that the rate at which f increases at P has its maximum value in the direction of v, and the corresponding rate of increase of f,
b) Determine a vector $\mathbf{w}$ such that the rate at which f decreases at P has its maximum value in the direction of w, and the corresponding rate of decrease of f.

7.

$$f(x,y) = \frac{1}{\sqrt{x^2 + y^2}}, \ P = (2,3)$$

8.

$$f(x,y) = \ln\left(\sqrt{x^2 + y^2}\right), \ P = (3,4)$$

In problems 9 and 10,
a) Compute

$$\left.\frac{d}{dt} f(\boldsymbol{\sigma}(t))\right|_{t=t_0}$$

by making use of the chain rule,
b) Compute the directional derivative of f in the tangential direction to the curve that is parametrized by σ at the point $\sigma(t_0)$.

9.

$$f(x,y) = e^{x^2 - y^2}, \ \boldsymbol{\sigma}(t) = (2\cos(t), 2\sin(t)), \ t_0 = \pi/6$$

10

$$f(x,y) = \arctan\left(\frac{y}{\sqrt{x^2 + y^2}}\right), \ \boldsymbol{\sigma}(t) = \left(\frac{1-t^2}{1+t^2}, \frac{2t}{1+t^2}\right), \ t_0 = 2.$$

In problems 11-13,
a) Find a vector that is orthogonal at the point P to the curve that is the graph of the given equation,
b) Determine an equation of the line that is tangent to that curve at P:

11.

$$2x^2 + 3y^2 = 35, \ (2,3)$$

12.

$$x^2 - y^2 = 4, \ \left(3, \sqrt{5}\right)$$

13.

$$e^{25 - x^2 - y^2} = 1, \ (3,4)$$

In problems 14-17,
a) Find a vector that is orthogonal at the point P to the surface that is the graph of the given equation
b) Determine an equation of the plane that is tangent to the given surface at P.

14.

$$z - x^2 + y^2 = 0, \ P = (4,3,7)$$

15.

$$x^2 - y^2 + z^2 = 1, \ P = (2,2,1)$$

16.

$$x^2 - y^2 - z^2 = 1, \; P = (3, 2, 2)$$

17.

$$x - \sin(y)\cos(z) = 0, \; P = (1, \pi/2, 0)$$

12.8 Local Maxima and Minima

The Definitions and Preliminary Examples

Definition 1 A function f has **a local maximum** at the point (x_0, y_0) if there exists an open disk D containing (x_0, y_0) such that $f(x, y) \leq f(x_0, y_0)$ for each $(x, y) \in D$. A function f has **a local minimum** at the point (x_0, y_0) if there exists an open disk D containing (x_0, y_0) such that $f(x, y) \leq f(x_0, y_0)$ for each $(x, y) \in D$.

Definition 2 **The absolute maximum** of a function f on the set $D \subset \mathbf{R}^2$ is $f(x_0, y_0)$ if $f(x, y) \leq f(x_0, y_0)$ for each $(x, y) \in D$. **The absolute minimum** of a function f on the set $D \subset R^2$ is $f(x_0, y_0)$ if $f(x, y) \geq f(x_0, y_0)$ for each $(x, y) \in D$.

Example 1 Let $Q(x, y) = x^2 + y^2$. Since $Q(x, y) \geq 0$ for each $(x, y) \in \mathbb{R}^2$ and $Q(0, 0) = 0$, Q attains its absolute minimum value 0 on $\mathbb{R}^2$ at $(0, 0)$. Of course, the only point at which Q has a local minimum is also $(0, 0)$. The function does not attain an absolute maximum on R^2 since it attains values of arbitrarily large magnitude. For example,

$$\lim_{x \to +\infty} Q(x, x) = \lim_{x \to +\infty} 2x^2 = +\infty.$$

$\square$

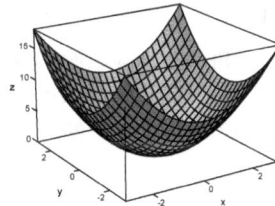

Figure 1

Example 2 Let $Q(x, y) = -x^2 - y^2$. Q attains its maximum value 0 at $(0, 0)$. The function does not attain an absolute minimum on $\mathbf{R}^2$ since it attains negative values of arbitrarily large magnitude. For example,

$$\lim_{x \to +\infty} Q(x, x) = \lim_{x \to +\infty} \left(-2x^2\right) = -\infty.$$

$\square$

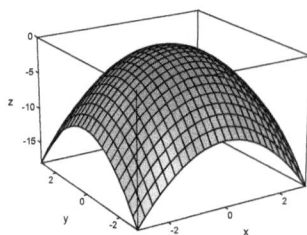

Figure 2

Example 3 Let $Q(x,y) = x^2 - y^2$. The function does not have a local extremum or an absolute maximum or minimum on $\mathbb{R}^2$. Indeed, the function attains positive and negative values of arbitrarily large magnitude since

$$\lim_{x \to +\infty} Q(x,0) = \lim_{x \to +\infty} x^2 = +\infty$$

and

$$\lim_{y \to +\infty} Q(0,y) = \lim_{y \to +\infty} -y^2 = -\infty.$$

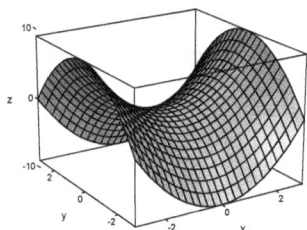

Fgiure 3

Figure 4 shows some level curves of Q. The darker color indicates a lower level on the graph of Q. We have $Q(x,0) = x^2$ and $Q(0,y) = -y^2$. Therefore, the restriction of Q to the x-axis has a minimum at $(0,0)$ and the restriction of Q to the y-axis has a maximum at $(0,0)$. The picture is consistent with these observations.$\square$

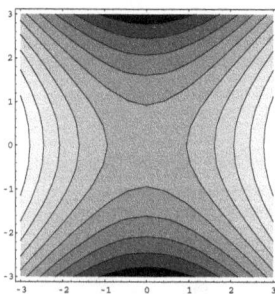

Figure 4

Definition 3 We say that (x_0, y_0) is a **saddle point of** Q if there exists a line that passes through (x_0, y_0) on which Q has a local maximum at (x_0, y_0), and a similar line on which Q has a local minimum at (x_0, y_0).

Thus, the function of Example 3 has a saddle point at $(0, 0)$.

The Second Derivative Test for Local Extrema

If f is a differentiable function of a single variable and has a local extremum at x_0, we must have $f'(x_0) = 0$. There is a similar condition for a function of two variables.

Proposition 1 If $f_x(x_0, y_0)$ **and** $f_y(x_0, y_0)$ **exist and** f **has a local maximum or local minimum at** (x_0, y_0)**, then**

$$\frac{\partial f}{\partial x}(x_0, y_0) = 0 \text{ and } \frac{\partial f}{\partial y}(x_0, y_0) = 0.$$

Proof

Set $g(x) = f(x, y_0)$. Thus, g is the restriction of f to the line $y = y_0$. Since g is a function of a single variable and has a local extremum at x_0, we have

$$\frac{dg}{dx}(x_0) = \frac{d}{dx} f(x, y_0) \bigg|_{x=x_0} = \frac{\partial f}{\partial x}(x_0, y_0) = 0.$$

Similarly, if we set $h(y) = f(x_0, y)$, so that h is the restriction of f to the line $x = x_0$, h has a local extremum at y_0. Therefore,

$$\frac{dh}{dy}(y_0) = \frac{d}{dy} f(x_0, y) \bigg|_{y=y_0} = \frac{\partial f}{\partial y}(x_0, y_0) = 0.$$

∎

Definition 4 We say that the point (x_0, y_0) is a **stationary point of** f if

$$\frac{\partial f}{\partial x}(x_0, y_0) = 0 \text{ and } \frac{\partial f}{\partial y}(x_0, y_0) = 0.$$

Thus, Proposition 1 asserts that **a necessary condition for a** f **to have a local maximum or minimum at** (x_0, y_0) **is that** (x_0, y_0) **is a stationary point of** f, provided that f has partial derivatives at (x_0, y_0). If we assume that f is differentiable at (x_0, y_0), the tangent plane to the graph of f at $(x_0, y_0, f(x_0, y_0))$ is the graph of the linear function

$$L(x, y) = f(x_0, y_0) + f_x(x_0, y_0)(x - x_0) + f_y(x_0, y_0)(y - y_0) = f(x_0, y_0),$$

so that it is parallel to the xy-plane.

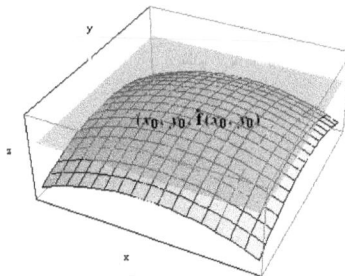

Figure 5

Remark 1 A function need not have a local extremum at a stationary point. For example, if $f(x,y) = x^2 - y^2$, as in Example 3, we have

$$\frac{\partial f}{\partial x}(x,y) = 2x \text{ and } \frac{\partial f}{\partial y}(x,y) = -2y,$$

so that the only stationary point of f is the origin $(0,0)$. We observed that f has a saddle point at $(0,0)$. In particular, f does not have a local maximum or minimum at $(0,0)$. ◊

Remark 2 A function can have a local extremum at a point even though its partial derivatives do not exist at that point. For example, let

$$f(x,y) = \sqrt{x^2 + y^2}.$$

We have

$$\frac{\partial f}{\partial x}(x,y) = \frac{1}{2\sqrt{x^2 + y^2}}(2x) = \frac{x}{\sqrt{x^2 + y^2}}$$

and

$$\frac{\partial f}{\partial y}(x,y) = \frac{y}{\sqrt{x^2 + y^2}}$$

if $(x,y) \neq (0,0)$. You can confirm that the partial derivatives of f do not exist at $(0,0)$. The function does have a local minimum at $(0,0)$, though. In fact,

$$f(x,y) = \sqrt{x^2 + y^2} \geq 0 = f(0,0)$$

for each $(x,y) \in \mathbf{R}^2$, so that the **absolute minimum** of f on R^2. The graph of f is a cone, as shown in Figure 6.◊

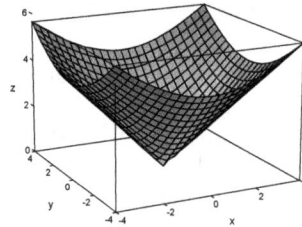

Figure 6

Remark 3 'Proposition 1 has a counterpart for a scalar function of any number of variables. For example, if f is a function of x, y and z, has partial derivatives and a local extremum at (x_0, y_0, z_0), then

$$\frac{\partial f}{\partial x}(x_0, y_0, z_0) = \frac{\partial f}{\partial y}(x_0, y_0, z_0) = \frac{\partial f}{\partial z}(x_0, y_0, z_0) = 0.$$

◊

Example 4 Let

$$f(x,y) = 2x^2 + 2xy + y^2$$

Determine the stationary points of f. Make use of your graphing utility to plot the graph of f. Does the picture suggest that f has a local maximum or minimum at each stationary point?

Solution

We have

$$\frac{\partial f}{\partial x}(x,y) = 4x + 2y \text{ and } \frac{\partial f}{\partial y}(x,y) = 2x + 2y.$$

Therefore, (x, y) is a stationary point of f if and only if

$$4x + 2y = 0,$$
$$2x + 2y = 0.$$

The only solution is $(0, 0)$. Thus, $(0, 0)$ is the only stationary point of f.

Figure 7 displays the graph of f. The picture suggests that f has a local (and absolute) minimum at $(0, 0)$. □

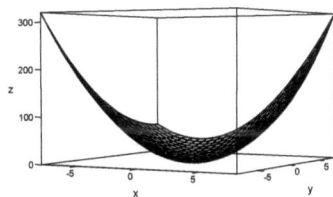

Figure 7

There is a counterpart of the second derivative test for the local extrema of a function of a single variable for functions of several variables. We will discuss the second derivative test only functions of two variables since the discussion for more than two variables requires machinery from linear algebra that you may not have at your disposal yet.

Theorem 1 (The Second Derivative Test for extrema) Assume that (x_0, y_0) is a stationary point f, and that f has continuous second-order partial derivatives in some open set that contains (x_0, y_0). Let

$$D(x_0, y_0) = f_{xx}(x_0, y_0) f_{yy}(x_0, y_0) - (f_{xy}(x_0, y_0))^2.$$

1. If $D(x_0, y_0) > 0$ and $f_{xx}(x_0, y_0) > 0$, f has a local minimum at (x_0, y_0),

2. If $D(x_0, y_0) > 0$ and $f_{xx}(x_0, y_0) < 0$, f has a local maximum at (x_0, y_0),
3. If $D(x_0, y_0) < 0$, (x_0, y_0) is a saddle point of f.

Remark 4 We will refer to

$$D(x,y) = f_{xx}(x,y) f_{yy}(x,y) - (f_{xy}(x,y))^2$$

as **the discriminant of f** at (x, y). The discriminant can be expressed as a determinant:

$$D(x,y) = \begin{vmatrix} f_{xx}(x,y) & f_{xy}(x,y) \\ f_{yx}(x,y) & f_{yy}(x,y) \end{vmatrix} = f_{xx}(x,y) f_{yy}(x,y) - (f_{xy}(x,y))^2,$$

since $f_{xy}(x, y) = f_{yx}(x, y)$, under the assumption that the second-order partial derivatives of f are continuous.

The second derivative test does not provide information about the nature of the stationary point (x_0, y_0) if the discriminant of f is 0 at (x_0, y_0). For example, let

$$f(x,y) = x^4 + y^4, \ g(x,y) = -x^4 - y^4 \text{ and } h(x,y) = x^4 - y^4.$$

Then $(0,0)$ is the only stationary point of each function. The discriminant vanishes at $(0,0)$ in each case. The function f attains its minimum value 0 at $(0,0)$, g attains its maximum value 0 at $(0,0)$ and h has a saddle point at $(0,0)$, as you can confirm easily. $\Diamond$

Example 5 Let

$$f(x,y) = 2x^2 + 2xy + y^2,$$

as in Example 4. Determine the nature of the stationary point of f.

Solution

As in Example 4,

$$\frac{\partial f}{\partial x}(x,y) = 4x + 2y \text{ and } \frac{\partial Q}{\partial y}(x,y) = 2x + 2y,$$

and $(0,0)$ is the only stationary point of Q
We have

$$f_{xx}(x,y) = 4, \ f_{yy}(x,y) = 2 \text{ and } f_{xy}(x,y) = 2$$

at each $(x,y) \in \mathbf{R}^2$. Therefore,

$$D(0,0) = f_{xx}(0,0) f_{yy}(0,0) - (f_{xy}(0,0))^2 = 4(2) - 2^2 = 4 > 0.$$

Since $f_{xx}(0,0) = 4 > 0$, f has a local minium at $(0,0)$, as we inferred in Example 4 from the graph of f. Figure 8 shows some level curves, and the darker color indicates a lower level on the graph of f. The picture is consistent with our assertion that f has a local minimum at $(0,0)$. $\square$

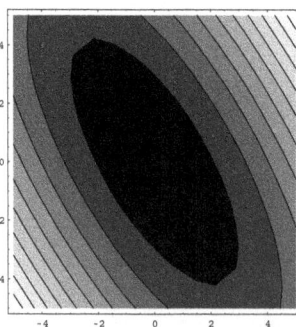

Figure 8

Remark 5 The function of Example 5 is a quadratic function. The nature of the stationary point (x_0, y_0) of a quadratic function can be determined by completing the squares in terms of the powers of $X = x - x_0$ and $Y = y - y_0$ (we translate the origin to (x_0, y_0)). For example, if

$$f(x,y) = 2x^2 + 2xy + y^2,$$

as in Example 5, the only stationary point is $(0,0)$. We have

$$
\begin{aligned}
f(x,y) &= 2\left(x^2 + xy\right) + y^2 \\
&= 2\left(x + \frac{y}{2}\right)^2 - \frac{y^2}{2} + y^2 = 2\left(x + \frac{y}{2}\right)^2 + \frac{y^2}{2}.
\end{aligned}
$$

Therefore,

$$f(x, y) \geq 0 = f(0, 0)$$

for each $(x, y) \in \mathbb{R}^2$. This shows that f attains its absolute minimum at $(0, 0)$. $\Diamond$

Example 6 Let $f(x, y) = x^2$. Determine the nature of the stationary points of f.

Solution

Since

$$\frac{\partial f}{\partial x}(x, y) = 2x \text{ and } \frac{\partial f}{\partial y}(x, y) = 0,$$

(x, y) is a stationary point of f if $x = 0$ and y is an arbitrary real number. Thus, any point $(0, y)$ on the y-axis is a stationary point of f. Since

$$\frac{\partial^2 f}{\partial x^2}(x, y) = 2, \cdot \frac{\partial^2 f}{\partial y \partial x}(x, y) = 0 \text{ and } \frac{\partial^2 f}{\partial x^2}(x, y) = 0,$$

the discriminant of f is

$$D(x, y) = \left(\frac{\partial^2 f}{\partial x^2}(x, y)\right)\left(\frac{\partial^2 f}{\partial x^2}(x, y)\right) - \left(\frac{\partial^2 f}{\partial y \partial x}(x, y)\right)^2 = 0.$$

This is another case where the second derivative test does not give any information about the nature of the stationary points of the function.
Note that

$$f(x, y) = x^2 \geq 0$$

and $f(0, y) = 0$ for each $y \in \mathbb{R}$, so that f attains its absolute minimum on each point of the y-axis. Figure 9 shows the graph of f which is a cylinder that has the same cross section (a parabola) on any plane that is parallel to the xz-plane. $\square$

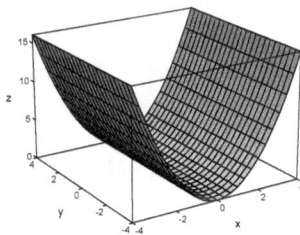

Figure 9

Example 7 Let

$$f(x, y) = x^2 + 4xy + y^2$$

Determine the nature of the stationary points of f.

Solution

We have

$$\frac{\partial f}{\partial x}(x, y) = 2x + 4y \text{ and } \frac{\partial f}{\partial y}(x, y) = 4x + 2y,$$

Therefore (x, y) is a stationary point of f if and only if

$$2x + 4y = 0,$$
$$4x + 2y = 0.$$

The only solution is $(0, 0)$. Thus, $(0, 0)$ is the only stationary point of f.
We have

$$f_{xx}(x, y) = 2, \ f_{xy}(x, y) = 4 \text{ and } f_{yy}(x, y) = 2,$$

so that

$$D(x, y) = (f_{xx}(x, y))(f_{yy}(x, y)) - (f_{xy}(x, y))^2 = (2)(2) - 16 = -12 < 0$$

Therefore, f has a saddle point $(0, 0)$. Figure 10 shows the graph of f and Figure 11 shows some level curves of f. The pictures are consistent with the fact that f has a saddle point at $(0, 0)$.
$\square$

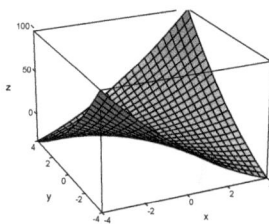

Figure 10

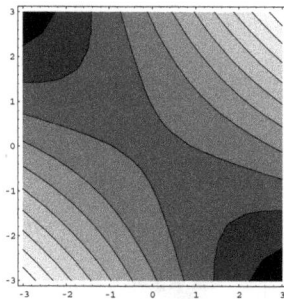

Figure 11

Remark 6 With reference to Example 7, we can determine directions in which $f(x, y)$ increases or decreases by completing the squares:

$$f(x, y) = x^2 + 4xy + y^2 = (x + 2y)^2 - 4y^2 + y^2$$
$$= (x + 2y)^2 - 3y^2.$$

Thus,

$$f(x, 0) = x^2,$$

so that the restriction of f to the x-axis has its minimum at $(0, 0)$. We also have

$$f(-2y, y) = -3y^2,$$

so that the restriction of to the line $x = -2y$ has its maximum value at $(0, 0)$. $\Diamond$

Example 8 Let

$$f(x,y) = \frac{1}{3}x^3 + \frac{1}{3}y^3 - xy.$$

Determine the nature of the stationary points of $f/$

Solution

Figure 12 shows the graph of f. The picture does not give a clear indication about the stationary points of f. In any case, our analysis will clarify matters.

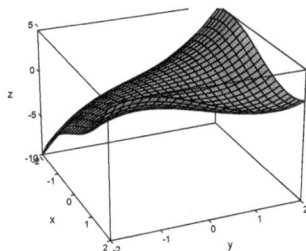

Figure 12

We have

$$f_x(x,y) = x^2 - y \text{ and } f_y(x,y) = y^2 - x$$

so that (x,y) is a stationary point of f if and only if

$$x^2 - y = 0,$$
$$y^2 - x = 0.$$

From the first equation, $y = x^2$. Therefore, the second equation leads to the following:

$$x^4 - x = 0 \Rightarrow x\left(x^3 - 1\right) = 0 \Rightarrow x = 0 \text{ or } x = 1.$$

If $x = 0$, then $y = 0$, Therefore, $(0,0)$ is a stationary point. If $x = 1$, then $y = 1$. Therefore, $(1,1)$ is a stationary point.
We have

$$f_{xx}(x,y) = 2x, \ f_{xy}(x,y) = -1 \text{ and } f_{yy}(x,y) = 2y.$$

Therefore

$$D(x,y) = (2x)(2y) - (-1)^2 = 4xy - 1.$$

Thus,

$$D(0,0) = -1 < 0$$

Therefore, f has a saddle point at $(0,0)$. Figure 13 shows some level curves of f near $(0,0)$. The picture is consistent with our assertion that $(0,0)$ is a saddle point of f. Clearly, f increases in some directions and decreases in some directions.

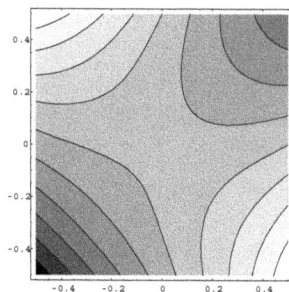

Figure 13

At $(1, 1)$,

$$f_{xx}(1, 1) = 2, \ f_{xy}(1, 1) = -1 \text{ and } f_{yy}(1, 1) = 2.$$

Therefore,

$$D = (2)(2) - 1 = 3 > 0.$$

We have $f_{xx}(1, 1) = 2 > 0$. Therefore, f has a local minimum at $(1, 1)$. The level curves in Figure 14 are consistent with our assertion.$\square$

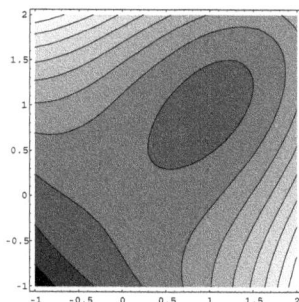

Figure 14

Example 9 Let

$$f(x, y) = \left(y^2 - x^2\right) e^{-x^2 - y^2}.$$

Determine the nature of the critical points of f.

Solution

Figure 15 shows the graph of f. The picture indicates that f has several local extrema and at least one saddle point. Our analysis will clarify the picture.

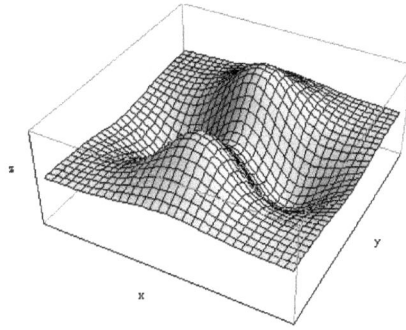

Figure 15

Then,

$$f_x(x,y) = \frac{\partial}{\partial x}\left(\left(y^2 - x^2\right)e^{-x^2-y^2}\right)$$
$$= -2xe^{-x^2-y^2} - \left(y^2 - x^2\right)2xe^{-x^2-y^2}$$
$$= -2xe^{-x^2-y^2}\left(1 + \left(y^2 - x^2\right)\right)$$

and

$$f_y(x,y) = \frac{\partial}{\partial y}\left(\left(y^2 - x^2\right)e^{-x^2-y^2}\right)$$
$$= 2ye^{-x^2-y^2} - 2y\left(y^2 - x^2\right)e^{-x^2-y^2}$$
$$= 2ye^{-x^2-y^2}\left(1 - \left(y^2 - x^2\right)\right).$$

Therefore,

$$-2x\left(1 + \left(y^2 - x^2\right)\right) = 0 \text{ and } 2y\left(1 - \left(y^2 - x^2\right)\right) = 0$$

If $x = 0$, then

$$2y\left(1 - y^2\right) = 0 \Leftrightarrow y = 0 \text{ or } y = \pm 1$$

Therefore, $(0,0)$, $(0,-1)$ and $(0,1)$ are stationary points.
If $y^2 - x^2 = -1$, then $4y = 0 \Rightarrow y = 0$, so that $x = \pm 1$. Therefore, $(-1,0)$ and $(1,0)$ are stationary points.
We have

$$\frac{\partial^2 f(x,y)}{\partial x^2} = -2e^{-x^2-y^2} + 10x^2 e^{-x^2-y^2} - 2e^{-x^2-y^2}y^2 + 4x^2 e^{-x^2-y^2}y^2 - 4x^4 e^{-x^2-y^2},$$

$$\frac{\partial^2 f(x,y)}{\partial y^2} = 2e^{-x^2-y^2} - 10e^{-x^2-y^2}y^2 + 2x^2 e^{-x^2-y^2} + 4e^{-x^2-y^2}y^4 - 4x^2 e^{-x^2-y^2}y^2,$$

$$\frac{\partial^2 f(x,y)}{\partial y \partial x} = -4\left(-y^2 + x^2\right)yxe^{-x^2-y^2}$$

The discriminant of f is

$$D(x,y) = \left(\frac{\partial^2 f(x,y)}{\partial x^2}\right)\left(\frac{\partial^2 f(x,y)}{\partial y^2}\right) - \left(\frac{\partial^2 f(x,y)}{\partial y \partial x}\right)^2$$

Table 1 displays the values the second-order partial derivatives and the discriminant of f at the stationary points, and the nature of the stationary points. Figure 16 shows some level curves of f. The picture is consistent with our conclusions.

(x,y)	f_{xx}	f_{yy}	f_{xy}	$D(x,y)$	
$(0,0)$	-2	2	0	-4	saddle point
$(0,1)$	$-4e^{-1}$	$-4e^{-1}$	0	$16e^{-2}$	maximum
$(0,-1)$	$-4e^{-1}$	$-4e^{-1}$	0	$16e^{-2}$	maximum
$(1,0)$	$4e^{-1}$	$4e^{-1}$	0	$16e^{-2}$	minimum
$(-1,0)$	$4e^{-1}$	$4e^{-1}$	0	$16e^{-2}$	minimum

Table 1

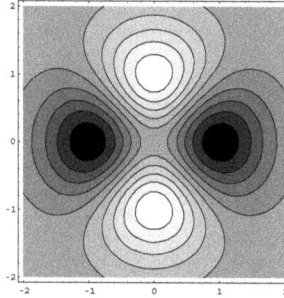

Figure 16

A Plausibility Argument for the Second Derivative Test

Proposition 2 Let $Q(x,y)$ be a quadratic function. Let (x_0, y_0) be an arbitrary point. Then,

$$Q(x,y) = Q(x_0, y_0) + \frac{\partial Q}{\partial x}(x_0, y_0)(x - x_0) + \frac{\partial Q}{\partial y}(x_0, y_0)(y - y_0)$$

$$+ \frac{1}{2}\frac{\partial^2 Q}{\partial x^2}(x - x_0)^2 + \frac{\partial^2 Q}{\partial x \partial y}(x - x_0)(y - y_0) + \frac{1}{2}\frac{\partial^2 Q}{\partial y^2}(y - y_0)^2$$

Proof

We can express $Q(x,y)$ as

$$Q(x,y) = A(x - x_0)^2 + 2B(x - x_0)(y - y_0) + C(y - y_0)^2 + D(x - x_0) + E(y - y_0) + F.$$

Then.

$$\frac{\partial Q}{\partial x} = 2A(x - x_0) + 2B(y - y_0) + D,$$

$$\frac{\partial Q}{\partial y} = 2B(x - x_0) + 2C(y - y_0) + E,$$

$$\frac{\partial^2 Q}{\partial x^2} = 2A, \ \frac{\partial^2 Q}{\partial y^2} = 2C, \ \frac{\partial^2 Q}{\partial y \partial x} = 2B.$$

Therefore,

$$\frac{\partial Q}{\partial x}(x_0, y_0) = D, \ \frac{\partial Q}{\partial y}(x_0, y_0) = E,$$

$$\frac{\partial^2 Q}{\partial x^2}(x_0, y_0) = 2A, \ \frac{\partial^2 Q}{\partial y^2}(x_0, y_0) = 2C, \ \frac{\partial^2 Q}{\partial y \partial x}(x_0, y_0) = 2B$$

■

As a result of Proposition 2, the quadratic function that has the same value as f at (x_0, y_0) and has the same first and second partial derivatives at (x_0, y_0) is uniquely determined:

Definition 5 Assume that f has continuous second-order partial derivatives in some open set that contains (x_0, y_0). **The quadratic approximation to f based at (x_0, y_0)** is

$$Q(x,y) = f(x_0, y_0) + \frac{\partial f}{\partial x}(x_0, y_0)(x - x_0) + \frac{\partial f}{\partial y}(x_0, y_0)(y - y_0) +$$

$$+ \frac{1}{2}\frac{\partial^2 f}{\partial x^2}(x_0, y_0)(x - x_0)^2 + \frac{\partial^2 f}{\partial y \partial x}(x_0, y_0)(x - x_0)(y - y_0) + \frac{1}{2}\frac{\partial^2 f}{\partial y^2}(x_0, y_0)(y - y_0)^2$$

We have $Q(x,y) \cong f(x,y)$ if (x,y) is close to (x_0, y_0) and the magnitude of the error is much smaller than $||(x - x_0, y - y_0)||^2$ if $||(x - x_0, y - y_0)||$ is small. In fact, the error $R(x - x_0, y - y_0)$ is such that

$$\lim_{(x,y) \to (x_0, y_0)} \frac{R(x - x_0, y - y_0)}{||(x - x_0, y - y_0)||^2} = 0.$$

We leave the proof of this fact to a course in advanced calculus. Note that **the linear approximation to f based at (x_0, y_0)** is

$$L(x,y) = f(x_0, y_0) + \frac{\partial f}{\partial x}(x_0, y_0)(x - x_0) + \frac{\partial f}{\partial y}(x_0, y_0)(y - y_0),$$

so that

$$Q(x,y) = L(x,y)$$
$$+ \frac{1}{2}\frac{\partial^2 f}{\partial x^2}(x_0, y_0)(x - x_0)^2 + \frac{\partial^2 f}{\partial y \partial x}(x_0, y_0)(x - x_0)(y - y_0) + \frac{1}{2}\frac{\partial^2 f}{\partial y^2}(x_0, y_0)(y - y_0)^2.$$

If (x_0, y_0) is a **stationary point of** f, we have

$$\frac{\partial f}{\partial x}(x_0, y_0) = 0 \text{ and } \frac{\partial f}{\partial y}(x_0, y_0) = 0,$$

so that the quadratic approximation to f based at (x_0, y_0) is

$$Q(x,y) = f(x_0, y_0) \qquad + \frac{1}{2}\frac{\partial^2 f}{\partial x^2}(x_0, y_0)(x - x_0)^2$$
$$+ \frac{\partial^2 f}{\partial y \partial x}(x_0, y_0)(x - x_0)(y - y_0) + \frac{1}{2}\frac{\partial^2 f}{\partial y^2}(x_0, y_0)(y - y_0)^2$$

Since $Q(x,y)$ approximates $f(x,y)$ very well near (x_0, y_0), it is reasonable to expect that the nature of the stationary point (x_0, y_0) of f will be revealed by analyzing the behavior of the quadratic function $Q(x,y)$.

We can set $X = x - x_0$, $Y = y - y_0$ and

$$\tilde{Q}(X,Y) = Q(X + x_0, Y + y_0) - f(x_0, y_0)$$
$$= + \frac{1}{2}\frac{\partial^2 f}{\partial x^2}(x_0, y_0)X^2 + \frac{\partial^2 f}{\partial y \partial x}(x_0, y_0)XY + \frac{1}{2}\frac{\partial^2 f}{\partial y^2}(x_0, y_0)Y^2,$$

so that $(0,0)$ is the stationary point of $\tilde{Q}$ and $\tilde{Q}(0,0) = 0$. We might as well assume that $(x_0, y_0) = (0,0)$ and $Q(0,0) = 0$ and simplify the notation. Thus, we assume that

$$Q(x,y) = \frac{1}{2}Q_{xx}x^2 + Q_{xy}xy + \frac{1}{2}Q_{yy}y^2.$$

Note the the second-order partial derivatives and the discriminant of a quadratic function are constants.

1. Let's assume that $D = Q_{xx}Q_{yy} - (Q_{xy})^2 > 0$ and $Q_{xx} > 0$. We complete the square:

$$Q(x,y) = \frac{1}{2}Q_{xx}\left(x^2 + \frac{2Q_{xy}xy}{Q_{xx}}\right) + \frac{1}{2}Q_{yy}y^2$$

$$= \frac{1}{2}Q_{xx}\left(x + \frac{Q_{xy}}{Q_{xx}}y\right)^2 - \frac{1}{2}\frac{Q_{xy}^2}{Q_{xx}}y^2 + \frac{1}{2}Q_{yy}y^2$$

$$= \frac{1}{2}Q_{xx}\left(x + \frac{Q_{xy}}{Q_{xx}}y\right)^2 + \frac{1}{2}\frac{Q_{xx}Q_{yy} - Q_{xy}^2}{Q_{xx}}y^2.$$

Therefore, $Q(x,y) \geq 0$ for each (x,y) and $Q(x,y) = 0$ iff $(x,y) = (0,0)$. Thus, Q attains its minimum value 0 at $(0,0)$ (there is no maximum).

2. Let's assume that $D = Q_{xx}Q_{yy} - (Q_{xy})^2 > 0$ and $Q_{xx} < 0$. Since

$$Q(x,y) = \frac{1}{2}Q_{xx}\left(x + \frac{Q_{xy}}{Q_{xx}}y\right)^2 + \frac{1}{2}\frac{Q_{xx}Q_{yy} - Q_{xy}^2}{Q_{xx}}y^2,$$

and $Q_{xx} < 0$ and

$$\frac{Q_{xx}Q_{yy} - Q_{xy}^2}{Q_{xx}} < 0,$$

we have

$$Q(x,y) \leq 0$$

for each (x,y) and $Q(x,y) = 0$ iff $(x,y) = (0,0)$. Thus, Q attains its maximum value 0 at $(0,0)$ (there is minimum).

3. Assume that $D = Q_{xx}Q_{yy} - (Q_{xy})^2 < 0$. If $Q_{xx} \neq 0$,

$$Q(x,y) = \frac{1}{2}Q_{xx}\left(x + \frac{Q_{xy}}{Q_{xx}}y\right)^2 + \frac{1}{2}\frac{Q_{xx}Q_{yy} - Q_{xy}^2}{Q_{xx}}y^2$$

If $Q_{xx} > 0$,

$$Q(x,0) = \frac{1}{2}Q_{xx}x^2 > 0 \text{ if } x \neq 0.$$

On the other hand, if

$$x + \frac{Q_{xy}}{Q_{xx}}y = 0,$$

then

$$Q(x,y) = \frac{1}{2}\frac{Q_{xx}Q_{yy} - Q_{xy}^2}{Q_{xx}}y^2 < 0$$

if $y \neq 0$. Therefore, Q has a saddle point at $(0,0)$.

If $Q_{xx} < 0$, then

$$Q(x,0) = \frac{1}{2}Q_{xx}x^2 < 0 \text{ if } x \neq 0,$$

whereas if

$$x + \frac{Q_{xy}}{Q_{xx}}y = 0$$

then

$$Q(x,y) = \frac{1}{2}\frac{Q_{xx}Q_{yy} - Q_{xy}^2}{Q_{xx}}y^2 > 0$$

if $y \neq 0$. Therefore Q has a saddle point at $(0,0)$.

There is a similar argument which shows that f has a saddle point at $(0,0)$ if $Q_{yy} \neq 0$.

If we have both Q_{xx} and Q_{yy} equal to 0, then

$$Q(x, y) = Q_{xy} xy,$$

and

$$D = -Q_{xy}^2 < 0 \Rightarrow Q_{xy} \neq 0,$$

so that

$$Q(x, x) = Q_{xy} x^2 \text{ and } Q(x, -x) = -Q_{xy} x^2.$$

Again, we have saddle a saddle point at $(0,0)$ (there are many lines along which Q increases as we move away from $(0,0)$ and many lines along which Q decreases.

Problems

In problems 1 - 8 determine the nature of the critical points of f (maximum minimum, or saddle point):

1.
$$f(x, y) = x^2 - y^2 + xy + x - y$$

2.
$$f(x, y) = x^2 + y^2 - xy + y$$

3.
$$f(x, y) = x^2 + y^2 + 2xy - 10$$

4.
$$f(x, y) = x^2 + y^2 + 3xy - 3y$$

5.
$$f(x, y) = x^2 - 3xy + 5x - 2y + 6y^2 + 8$$

6.
$$f(x, y) = 3x - x^3 - 3xy^2$$

7.
$$f(x, y) = -x^4 - y^4 - 4xy + \frac{1}{16}.$$

8.
$$f(x, y) = x^3 - 12xy + 8y^3$$

12.9   Absolute Extrema and Lagrange Multipliers

Absolute Extrema

As we saw in Section 13.6, a function need not have an absolute maximum or minimum on a given set. For example, if $f(x, y) = x^2 + y^2$ then f does not have an absolute maximum on $\mathbf{R}^2$ since f attains arbitrarily large values on $\mathbf{R}^2$. Indeed,

$$\lim_{x \to +\infty} f(x, x) = \lim_{x \to +\infty} 2x^2 = +\infty.$$

The function does have an absolute minimum:

$$f(x, y) = x^2 + y^2 \geq 0 = f(0, 0).$$

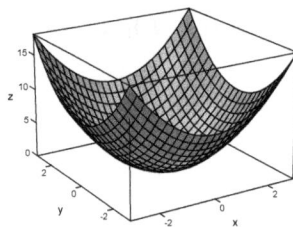

Figure 1: $f(x, y) = x^2 + y^2$

On the other hand, if we set $g(x) = -f(x) = -x^2 - y^2$, then g does not have an absolute minimum, since g attains negative values of arbitrarily large magnitude. The function has an absolute maximum on $\mathbb{R}^2$ since

$$g(x, y) = -x^2 - y^2 \leq 0 = g(0, 0).$$

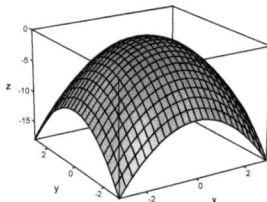

Figure 2: $g(x, y) = -x^2 - y^2$

In the above cases, the underlying set ($\mathbb{R}^2$) is unbounded. A function need not attain an absolute extremum on a bounded set either. For example, if $f(x, y) = x^2 + y^2$ and D is the open unit disk $\{(x, y) : x^2 + y^2 < 1\}$, f does not attain an absolute maximum on D since

$$f(x, y) = x^2 + y^2 < 1$$

for each $(x, y) \in D$ and f has values that are arbitrarily close to 1. Indeed,

$$\lim_{\substack{x^2 + y^2 = r^2 \\ r \to 1}} f(x, y) = \lim_{r \to 1} r^2 = 1.$$

Thus, the only candidate for the absolute maximum of f on D is 1, but f does not attain the value 1 in D.

Just as in the case of functions of a single variable, a continuous function of several variables attains its absolute extrema on a closed and bounded set. Let's state the theorem for a function of two variables:

Theorem 1 Assume that $D \subset \mathbb{R}^2$ is closed and bounded, and f is a scalar function that is continuous on D. Then f attains its absolute maximum and absolute minimum on D.

We leave the proof of Theorem 1 to a course in advanced calculus.

Just as in the case of functions of a single variable, we can implement the following strategy in order to find the absolute extrema of a function f that is continuous on a closed and bounded set $D \subset \mathbb{R}^2$:

1. Determine the stationary points of f in the interior of D and the points in the interior of D where f does not have partial derivatives. Calculate the corresponding values of f.

2. Determine the maximum and minimum values of f on the boundary of D.

3. Determine the absolute extrema of f on D by comparing the values that have been computed in steps 1 and 2.

Example 1 Let $f(x,y) = x^2 + y^2$. Determine the absolute maximum and minimum of f on the square $D = [-2, 2] \times [-2, 2]$.

Solution

The only stationary point of f is $(0, 0)$ and $f(0, 0) = 0$.
The restriction of f to the side $L_1 = \{(x, -2) : -2 \leq x \leq 2\}$ of D is

$$f(x, -2) = x^2 + 4.$$

Therefore, the maximum on f on L_1 is $f(2, -2) = 8$ and the minimum of f on L_1 is $f(0, -2) = 4$.
Since $f(x, 2) = x^2 + 4 = f(x, -2)$, the maximum on f on $L_2 = \{(x, 2) : -2 \leq x \leq 2\}$ is also 8 and the minimum of f on L_2 is 4.
Similarly, the maximum of f on the other sides of D is 8 and the minimum of f on the other sides of D is 4.
Thus, the absolute maximum of f on D is 8 and the absolute minimum of f is 0. $\square$

Lagrange Multipliers

There is a useful necessary condition for a function to have a local extremum on a curve:

Theorem 2 (Lagrange Multipliers) Assume that C is a smooth curve in R^2 that is the graph of the equation $g(x, y) = c$, where c is a constant. If the restriction of the differentiable function f to C has a local extremum at the point (x_0, y_0) then there exists a constant λ such that

$$\nabla f(x_0, y_0) = \lambda \nabla g(x_0, y_0),$$

i.e.,

$$\frac{\partial f}{\partial x}(x_0, y_0) = \lambda \frac{\partial g}{\partial x}(x_0, y_0),$$
$$\frac{\partial f}{\partial y}(x_0, y_0) = \lambda \frac{\partial g}{\partial y}(x_0, y_0)$$

Proof

Since the curve C is assumed to be smooth, we can assume that a segment of C that contains (x_0, y_0) in its interior can be parametrized by a function $\sigma : J \to \mathbb{R}^2$ such that $\sigma(t_0) = (x_0, y_0)$, where t_0 is in the interior of the interval J. Thus, $f \circ \sigma$ has a local extremum at t_0, so that

$$\left. \frac{df(\sigma(t))}{dt} \right|_{t=t_0} = 0.$$

By the chain rule,

$$\left. \frac{df(\sigma(t))}{dt} \right|_{t=t_0} = \boldsymbol{\nabla} f(\sigma(t_0)) \cdot \frac{d\boldsymbol{\sigma}}{dt}(t_0) = \boldsymbol{\nabla} f(x_0, y_0) \cdot \frac{d\boldsymbol{\sigma}}{dt}(t_0) = 0$$

Therefore, $\nabla f(x_0, y_0)$ is orthogonal to the vector $\boldsymbol{\sigma}'(t_0)$ that is tangent to C at (x_0, y_0). Since C is the graph of $g(x, y) = c$, we know that $\nabla g(x_0, y_0)$ is also orthogonal to C at (x_0, y_0). Therefore, $\nabla f(x_0, y_0)$ is parallel to $\nabla g(x_0, y_0)$. Thus, there is a scalar λ such that

$$\nabla f(x_0, y_0) = \lambda \nabla g(x_0, y_0).$$

In terms of the components of the vectors,

$$\frac{\partial f}{\partial x}(x_0, y_0) = \lambda \frac{\partial g}{\partial x}(x_0, y_0),$$
$$\frac{\partial f}{\partial y}(x_0, y_0) = \lambda \frac{\partial g}{\partial y}(x_0, y_0)$$

■

Remark 1 The fact that $\nabla f(x_0, y_0)$ is parallel to $\nabla g(x_0, y_0)$ implies that the level curve of f that passes through (x_0, y_0) and the curve $g(x, y) = 0$ have a common tangent at (x_0, y_0).◊

With reference to the statement of Theorem 2, the number λ is called a **Lagrange multiplier.** Note that we have to solve the three equations

$$\begin{aligned}
\frac{\partial f}{\partial x}(x, y) &= \lambda \frac{\partial g}{\partial x}(x, y). \\
\frac{\partial f}{\partial y}(x, y) &= \lambda \frac{\partial g}{\partial y}(x, y), \\
g(x, y) &= c,
\end{aligned}$$

in the three unknowns x, y and λ.

Example 2 Determine the extrema of $f(x, y) = y^2 - x^2$ on the unit circle $x^2 + y^2 = 1$.

Solution
We set $g(x, y) = x^2 + y^2$, so that the constraint is $g(x, y) = 1$. We have

$$\nabla f(x, y) = (-2x, 2y) \text{ and } \nabla g(x, y) = (2x, 2y).$$

Therefore, $\nabla f(x, y) = \lambda \nabla g(x, y)$ iff

$$-2x = 2\lambda x$$
$$2y = 2\lambda y$$

i.e.,

$$(1 + \lambda) x = 0$$
$$(1 - \lambda) y = 0$$

and

$$x^2 + y^2 = 1$$

From the first equation, $x = 0$ or $\lambda = -1$. If $x = 0$, then $y = \pm 1$. Therefore, the candidates are $(0, -1)$ and $(0, 1)$.
If $\lambda = -1$, then $y = 0$, so that $x = \pm 1$. Therefore, the candidates are $(-1, 0)$ and $(0, 1)$. We have

$$f(0, \pm 1) = 1 \text{and } f(\pm 1, 0) = -1$$

Therefore, the minimum of f on the unit circle is -1 and the maximum is 1.

Figure 3 shows some level curves of f and the unit circle. Note that the level curves of f that pass through the points determined with the help of Theorem 2.are tangent to the "constraint curve" $x^2 + y^2 = 1$ at those points (Remark 1. $\square$

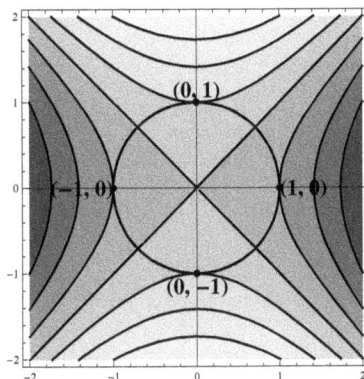

Figure 3

Remark 2 We could have obtained the result directly:

On the circle,

$$f(x,y) = y^2 - x^2 = (1 - x^2) - x^2 = 1 - 2x^2,$$

We have

$$\frac{d}{dx}(1 - 2x^2) = -4x = 0 \Rightarrow x = 0,$$

so that we obtain the points $(0, \pm 1)$. We must also take into account the endpoints ± 1, so that the corresponding points are $(-1, 0)$ and $(1, 0)$. $\lozenge$

Example 3 Let $f(x,y) = y^2 - x^2$. Determine the maximum and minimum values of f inside the unit disk

$$G = \left\{ (x,y) : x^2 + y^2 \leq 1 \right\}$$

Solution

In Example 2 we determined the maximum and minimum values of f on the boundary:

$$f(\pm 1, 0) = -1 \text{ and } f(0, \pm 1) = 1$$

The critical point in the interior is $(0, 0)$. We have $f(0, 0) = 0.$.Therefore, the maximum value of f in G is 1 and the minimum value is -1. $\square$

The theorem on Lagrange multipliers has counterparts for functions of any number of variables. For example, assume that the surface S is the graph of the equation $g(x, y, z) = c$, where c is a constant. Thus, S is a level surface of g. Also assume that f attains a local extremum at $P_0 = (x_0, y_0, z_0)$. We know that $\nabla g(P_0)$ is orthogonal at P_0 to any curve on S that passes through P_0. Assume that C is a curve on S such that $\sigma(t_0) = P_0$. The function $f(\sigma(t))$ attains a local extremum at t_0. Therefore,

$$0 = \left. \frac{df(\sigma(t))}{dt} \right|_{t=t_0} = \nabla f(\sigma(t_0)) \cdot \frac{d\sigma}{dt}(t_0)$$

Therefore, $\nabla f(P_0)$ is also orthogonal to such a curve at P_0. Thus, $\nabla f(P_0)$ is parallel to $\nabla g(P_0)$. Therefore there exists a number λ such that

$$\nabla f(P_0) = \lambda \nabla g(P_0).$$

Therefore,

$$\frac{\partial f}{\partial x}(x_0, y_0, z_0) = \lambda \frac{\partial g}{\partial x}(x_0, y_0, z_0)$$

$$\frac{\partial f}{\partial y}(x_0, y_0, z_0) = \lambda \frac{\partial g}{\partial y}(x_0, y_0, z_0)$$

$$\frac{\partial f}{\partial z}(x_0, y_0, z_0) = \lambda \frac{\partial g}{\partial z}(x_0, y_0, z_0)$$

and

$$g(x_0, y_0, z_0) = c$$

Example 4 Let $f(x, y, z) = x + y + z$. Determine the maximum of f on the unit sphere $x^2 + y^2 + z^2 = 1$.

Solution

If we set $g(x, y, z) = x^2 + y^2 + z^2$, the constraint equation is

$$g(x, y, z) = 1.$$

Therefore, we need to have $\nabla f(x, y, z) = \lambda \nabla g(x, y, z)$ for some λ. Since

$$\nabla f(x, y, z) = (1, 1, 1) \text{ and } \nabla g(x, y, z) = (2x, 2y, 2z),$$

we obtain the equations

$$1 = 2\lambda x$$
$$1 = 2\lambda y$$
$$1 = 2\lambda z,$$

and

$$x^2 + y^2 + z^2 = 1.$$

Since $2\lambda x = 1$ we have $\lambda \neq 0$. Thus, the first three equations lead to the following equalities:

$$x = y = z = \frac{1}{2\lambda}$$

We substitute these equalities in the last equation:

$$\frac{3}{4\lambda^2} = 1 \Rightarrow \lambda = \pm \frac{\sqrt{3}}{2}.$$

If $\lambda = \sqrt{3}/2$, we have

$$x = y = z = \frac{1}{2\lambda} = \frac{1}{\sqrt{3}},$$

if $\lambda = -\sqrt{3}/2$, we have

$$x = y = z = \frac{1}{2\lambda} = -\frac{1}{\sqrt{3}}$$

We have

$$f\left(\frac{1}{\sqrt{3}}\right) = \frac{3}{\sqrt{3}} = \sqrt{3},$$

and

$$f\left(-\frac{1}{\sqrt{3}}\right) = -\sqrt{3}.$$

Therefore, the maximum value of f on the unit sphere is $\sqrt{3}$ and the minimum value is $-\sqrt{3}$. Note that

$$\nabla f = (1, 1, 1)$$

and

$$\nabla g\left(x, y, z\right) = (2x, 2y, 2z),$$

so that

$$\nabla g\left(\frac{1}{\sqrt{3}}, \frac{1}{\sqrt{3}}, \frac{1}{\sqrt{3}}\right) = \frac{2}{\sqrt{3}}(1, 1, 1).$$

Therefore

$$\nabla f = \frac{\sqrt{3}}{2}\nabla g\left(\frac{1}{\sqrt{3}}, \frac{1}{\sqrt{3}}, \frac{1}{\sqrt{3}}\right)$$

Thus, the graph of f (a plane) is "tangential" to the graph of the equation $g\left(x, y, z\right) = 1$ (a sphere) at the point was determined. Figure 4 is consistent with this observation. $\square$

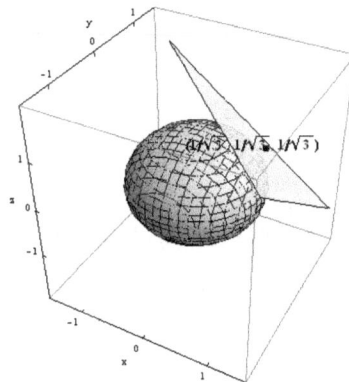

Figure 4

Maximizing Output under Budgetary Constraints

Assume that the total value of a certain **product** produced by a firm depends on the **capital** and the cost of **labor**. Thus, if we denote the value of the product by z, the capital by x and the cost of labor by y, there exists a function f such that $z = f\left(x, y\right)$. The firm would like to maximize f subject to a constraint $g\left(x, y\right) = qx + py = B$, where p, q and B are constants. Thus, the necessary conditions provided by the technique of Lagrange multipliers are

$$\frac{\partial f}{\partial x} = \lambda q, \ \frac{\partial f}{\partial y} = \lambda p \text{ and } py + qx = B.$$

Let $X = qx$ and $Y = py$, so that X is the dollar value of capital and Y is the dollar value of labor. Then,

$$\frac{\partial f}{\partial X} = \frac{1}{q}\frac{\partial f}{\partial x} = \lambda \text{ and } \frac{\partial f}{\partial Y} = \frac{1}{p}\frac{\partial f}{\partial y} = \lambda.$$

Thus,

$$\frac{\partial f}{\partial X} = \frac{\partial f}{\partial Y} = \lambda$$

Thus, at the optimum production level, the marginal change in output per dollar's worth of additional capital investment is equal to the marginal change in output per dollar's worth of additional labor, and λ is the common value. Thus, at the optimum production, the exchange of a dollar's worth of capital for a dollar's worth of labor does not change the output.

Another interpretation: Consider the optimal value as a function of the budget B. Set $x(B)$ and $y(B)$ denote the optimal capital and labor, respectively corresponding to $g(x,y) = qx + py = B$. We have

$$\frac{d}{dB} f(x(B), y(B)) = \frac{\partial f}{\partial x}\frac{dx}{dB} + \frac{\partial f}{\partial y}\frac{dy}{dB}$$

We have

$$f_x = \lambda g_x \text{ and } f_y = \lambda g_y$$

and $(x, y) = (x(B), y(B))$. Therefore,

$$\begin{aligned}
\frac{d}{dB} f(x(B), y(B)) &= \frac{\partial f}{\partial x}\frac{dx}{dB} + \frac{\partial f}{\partial y}\frac{dy}{dB} \\
&= \lambda\left(\frac{\partial g}{\partial x}\frac{dx}{dB} + \frac{\partial g}{\partial y}\frac{dy}{dB}\right) \\
&= \lambda\frac{g(x(B), y(B))}{dB}
\end{aligned}$$

Since $g(x(B), y(B)) = B$, we have

$$\frac{g(x(B), y(B))}{dB} = 1.$$

Therefore,

$$\frac{d}{dB} f(x(B), y(B)) = \lambda$$

Thus, **the value of** λ is the rate of the maximum value of f with respect to the budgetary allocation.

Example 5 Assume that

$$z = f(x, y) = Ax^\alpha y^{1-\alpha},$$

where $A > 0$, $\alpha > 0$ and $\alpha < 1$. This is a **Cobb-Douglas production function**. The **price of capital** is q, the **price of labor** is p, and we have a **budgetary constraint** $g(x, y) = qx + py = B$, where B is a positive constant. We would like to maximize the value of the product subject to the budgetary constraint.

Solution

We apply the technique of Lagrange multipliers. We need to have $\nabla f = \lambda \nabla g$. Thus,

$$\begin{aligned}
\alpha Ax^{\alpha-1}y^{1-\alpha} &= \lambda q, \\
(1-\alpha) Ax^\alpha y^{-\alpha} &= \lambda p, \\
qx + py &= B.
\end{aligned}$$

Eliminating λ from the first two equations,

$$\frac{\alpha Ax^{\alpha-1}y^{1-\alpha}}{q} = \frac{(1-\alpha) Ax^\alpha y^{-\alpha}}{p},$$

so that

$$\alpha p y = (1 - \alpha) q x.$$

Therefore,

$$q x + \frac{1 - \alpha}{\alpha} q x = B \Rightarrow \frac{1}{\alpha} q x = B \Rightarrow x = \frac{\alpha B}{q},$$

and

$$p y = B - q x = B - \alpha B = (1 - \alpha) B \Rightarrow y = \frac{1 - \alpha}{p} B.$$

Therefore, the candidate for (x, y) in order to maximize the value of the output is

$$\left(\frac{\alpha B}{q}, \frac{1 - \alpha}{p} B \right)$$

the corresponding value of λ is

$$
\lambda = \frac{\alpha A x^{\alpha - 1} y^{1 - \alpha}}{q} = \frac{\alpha A \left(\frac{\alpha B}{q} \right)^{\alpha - 1} \left(\frac{1 - \alpha}{p} B \right)^{1 - \alpha}}{q}
$$

$$
= \frac{\alpha A \alpha^{\alpha - 1} B^{\alpha - 1} (1 - \alpha)^{1 - \alpha} B^{1 - \alpha}}{q^{\alpha - 1} p^{\alpha - 1} q}
$$

$$
= \frac{A \alpha^{\alpha} (1 - \alpha)^{1 - \alpha}}{q^{\alpha} p^{\alpha - 1}}
$$

The corresponding value of f is

$$
f(x(B), y(B)) = A x^{\alpha} y^{1 - \alpha} = A \left(\frac{\alpha B}{q} \right)^{\alpha} \left(\frac{1 - \alpha}{p} B \right)^{1 - \alpha}
$$

$$
= \frac{A B \alpha^{\alpha} (1 - \alpha)^{1 - \alpha}}{q^{\alpha} p^{\alpha - 1}}
$$

We do have

$$\lambda(B) = \frac{d}{dB} f(x(B), y(B))$$

For example let $A = 1$, $B = 1$, $\alpha = 1/3$. Then,

$$z = f(x, y) = x^{1/3} y^{2/3}.$$

subject to the constraint $q x + p y = 1$. Let $q = 1$ and $p = 2$. Then, $x + 2y = 1$ is the constraint. The optimal values are

$$x = \frac{\alpha B}{q} = \frac{\frac{1}{3}}{1} = \frac{1}{3} \text{ and } y = \frac{\frac{2}{3}}{2} = \frac{1}{3}.$$

We have

$$z = x^{1/3} y^{2/3} = \frac{1}{3^{1/3}} \left(\frac{1}{3} \right)^{2/3} = \frac{1}{3}$$

Figure 5 shows the relevant level curve of f and the graph of the constraint $x + 2y = 1$. $\square$

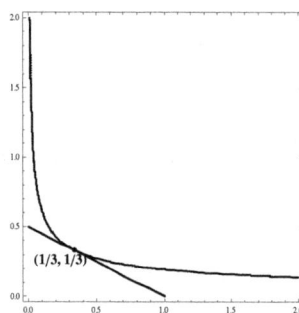

Figure 5

Problems

In problems 1-6, make use of Lagrange multipliers to determine the maximum and minimum values of f on the set D:

1.
$$f(x,y) = x + y; \ D = \{(x,y) : x^2 + y^2 = 4\}.$$

2.
$$f(x,y) = 3x - 4y, \ D = \{(x,y) : x^2 + y^2 = 9\}$$

3.
$$f(x,y) = xy, \ D = \{(x,y) : 4x^2 + y^2 = 8\}$$

4.
$$f(x,y) = x + y^2, \ D = \{(x,y) : 2x^2 + y^2 = 1\}$$

5.
$$f(x,y) = x^2 + 2y^2, \ D = \{(x,y) : x^2 + y^2 \le 4\}$$

6.
$$f(x,y) = xy, \ D = \{(x,y) : x^2 + 2y^2 \le 1\}$$

Chapter 13

Multiple Integrals

In this chapter we will discuss the integrals of real-valued functions of several variables. Integration of functions of two variables involves **double integrals**. In certain cases it is useful to make use of **polar coordinates**. Integration of functions of three variables involves **triple integrals**. In certain cases it is useful to use **cylindrical** or **spherical coordinates**. Double and triple integrals are needed for the calculation of **volumes** and **mass**. They are also essential for the understanding of the major facts of vector analysis that will be taken up in the next chapter.

13.1   Double Integrals over Rectangles

Let D be the rectangle $[a, b] \times [c, d]$ so that

$$D = \left\{ (x, y) \in \mathbb{R}^2 : a \leq x \leq b \text{ and } c \leq y \leq d \right\}.$$

Assume that f is a real-valued function that is continuous on D and $f(x, y) \geq 0$ for each $(x, y) \in D$. We would like to determine the volume of the region G in the xyz-space that is between the graph of $z = f(x, y)$ and D.

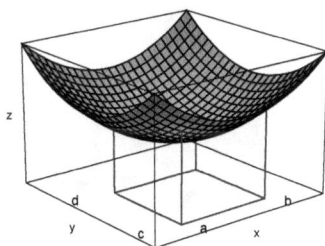

Figure 1: The integral of a positive-valued function is a volume

As in the case of the area problem in connection with single-variable functions, we will begin with approximations. Let

$$a = x_0 < x_1 < \cdots < x_{j-1} < x_j < \cdots < x_{m-1} < x_m = b,$$

and

$$c = y_0 < y_1 < \cdots < y_{k-1} < y_k < \cdots < y_{n-1} < y_n = d,$$

be partitions of $[a, b]$ and $[c, d]$, respectively. Let

$$\Delta x_j = x_j - x_{j-1} \text{ and } \Delta y_k = y_k - y_{k-1}$$

denote the lengths of the corresponding intervals. The grid that consists of the lines $x = x_j$, $j = 0, 1, 2, \ldots, m$ and $y = y_k$, $k = 1, 2, \ldots, n$, forms a collection of rectangles

$$R_{jk} = [x_{j-1}, x_j] \times [y_{k-1}, y_k]$$

that covers the rectangle D.

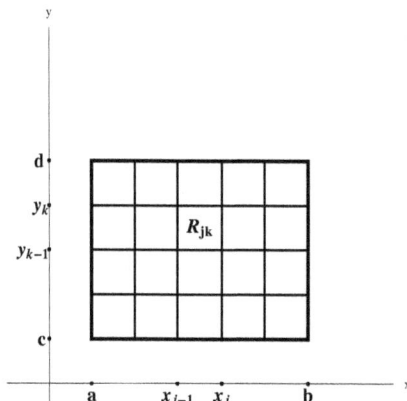

Figure 2: A rectangle is covered by smaller rectangles

We denote the area of R_{jk} by ΔA_{jk}, so that

$$\Delta A_{jk} = \Delta x_j \, \Delta y_k,$$

We sample a point $\left(x_j^*, y_k^*\right)$ from the rectangle R_{jk} and approximate the volume of the part of the region G between the graph of f and R_{jk} by the volume of the rectangular box with base R_{jk} and height $f\left(x_j^*, y_k^*\right)$, namely,

$$f\left(x_j^*, y_k^*\right) \Delta A_{jk} = f\left(x_j^*, y_k^*\right) \Delta x_j \, \Delta y_k$$

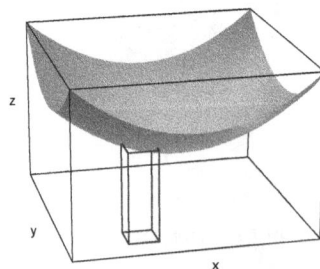

Figure 3: The volume over R_{jk} is approximated by the volume of a box

The volume of G is approximated by the **Riemann sum**

$$\sum_{k=1}^{n}\sum_{j=1}^{m} f\left(x_j^*, y_k^*\right) \Delta A_{jk} = \sum_{k=1}^{n}\sum_{j=1}^{m} f\left(x_j^*, y_k^*\right) \Delta x_j \, \Delta y_k$$

Under the assumption that f is continuous on D, there is number that is approximated by such a sum with desired accuracy provided that the rectangles R_{jk} are small enough. We will identify the volume of G with that number and denote it by the symbol

$$\iint_D f(x,y)\, dA = \iint_D f(x,y)\, dxdy.$$

Thus,

$$\left| \sum_{k=1}^{n} \sum_{j=1}^{m} f\left(x_k^*, y_j^*\right) \Delta A_{jk} - \iint_D f(x,y)\, dA \right|$$

is as small as we please provided that the maximum of ΔA_{jk} is small enough. The sums and the approximation property make sense even if f is of variable sign over D:

Definition 1 Let f be defined on the rectangle $D = [a,b] \times [c,d]$. We say that f is integrable on D and its **double integral on D** is

$$\iint_D f(x,y)\, dA$$

if

$$\left| \sum_{k=1}^{n} \sum_{j=1}^{m} f\left(x_k^*, y_j^*\right) \Delta A_{jk} - \iint_D f(x,y)\, dA \right|$$

as small as required provided that the maximum of ΔA_{jk} is small enough.

Even though we motivated the definition of the double integral as the volume of the region under the graph of a positive-valued function, the definition makes sense when the integrand has variable sign. Just as in the case of the integral of a function of a single variable, **we can interpret a double integral as "signed volume"**: If f is negative-valued on D, the double integral of f on D is $(-1) \times$ (volume of the region between the graph of f and D). We may refer to a double integral simply as an integral.

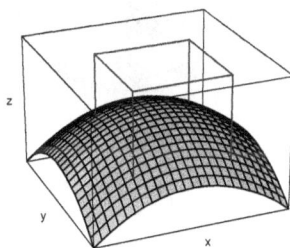

Figure 4: The integral of a negative-valued function is signed volume

Remark   It is known that the (double) integral

$$\iint_D f(x,y)\, dA$$

exists if f is continuous on D. We leave the proof of this fact to a course in advanced calculus. $\Diamond$

The following theorem enables us to reduce the calculation of a double integral to the calculation of integrals in a single variable:

Theorem 1 (Fubini's Theorem) Let f be continuous on $D = [a, b] \times [c, d]$. We have

$$\int \int_D f(x, y) \, dA = \int_{x=a}^{x=b} \left(\int_{y=c}^{y=d} f(x, y) \, dy \right) dx = \int_{y=c}^{y=d} \left(\int_{x=a}^{x=b} f(x, y) \, dx \right) dy.$$

The proof of Fubini's Theorem is left to a course in advanced calculus. Intuitively, the theorem corresponds to rearranging the double sum

$$\sum_{k=1}^{n} \sum_{j=1}^{m} f\left(x_j^*, y_k^*\right) \Delta A_{jk} = \sum_{k=1}^{n} \sum_{j=1}^{m} f\left(x_j^*, y_k^*\right) \Delta x_j \, \Delta y_k$$

as

$$\sum_{j=1}^{m} \left(\sum_{k=1}^{n} f\left(x_j^*, y_k^*\right) \Delta y_k \right) \Delta x_j \text{ or } \sum_{k=1}^{n} \left(\sum_{j=1}^{m} f\left(x_j^*, y_k^*\right) \Delta x_j \right) \Delta y_k.$$

You can also make a connection with the method of slices that was used for the computation certain volumes. If f is positive-valued on D, the integral

$$A(x) = \int_{y=c}^{y=d} f(x, y) \, dy$$

is the area of a cross section of the region between the graph of f and the rectangle D. The volume of that region can be calculated as

$$\int_{x=a}^{x=b} A(x) \, dx = \int_{x=a}^{x=b} \left(\int_{y=c}^{y=d} f(x, y) \, dy \right) dx.$$

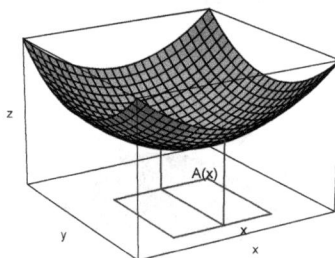

Figure 5: Fubini's Theorem is related to the method of slices

We can reverse the roles of x and y and obtain the equality

$$\int \int_D f(x, y) \, dA = \int_{y=c}^{y=d} \left(\int_{x=a}^{x=b} f(x, y) \, dx \right) dy$$

Remark 1 We may use the notation

$$\int \int_D f(x, y) \, dx dy$$

to denote

$$\int\int_D f(x,y)\, dA.$$

Thus, $dxdy$ does not mean that we integrate with respect to x first, unless we indicate the limits for x and y.

If there is no danger of confusion, we may dispense with the parentheses and set

$$\int_{x=a}^{x=b}\int_{y=c}^{y=d} f(x,y)\, dydx = \int_{x=a}^{x=b}\left(\int_{y=c}^{y=d} f(x,y)\, dy\right) dx,$$

and

$$\int_{y=c}^{y=d}\int_{x=a}^{x=b} f(x,y)\, dxdy = \int_{y=c}^{y=d}\left(\int_{x=a}^{x=b} f(x,y)\, dx\right) dy.$$

$\Diamond$

Example 1 Let $f(x,y) = x^2 + y^2$.Determine the volume of the region bounded by the graph of $z = f(x,y)$, the xy-plane, and the planes $x = \pm 2$, $y = \pm 2$.

Solution

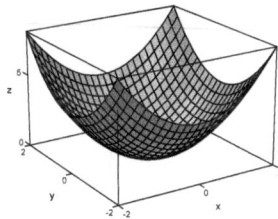

Figure 6

The volume is

$$\int\int_D \left(x^2 + y^2\right) dxdy,$$

where D is the rectangle $[-2, 2] \times [-2, 2]$. Therefore,

$$\int\int_D \left(x^2 + y^2\right) dxdy = \int_{y=-2}^{y=2}\left(\int_{x=-2}^{x=2}\left(x^2 + y^2\right) dx\right) dy$$

$$= \int_{y=-2}^{y=2}\left(\frac{x^3}{3} + y^2 x\Bigg|_{x=-2}^{x=2}\right) dy$$

$$= \int_{y=-2}^{y=2}\left(\left(\frac{8}{3} + 2y^2\right) - \left(-\frac{8}{3} - 2y^2\right)\right) dy$$

$$= \int_{y=-2}^{y=2}\left(4y^2 + \frac{16}{3}\right) dy$$

$$= \frac{4}{3}y^3 + \frac{16}{3}y\Bigg|_{y=-2}^{2} = \frac{128}{3}.$$

$\square$

Example 2 Determine

$$\int\int_D x \sin(xy)\,dxdy,$$

where D is the rectangle $[-\pi, \pi] \times [0, 1]$.

Solution

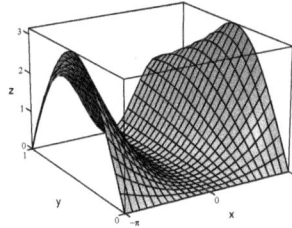

Figure 7

We have

$$\int\int_D x \sin(xy)\,dxdy = \int_{y=0}^{y=1} \left(\int_{x=-\pi}^{x=\pi} x \sin(xy)\,dx \right) dy.$$

The inner integral requires integration by parts. Let's reverse the order of integration:

$$\int\int_D x \sin(xy)\,dxdy = \int_{x=-\pi}^{x=\pi} \left(\int_{y=0}^{y=1} x \sin(xy)\,dy \right) dx..$$

The inner integral is

$$\int_{y=0}^{y=1} x \sin(xy)\,dy = -\cos(xy)|_0^1 = -\cos(x) + 1.$$

Therefore,

$$\int\int_D x \sin(xy)\,dxdy = \int_{x=-\pi}^{x=\pi} (-\cos(x) + 1)\,dx = -\sin(x) + x|_{-\pi}^{\pi} = 2\pi.$$

$\square$

Remark 2 If D is a rectangular region in $\mathbb{R}^2$ so that $D = [a, b] \times [c, d]$, an integral of the form

$$\int\int_D f(x)\,g(y)\,dxdy$$

can be expressed as a product of single-variable integrals:

$$\int\int_D f(x)\,g(y)\,dxdy = \left(\int_a^b f(x)\,dx \right) \left(\int_c^d g(y)\,dy \right).$$

Indeed,

$$
\begin{aligned}
\int\int_D f(x)\,g(y)\,dxdy &= \int_{x=a}^{x=b}\left(\int_{y=c}^{y=d} f(x)\,g(y)\,dy\right) dx \\
&= \int_{x=a}^{x=b} f(x)\left(\int_{y=c}^{y=d} g(y)\,dy\right) dx \\
&= \left(\int_{y=c}^{y=d} g(y)\,dy\right)\int_{x=a}^{x=b} f(x)\,dx \\
&= \left(\int_a^b f(x)\,dx\right)\left(\int_c^d g(y)\,dy\right)
\end{aligned}
$$

◊

Example 3 Evaluate

$$
\int\int_D \sin(x)\cos(y)\,dxdy,
$$

where D is the rectangle $[-\pi/2, 0] \times [0, \pi/2]$.

Solution

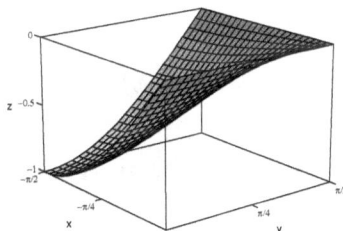

Figure 8

$$
\begin{aligned}
\int\int_D \sin(x)\cos(y)\,dxdy &= \int_{y=0}^{\pi/2}\int_{x=-\pi/2}^{0} \sin(x)\cos(y)\,dxdy \\
&= \left(\int_{x=-\pi/2}^{0} \sin(x)\,dx\right)\left(\int_{y=0}^{\pi/2} \cos(y)\,dxdy\right) \\
&\quad \left(-\cos(x)\big|_{x=-\pi/2}^{x=0}\right)\left(\sin(y)\big|_{y=0}^{\pi/2}\right) \\
&= (-1)(1) = -1.
\end{aligned}
$$

Note that the integral is negative, consistent with the fact that the integrand is negative on D.
□

Problems

In problems 1-4, calculate the given iterated integral:

1.

$$\int_{y=1}^{y=2}\left(\int_{x=0}^{x=3}\left(x^2+3xy^2\right)dx\right)dy$$

2.

$$\int_{x=0}^{x=1}\left(\int_{y=\pi/4}^{y=\pi/3}x\sin\left(y\right)dy\right)dx$$

3.

$$\int_{y=2}^{y=4}\left(\int_{x==1}^{x=3}e^{xy}ydx\right)dy$$

4.

$$\int_{x=0}^{x=2}\left(\int_{y=1}^{y=3}xy\sqrt{1+y^2}dy\right)dx$$

In problems 5-8, calculate the double integral:

5.

$$\iint_D\frac{xy^2}{x^2+1}dA \text{ where } D=\{(x,y):0\le x\le 1,\ -2\le y\le 2\}$$

6.

$$\iint_D xye^{x^2y}dA \text{ where } D=\{(x,y):0\le x\le 1,\ 0\le y\le 2\}$$

7.

$$\iint_D\frac{x^2}{1+y^2}dA \text{ where } D=\{(x,y):0\le x\le 1,\ 0\le y\le 1\}$$

8.

$$\iint_D y\cos\left(x+y\right)dA,\ \ D=\{(x,y):0\le x\le\frac{\pi}{3},\ 0\le y\le\frac{\pi}{2}\}$$

13.2   Double Integrals over Non-Rectangular Regions

In this section we will discuss double integrals on regions that are not rectangular. Let D be a bounded region in the Cartesian coordinate plane that has a boundary consisting of smooth segments. As in the case of a rectangle, let's begin with a function f that is continuous and nonnegative on D. In order to approximate the volume of the region G in the xyz-space that is between the graph of $z=f\left(x,y\right)$ and D we form a mesh in the xy-plane that consists of vertical and horizontal lines. In this case D is not necessarily the union of rectangles. Let's consider only those rectangles that have nonempty intersection with D and label them as

$$R_{jk}=[x_{j-1},x_j]\times[y_{k-1},y_k],\text{ where }j=0,1,2,\ldots,m\text{ and }k=0,1,2,\ldots,n.$$

We set

$$\Delta x_j=x_j-x_{j-1}\text{ and }\Delta y_k=y_k-y_{k-1}.$$

We can approximate the volume of G by a sum

$$\sum_{k=1}^{n}\sum_{j=1}^{m}f\left(x_j^*,y_k^*\right)\Delta A_{jk}=\sum_{k=1}^{n}\sum_{j=1}^{m}f\left(x_j^*,y_k^*\right)\Delta x_j\ \Delta y_k,$$

where $\left(x_j^*,y_k^*\right)\in R_{jk}.$.The double integral of f on D is the number

$$\iint_D f\left(x,y\right)dA\ =\int\int_D f\left(x,y\right)dxdy$$

that can be approximated by such sums as accurately as desired provided that the maximum of ΔA_{jk} is small enough. Such a number exists for any continuous function f on D even if the sign of f may vary.

There are counterparts of the Fubini Theorem that allowed us to calculated a double integral on a rectangle as an iterated integral. Assume that the region D in the xy-plane is bounded by the lines $x = a$, $x = b$ and the graphs of functions $f(x)$ and $g(x)$, as in Figure 1. We will refer to such a region as an **x-simple region**, or as **a region of type 1**.

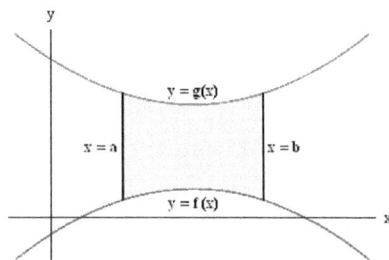

Figure 1

Assume that F is continuous on D. We can evaluate the double integral as an iterated integral:

$$\int\int_D F(x,y)\,dA = \int_{x=a}^{x=b}\left(\int_{y=f(x)}^{y=g(x)} F(x,y)\,dy\right)dx.$$

We will leave the proof to a course in advanced calculus. The statement should be plausible, as in the case of rectangular regions: As x varies from a to b,.the region is swept by rectangles of "infinitesimal" thickness dx that extend from $y = g(x)$ to $y = f(x)$.

The roles of x and y may be reversed: A region D is a **y-simple region, or a region of type 2**, if it is bounded by the lines $y = a$, $y = b$ and the graphs of $x = f(y)$ and $x = g(y)$, as in Figure 2. We have

$$\int\int_D F(x,y)\,dA = \int_{y=a}^{y=b}\left(\int_{x=f(y)}^{x=g(y)} F(x,y)\,dx\right)dxy.$$

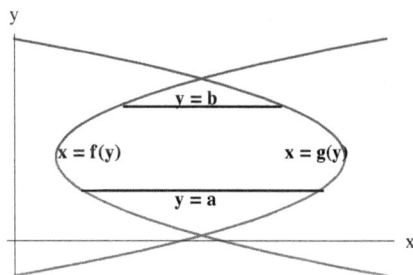

Figure 2

Example 1 Evaluate

$$\int\int_D \sqrt{1+x^2}\,dxdy,$$

where D is the triangle that is bounded by $y = x$, $x = 1$ and the x-axis.

Solution

Figure 3 shows the triangle D. The integral is the volume of the region between the graph of $z = \sqrt{1+x^2}$ and the triangle D in the xy-plane, as shown in Figure 4.

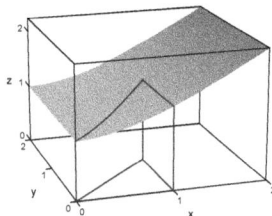

Figure 4

The region D is both x-simple and y-simple. We will treat it as an x-simple region. Thus,

$$\int\int_D \sqrt{1+x^2}dxdy = \int_{x=0}^{1}\left(\int_{y=0}^{x}\sqrt{1+x^2}dy\right)dx = \int_{x=0}^{1}\sqrt{1+x^2}\left(\int_{y=0}^{x}1dy\right)dx$$

$$= \int_{x=0}^{1}\sqrt{1+x^2}\left(y|_0^x\right)dx$$

$$= \int_{x=0}^{x=1}\sqrt{1+x^2}xdx.$$

We set $u = 1+x^2$, so that $du = 2xdx$. Therefore,

$$\int_{x=0}^{x=1}\sqrt{1+x^2}xdx = \frac{1}{2}\int_{u=1}^{2}\sqrt{u}du = \frac{1}{2}\left(\frac{u^{3/2}}{3/2}\bigg|_{u=1}^{2}\right) = \frac{2^{3/2}-1}{3}.$$

□

Example 2 Evaluate

$$\int\int_D x\sin(y)\,dA,$$

where D is the region bounded by $y = x$, $y = x^2$, $x = 0$ and $x = 1$.

Solution

Figure 5 shows the region D.

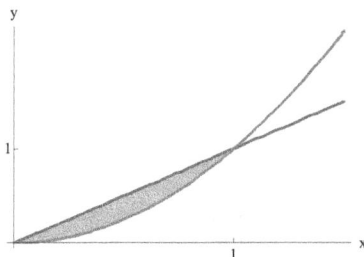

Figure 5

This is a region of type 1 and of type 2. We will treat it as a region of type I, and integrate with respect to y first:

$$\int\int_D x\sin(y)\,dA = \int_{x=0}^{x=1}\left(\int_{y=x^2}^{y=x} x\sin(y)\,dy\right)dx = \int_{x=0}^{x=1} x\left(\int_{y=x^2}^{y=x}\sin(y)\,dy\right)dx$$

$$= \int_{x=0}^{x=1} x\left(-\cos(y)\big|_{y=x^2}^{y=x}\right)dx$$

$$= \int_{x=0}^{x=1} x\left(-\cos(x)+\cos\left(x^2\right)\right)dx$$

$$= -\int_0^1 x\cos(x)\,dx + \int_0^1 x\cos\left(x^2\right)dx.$$

The first integral requires integration by parts: If we set $u = x$ and $dv = \cos(x)$, then

$$du = dx \text{ and } v = \int\cos(x)\,dx = \sin(x).$$

Thus,

$$\int x\cos(x)\,dx = \int u\,dv = uv - \int v\,du$$

$$= x\sin(x) - \int\sin(x)\,dx$$

$$= x\sin(x) + \cos(x).$$

Therefore,

$$\int_0^1 x\cos(x)\,dx = \sin(1) + \cos(1) - \cos(0) = \sin(1) + \cos(1) - 1.$$

As for the second integral, we set $u = x^2$ so that $du = 2x\,dx$. Thus,

$$\int x\cos\left(x^2\right)dx = \frac{1}{2}\int\cos(u)\,du = \frac{1}{2}\sin\left(x^2\right).$$

Therefore,

$$\int_0^1 x\cos\left(x^2\right)dx = \frac{1}{2}\sin(1).$$

Thus,

$$\int\int_D x\sin(y)\,dA = -\int_0^1 x\cos(x)\,dx + \int_0^1 x\cos\left(x^2\right)dx$$

$$= -\sin(1) - \cos(1) + 1 + \frac{1}{2}\sin(1)$$

$$= 1 - \cos(1) - \frac{1}{2}\sin(1).$$

$\square$

Example 3 Evaluate

$$\int\int_D ye^x\,dA$$

where D is the region bounded by $x = y^2/4$, $x = 1$, $y = 1$ and $y = 2$.

Solution

Figure 6 shows the region D.

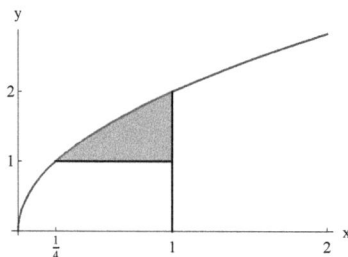

Figure 6

This is a region of type 1 and of type 2. We will treat it as a region of type 2, and integrate with respect to x first:

$$
\begin{aligned}
\int\int_D y e^{x^2}\, dA = \int_{y=1}^{y=2} \left(\int_{x=y^2/4}^{x=1} y e^x\, dx \right) dy &= \int_{y=1}^{y=2} y \left(\int_{x=y^2/4}^{x=1} e^x\, dx \right) dy \\
&= \int_{y=1}^{y=2} y \left(e^x \big|_{y^2/4}^{1} \right) dy \\
&= \int_{y=1}^{y=2} y \left(e - e^{y^2/4} \right) dy \\
&= \frac{e}{2} y^2 - 2 e^{y^2/4} \Big|_1^2 \\
&= 2e - 2e - \frac{e}{2} + 2 e^{1/4} \\
&= -\frac{e}{2} + 2 e^{1/4}.
\end{aligned}
$$

$\square$

In some cases, changing the order of integration enables us to evaluate an integral that is intractable in its original form, as in the following example.

Example 4 Evaluate the iterated integral

$$
\int_{x=0}^{1} \left(\int_{y=2x}^{y=2} e^{-y^2}\, dy \right) dx.
$$

Solution

We cannot evaluate the given iterated integral in terms of familiar antiderivatives. The first thing to do is to sketch the region of integration:

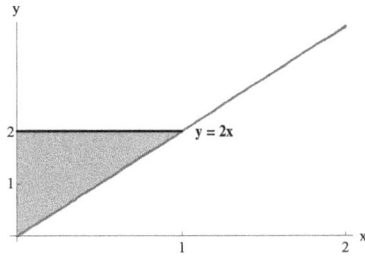

Figure 7

We will reverse the order of integration:

$$\int_{x=0}^{1} \left(\int_{y=2x}^{2} e^{-y^2} dy \right) dx = \int_{y=0}^{y=2} \left(\int_{x=0}^{x=y/2} e^{-y^2} dx \right) dy = \int_{y=0}^{y=2} e^{-y^2} \left(\int_{x=0}^{x=y/2} dx \right) dy$$

$$= \int_{y=0}^{y=2} e^{-y^2} \left(x \big|_{0}^{y/2} \right) dy$$

$$= \int_{y=0}^{y=2} e^{-y^2} \frac{y}{2} dy$$

$$= -\frac{1}{4} e^{-y^2} \bigg|_{0}^{2} = -\frac{1}{4} e^{-4} + \frac{1}{4}.$$

□

Problems

In problems 1 and 2 evaluate the given iterated integral:

1.

$$\int_{y=0}^{y=4} \int_{x=0}^{x=\sqrt{y}} xy^2 dxdy$$

2.

$$\int_{\theta=0}^{\theta=\pi/2} \int_{r=0}^{r=\cos(\theta)} e^{\sin(\theta)} drd\theta$$

In problems 3-6, sketch the region D and evaluate the given double integral:

3.

$$\int \int_{D} 4x^3 y dA,$$

where D is the region bounded by the graphs of $y = x^2$ and $y = 2x$.

4.

$$\int \int_{D} y\sqrt{x} dA,$$

where D is the region bounded by the graphs of $y = x^2$, $y = 0$, $x = 0$ and $x = 4$.

5.

$$\int \int_{D} y^2 e^{xy} dA,$$

where D is the region bounded by the graphs of $y = x$, $y = 1$ and $x = 0$.

6.

$$\int \int_{D} y^2 \sin\left(x^2\right) dA,$$

where D is the region bounded by the graphs of $y = -x^{1/3}$, $y = x^{1/3}$, $x = 0$ and $x = \sqrt{\pi}$.

In problems 7-9 determine the volume of the region D in $\mathbb{R}^3$:

7. D is bounded by the paraboloid $z = x^2 + y^2 + 1$, the xy-plane, the surfaces $y = x^2$ and $y = 1$.

8. D is bounded by the graphs of $z = e^x$, $z = -e^y$, $x = 0$, $x = 1$, $y = 0$ and $y = 1$.

9. D is the tetrahedron bounded by the plane $3x + 2y + z = 6$ and the coordinate planes.

In problems 10-12,
a) Sketch the region of integration,
b) Evaluate the integral by reversing the order of integration:

10.

$$\int_{y=0}^{y=1} \int_{x=3y}^{3} e^{x^2} \, dx \, dy$$

11.

$$\int_{y=0}^{y=\sqrt{\pi}/2} \int_{x=y}^{x=\sqrt{\pi}/2} \cos\left(x^2\right) \, dx \, dy.$$

12.

$$\int_{y=0}^{y=1} \int_{x=\arcsin(y)}^{\pi/2} \cos\left(x\right) \sqrt{1 + \cos^2\left(x\right)} \, dx \, dy.$$

13.3   Double Integrals in Polar Coordinates

In some cases it is convenient to evaluate a double integral by converting it to an integral in polar coordinates.

Assume that D is the set of points in the plane with polar coordinates (r, θ) such that $\alpha \le \theta \le \beta$ and $a \le r \le b$, and we would like to approximate

$$\int \int_D f\left(x, y\right) dA$$

by a sum that will be constructed as follows: Let

$$\alpha = \theta_0 < \theta_1 < \theta_2 < \cdots < \theta_{j-1} < \theta_j < \cdots < \theta_m = \beta,$$

be a partition of the interval $[\alpha, \beta]$, and let

$$a = r_0 < r_1 < r_2 < \cdots < r_{k-1} < r_k < \cdots < r_n = b$$

be a partition of $[a, b]$. We set

$$\Delta\theta_j = \theta_j - \theta_{j-1} \text{ and } \Delta r_k = r_k - r_{k-1}.$$

The rays $\theta = \theta_j$ and the arcs $r = r_k$ form a grid that covers D, as illustrated in Figure 1.

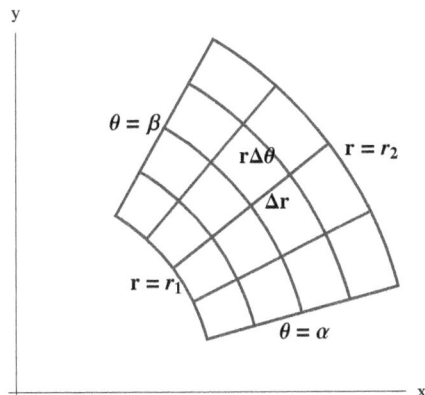

Figure 1

The area of the "polar rectangle that consists of the points with polar coordinates (r, θ), where $r_{k-1} < r < r_k$ and $\theta_{j-1} < \theta < \theta_j$ is

$$\frac{1}{2} r_k^2 \Delta\theta_j - \frac{1}{2} r_{k-1}^2 \Delta\theta_j = \frac{1}{2} \left(r_k^2 - r_{k-1}^2 \right) \Delta\theta_j = \frac{1}{2} \left(r_k + r_{k-1} \right) \left(r_k - r_{k-1} \right) \Delta\theta_j$$
$$= \tilde{r}_k \Delta r_k \Delta\theta_j,$$

where

$$\tilde{r}_k = \frac{1}{2} \left(r_k + r_{k-1} \right).$$

Therefore, we can approximate the part of

$$\int \int_D f(x, y) \, dA$$

that corresponds to that polar rectangle by

$$f \left(\tilde{r}_k \cos \left(\tilde{\theta}_j \right), \tilde{r} \sin \left(\tilde{\theta}_j \right) \right) \tilde{r}_k \Delta r_k \Delta\theta_j,$$

where

$$\tilde{\theta}_j = \frac{1}{2} \left(\theta_j + \theta_{j-1} \right)$$

and the total integral by the sum

$$\sum_{j=1}^{m} \sum_{k=1}^{n} f \left(\tilde{r}_k \cos \left(\tilde{\theta}_j \right), \tilde{r} \sin \left(\tilde{\theta}_j \right) \right) \tilde{r}_k \Delta r_k \Delta\theta_j.$$

This is a Riemann sum for the integral

$$\int_{\theta=\alpha}^{\theta=\beta} \int_{r=r_1}^{r=r_2} f \left(r \cos \left(\theta \right), r \sin \left(\theta \right) \right) r \, dr \, d\theta.$$

Thus, it is reasonable to conjecture that

$$\int \int_D f(x, y) \, dA = \int_{\theta=\alpha}^{\theta=\beta} \int_{r=r_1}^{r=r_2} f \left(r \cos \left(\theta \right), r \sin \left(\theta \right) \right) r \, dr \, d\theta.$$

This is indeed true. As a matter of fact the following more general fact valid:

Theorem 1 Assume that D is the set of points in the plane with polar coordinates (r, θ) such that $\alpha \leq \theta \leq \beta$ and

$$r_1(\theta) \leq r \leq r_2(\theta),$$

where the functions $r_1(\theta)$ and $r_2(\theta)$ are continuous. If f is continuous on D then

$$\int\int_D f(x, y)\, dA = \int_{\theta=\alpha}^{\theta=\beta} \int_{r=r_1(\theta)}^{r_2(\theta)} f(r\cos(\theta), r\sin(\theta))\, r\, dr\, d\theta.$$

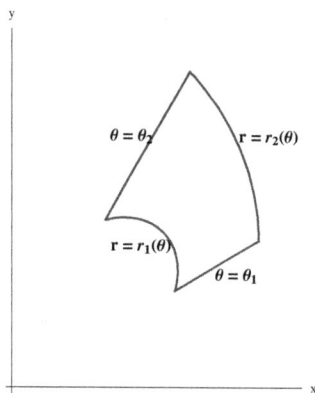

Figure 2

We leave the proof of Theorem 1 to a course in advanced calculus. We can view $r\, dr\, d\theta$ as the area of a "polar rectangle" with sides $r\, d\theta$ and dr.

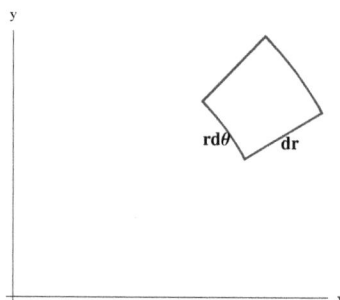

Figure 3

The conversion of a double integral from Cartesian to polar coordinates is useful if the region of integration and the integrand can be expressed conveniently in polar coordinates, as in the following examples.

Example 1 Let D be the annular region between the circles of radius $\pi/2$ and π centered at the origin. Evaluate

$$\int\int_D \frac{\sin\left(\sqrt{x^2 + y^2}\right)}{\sqrt{x^2 + y^2}}\, dx\, dy,$$

by expressing the integral in polar coordinates.

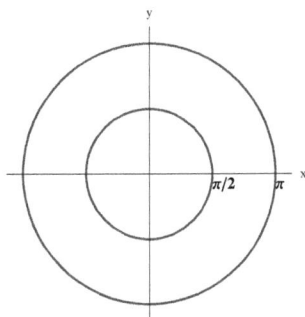

Figure 4

Solution

$$\int\int_D \frac{\sin\left(\sqrt{x^2+y^2}\right)}{\sqrt{x^2+y^2}}dxdy = \int_{\theta=0}^{\theta=2\pi}\int_{r=\pi/2}^{r=\pi}\frac{\sin(r)}{r}rdrd\theta$$

$$= \int_{\theta=0}^{\theta=2\pi}\int_{r=\pi/2}^{r=\pi}\sin(r)\,drd\theta$$

$$= \int_{\theta=0}^{\theta=2\pi}\left(-\cos(r)|_{\pi/2}^{\pi}\right)d\theta$$

$$= \int_{\theta=0}^{\theta=2\pi}\left(-\cos(\pi)+\cos(\pi/2)\right)d\theta$$

$$= \int_{\theta=0}^{\theta=2\pi}d\theta = 2\pi.$$

Remark 1 if D is the set of points in the plane with polar coordinates (r,θ) such that $\alpha \leq \theta \leq \beta$ and $0 \leq r \leq r(\theta)$, the area of D is

$$\frac{1}{2}\int_{\theta=\alpha}^{\theta=\beta}r^2(\theta)\,d\theta.$$

Indeed,

$$\text{Area of } D = \int\int_D dxdy = \int_{\theta=\alpha}^{\theta=\beta}\int_{r=0}^{r(\theta)}rdrd\theta = \int_{\theta=\alpha}^{\theta=\beta}\frac{1}{2}r^2(\theta)\,d\theta.$$

◊

Example 2 Compute the area inside one loop of the curve C that has the equation

$$r = \cos(3\theta)$$

in polar coordinates.

Solution

Figure 6 shows the graph of $r = \cos(3\theta)$ in the Cartesian θr-plane and the curve C (in the xy-plane).

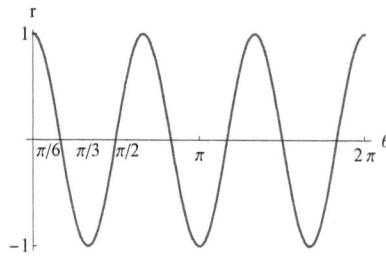

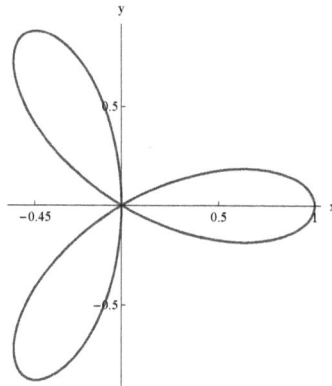

Figure 6

Note that

$$\cos(3\theta) = 0 \text{ if } 3\theta = \frac{\pi}{2} \Leftrightarrow \theta = \frac{\pi}{6}.$$

The part of the curve in the first quadrant is traced when θ increases from 0 to $\pi/6$. By symmetry, the area inside one loop is

$$2\left(\frac{1}{2}\int_{\theta=0}^{\theta=\pi/6} r^2(\theta)\,d\theta\right) = \int_0^{\pi/6}\cos^2(3\theta)\,d\theta = \frac{1}{3}\int_{u=0}^{u=\pi/2}\cos^2(u)\,du$$

$$= \frac{1}{3}\left(\frac{u}{2} + \frac{\cos(u)\sin(u)}{2}\bigg|_0^{\pi/2}\right)$$

$$= \frac{1}{3}\left(\frac{\pi}{4}\right) = \frac{\pi}{12}.$$

□

Example 3 Show that

$$\int_0^\infty e^{-x^2}\,dx = \frac{1}{2}\sqrt{\pi}.$$

Solution

We have shown that the improper integral

$$\int_0^\infty e^{-x^2}\,dx$$

converges. Since e^{-x^2} defines an even function,

$$\int_0^\infty e^{-x^2}\,dx = \frac{1}{2}\int_{-\infty}^\infty e^{-x^2}\,dx.$$

We have

$$\int_{-\infty}^\infty e^{-x^2}\,dx = \lim_{A\to\infty}\int_{-A}^A e^{-x^2}\,dx.$$

Now,

$$\left(\int_{-A}^A e^{-x^2}\,dx\right)^2 = \left(\int_{-A}^A e^{-x^2}\,dx\right)\left(\int_{-A}^A e^{-x^2}\,dx\right)$$

$$= \left(\int_{-A}^A e^{-x^2}\,dx\right)\left(\int_{-A}^A e^{-y^2}\,dy\right)$$

since the variable of integration is a dummy variable. Thus,

$$\left(\int_{-A}^A e^{-x^2}\,dx\right)^2 = \int_{y=-A}^{y=A}\int_{x=-A}^{x=A} e^{-x^2}e^{-y^2}\,dx\,dy = \int\int_{R_A} e^{-x^2}e^{-y^2}\,dx\,dy$$

$$= \int\int_{R_A} e^{-x^2-y^2}\,dx\,dy$$

where R_A is the square $[-A, A]\times[-A, A]$. Therefore,

$$\left(\int_{-\infty}^\infty e^{-x^2}\,dx\right)^2 = \lim_{A\to\infty}\left(\int_{-A}^A e^{-x^2}\,dx\right)^2 = \lim_{A\to\infty}\int\int_{R_A} e^{-x^2-y^2}\,dx\,dy.$$

We can evaluate the limit as

$$\lim_{R\to\infty}\int\int_{D_R} e^{-x^2-y^2}\,dx\,dy,$$

where D_R is the disk of radius R centered at the origin. We transform the integral to polar coordinates:

$$\int\int_{D_R} e^{-x^2-y^2}\,dx\,dy = \int_{\theta=0}^{\theta=2\pi}\int_{r=0}^{r=R} e^{-r^2} r\,dr\,d\theta$$

$$= \left(\int_{\theta=0}^{\theta=2\pi} d\theta\right)\left(\int_{r=0}^{r=R} e^{-r^2} r\,dr\right)$$

$$= 2\pi\left(-\frac{1}{2}e^{-r^2}\Big|_0^R\right) = 2\pi\left(-\frac{1}{2}e^{-R^2}+\frac{1}{2}\right) = -\pi e^{-R^2}+\pi.$$

Therefore,

$$\left(\int_{-\infty}^\infty e^{-x^2}\,dx\right)^2 = \lim_{R\to\infty}\int\int_{D_R} e^{-x^2-y^2}\,dx\,dy$$

$$= \lim_{R\to\infty}\left(-\pi e^{-R^2}+\pi\right) = \pi,$$

so that

$$\int_{-\infty}^\infty e^{-x^2}\,dx = \sqrt{\pi}.$$

Thus,

$$\int_0^\infty e^{-x^2}\,dx = \frac{1}{2}\int_{-\infty}^\infty e^{-x^2}\,dx = \frac{1}{2}\sqrt{\pi},$$

as claimed. $\square$

Example 4 Determine the volume of the region between the paraboloid $z = 4 - x^2 - y^2$ and the xy-plane.

Solution

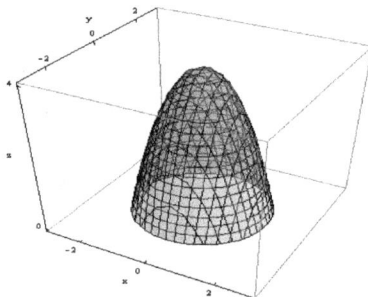

Figure 7

The given paraboloid intersects the xy-plane along the circle

$$4 - x^2 - y^2 = 0 \Leftrightarrow x^2 + y^2 = 4.$$

This a circle of radius 2 centered at the origin. The required volume is

$$\int\int_{x^2+y^2\leq 4}\left(4 - x^2 - y^2\right)dxdy = \int_{\theta=0}^{2\pi}\int_{r=0}^{r=2}\left(4 - r^2\right)r\,dr\,d\theta = 2\pi\int_{r=0}^{r=2}\left(4r - r^3\right)dr$$

$$= 2\pi\left(2r^2 - \frac{1}{4}r^4\Big|_0^2\right)$$

$$= 2\pi\left(4\right) = 8\pi.$$

Problems

In problems 1-5,
a) Sketch the region D,
b) Evaluate the given double integral by transforming the integral to an integral in polar coordinates:

1.

$$\int\int_D xy\,dA,$$

where

$$D = \{(x,y) : x^2 + y^2 \leq 16 \text{ and } x \geq 0,\ y \geq 0\}$$

(i.e., D is the part of the disk of radius 4 centered at the origin that is in the first quadrant).

2.

$$\int\int_D \sin\left(x^2 + y^2\right)dA,$$

where
$$D = \{(x,y) : x^2 + y^2 \leq 4 \text{ and } y \geq 0\}$$
(i.e., D is the part of the disk of radius 2 centered at the origin that is in the upper half-plane).

3.
$$\int\int_D \sqrt{9 - x^2 - y^2}\, dA,$$

where
$$D = \{(x,y) : x^2 + y^2 \leq 9 \text{ and } x \geq 0\}$$
(i.e., D is the part of the disk of radius 3 centered at the origin that is in the right half-plane).

4.
$$\int\int_D e^{-x^2 - y^2}\, dA,$$

where
$$D = \{(x,y) : 1 \leq x^2 + y^2 \leq 4 \text{ and } y \geq 0\}$$
(i.e., D is the annular region in the upper half-plane between the circles of radius 1 and 2 centered at the origin).

5.
$$\int\int_D \arctan\left(\frac{y}{x}\right) dA,$$

where
$$D = \{(x,y) : 1 \leq x^2 + y^2 \leq 4 \text{ and } 0 \leq y \leq x\}.$$

In problems 6 and 7, sketch the region D and determine the area of D by using a double integral in polar coordinates:

6. D is the region inside one loop of the graph of $r = \sin(2\theta)$.

7. D is the region inside the cardioid $r = 1 + \cos(\theta)$ and ou tside the circle $r = 3\cos(\theta)$.

In problems 8 and 9, make use of polar coordinates to compute the volume of the region D in $\mathbb{R}^3$:

8. D is between the cone $z = \sqrt{x^2 + y^2}$ and the cylinder $x^2 + y^2 = 4$, and above the xy-plane, as in the picture:

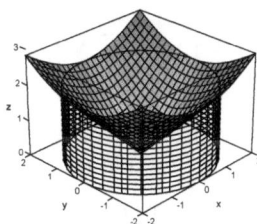

9. D is between the sphere $x^2 + y^2 + z^2 = 16$ and the cylinder $x^2 + y^2 = 4$, as in the picture:

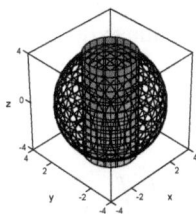

In problems 10 and 11,

a) Sketch the region that is relevant to the given iterated integral,

b) Make use of polar coordinates to evaluate the integral:

10.

$$\int_{x=-3}^{x=3} \int_{y=0}^{y=\sqrt{9-x^2}} \sin\left(x^2 + y^2\right) dy dx.$$

11.

$$\int_{y=0}^{y=1} \int_{x=y}^{x=\sqrt{2-y^2}} x dx dy$$

13.4   Applications of Double Integrals

In this section we will discuss some physical applications of double integrals and some applications to probability. In physical applications you may assume that the units are mks-units (meter-kilogram-second).

Density and Mass

Assume that a thin plate is modeled as a region D in the xy-plane. If the plate has area A and constant **density** ρ (per unit area) its **mass** is

$$m = \rho A.$$

How about the mass of such a plate if its density is not constant but is a function $\rho(x, y)$? A reasonable approach is to cover D with a mesh of rectangles, as in the general definition of a double integral on D. Assuming that $\rho(x, y)$ is continuous, we can assume that density has the constant value $\rho\left(x_j^*, y_k^*\right)$ on the rectangle $R_{jk} = [x_{j-1}, x_j] \times [y_{k-1}, y_k]$ (we are using the same notation as in the definition of a double integral). The mass of the portion of the plate on R_{jk} is approximated by

$$\rho\left(x_j^*, y_k^*\right) \Delta A_{jk} = \rho\left(x_j^*, y_k^*\right) \Delta x_j \Delta y_k,$$

and the total mass is approximated by the sum

$$\sum_{k=1}^{n}\sum_{j=1}^{n} \rho\left(x_j^*, y_k^*\right) \Delta A_{jk} = \sum_{k=1}^{n}\sum_{j=1}^{n} \rho\left(x_j^*, y_k^*\right) \Delta x_j \Delta y_k,$$

where the sum is over those rectangles R_{jk} that have nonempty intersection with the region D. Since such Riemann sums approximate the double integral

$$m(D) = \int\int_D \rho(x, y)\, dA$$

as accurately as required provided that the mesh is sufficiently fine, we will calculate the mass $m(D)$ of the plate as the above double integral.

Example 1 Assume that a thin plate of mass density $\sqrt{x^2 + y^2}$ (kg/m^2) is in the shape of a disk of radius 2 meters centered at the origin. Calculate the mass of the plate.

Solution

The mass is

$$m(D) = \int\int_{D_2} \sqrt{x^2 + y^2} dx dy$$

where D denotes the disk of radius 2 meters centered at the origin. We transform to polar coordinates:

$$m(D) = \int\int_D \sqrt{x^2 + y^2}\, dxdy \quad = \quad \int_{\theta=0}^{\theta=2\pi} \int_{r=0}^{2} r^2 dr d\theta$$

$$= \quad 2\pi \left(\frac{1}{3} r^3 \Big|_0^2 \right) = 2\pi \left(\frac{8}{3} \right) = \frac{16\pi}{3} \cong 16.755\,2 \text{ kg.}$$

$\square$

Moments and Center of Mass

The moment of a point mass m at (x, y) with respect to the x-axis is defined as my and its moment with respect to the y-axis is defined as mx. If the region D represents a thin plate that has mass density $\rho(x, y)$, as in our discussion of mass, we arrive at the definitions of moments as follows: As before, we will cover D with a mesh of rectangles and approximate the mass of the part of the plate corresponding to the small rectangle $R_{jk} = [x_{j-1}, x_j] \times [y_{k-1}, y_k]$ by $\rho\left(x_j^*, y_k^*\right) \Delta x_j \Delta y_k$, where $\left(x_j^*, y_k^*\right)$ is a point inside that rectangle. Let's assume that this mass is concentrated at a point $\left(x_j^*, y_k^*\right)$, so that its moment with respect to the x-axis is

$$y_k^* \rho\left(x_j^*, y_k^*\right) \Delta x_j \Delta y_k$$

and its moment with respect to the y-axis is

$$x_j^* \rho\left(x_j^*, y_k^*\right) \Delta x_j \Delta y_k.$$

The sum of the moments with respect to the x-axis is

$$\sum_{k=1}^{n} \sum_{j=1}^{n} y_k^* \rho\left(x_j^*, y_k^*\right) \Delta x_j \Delta y_k = \sum_{k=1}^{n} \sum_{j=1}^{n} y_k^* \rho\left(x_j^*, y_k^*\right) \Delta A_{jk},$$

and the sum of the moments with respect to the y-axis is

$$\sum_{k=1}^{n} \sum_{j=1}^{n} x_j^* \rho\left(x_j^*, y_k^*\right) \Delta x_j \Delta y_k = \sum_{k=1}^{n} \sum_{j=1}^{n} x_j^* \rho\left(x_j^*, y_k^*\right) \Delta A_{jk}.$$

These are Riemann sums that approximate the integrals

$$\int\int_D y\rho(x, y)\, dxdy \text{ and } \int\int_D x\rho(x, y)\, dxdy,$$

respectively. We will define **the moment of the plate with respect to the x-axis** as

$$M_x(D) = \int\int_D y\rho(x, y)\, dxdy = \int\int_D y\rho(x, y)\, dA,$$

and **the moment of the plate with respect to the y-axis** as

$$M_y(D) = \int\int_D x\rho(x, y)\, dxdy = \int\int_D x\rho(x, y)\, dA.$$

The center of mass of plate with mass $m(D)$ is the point $(\bar{x}, \bar{y})$, where

$$\bar{x} = \frac{M_y(D)}{m(D)} \text{ and } \bar{y} = \frac{M_x(D)}{m(D)}.$$

Physically, if we assume that the plate is supported only at the center of mass it would remain horizontal.

Example 2 Assume that a plate is represented as a half-disk

$$D = \left\{ (x, y) : x^2 + y^2 \leq 9, \ x \geq 0 \right\},$$

and has density

$$\rho\left(x, y\right) = x^2 + y^2.$$

Calculate the center of mass of the plate.

Solution

The mass of the plate is

$$m\left(D\right) = \int\int_D \left(x^2 + y^2\right) dxdy.$$

Let's transform to polar coordinates:

$$m\left(D\right) = \int\int_D \left(x^2 + y^2\right) dxdy. \quad = \int_{\theta=-\pi/2}^{\theta=\pi/2} \int_{r=0}^{r=3} r^3 dr d\theta$$

$$= \pi \left(\frac{1}{4} r^4 \Big|_0^3\right) = \frac{3^4 \pi}{4} = \frac{81\pi}{4} \cong 63.617\,3.$$

The moment of the plate with respect to the y-axis is

$$M_y\left(D\right) = \int\int_D x\rho\left(x, y\right) dxdy = \int\int_D x\left(x^2 + y^2\right) dxdy.$$

We will transform to polar coordinates again:

$$M_y\left(D\right) = \int\int_D x\left(x^2 + y^2\right) dxdy \quad = \int_{\theta=-\pi/2}^{\theta=\pi/2} \int_{r=0}^{r=3} r\cos\left(\theta\right) r^3 dr d\theta$$

$$= \int_{\theta=-\pi/2}^{\theta=\pi/2} \sin\left(\theta\right) \left(\int_{r=0}^{r=3} r^4 dr\right) d\theta$$

$$= \left(\int_{\theta=-\pi/2}^{\theta=\pi/2} \sin\left(\theta\right) d\theta\right) \left(\int_{r=0}^{r=3} r^4 dr\right)$$

$$= \left(\sin\Big|_{-\pi/2}^{\pi/2}\right) \left(\frac{r^5}{5}\Big|_0^3\right)$$

$$= (2) \left(\frac{3^5}{5}\right) = \frac{486}{5}$$

Therefore, the x-coordinate of the center of mass is

$$\bar{x} = \frac{M_y\left(D\right)}{m\left(D\right)} = \frac{\dfrac{486}{5}}{\dfrac{81\pi}{4}} = \frac{24}{5\pi} \cong 1.527\,89$$

The moment of the plate with respect to the x-axis is

$$M_x\left(D\right) = \int\int_D y\rho\left(x, y\right) dxdy = \int\int_D y\left(x^2 + y^2\right) dxdy.$$

We will transform to polar coordinates:

$$
\begin{aligned}
M_x(D) = \int\int_D y\left(x^2 + y^2\right) dx dy &= \int_{\theta=-\pi/2}^{\theta=\pi/2}\int_{r=0}^{r=3} r\sin(\theta)\, r^3 dr d\theta \\
&= \int_{\theta=-\pi/2}^{\theta=\pi/2} \sin(\theta)\left(\int_{r=0}^{r=3} r^4 dr\right) d\theta \\
&= \left(\int_{\theta=-\pi/2}^{\theta=\pi/2} \sin(\theta)\, d\theta\right)\left(\int_{r=0}^{r=3} r^4 dr\right) \\
&= \left(-\cos(\theta)\big|_{-\pi/2}^{\pi/2}\right)\left(\frac{r^5}{5}\bigg|_0^3\right) = 0
\end{aligned}
$$

Therefore, the y-cordinate of the center of mass is 0. The center of mass is the point $(24/5\pi, 0) \cong (1.527\,89, 0)$. $\square$

Probability

We discussed the probability density function f that corresponds to a single continuous random variable X: The probability that X has values between a and b is

$$
P(a \le X \le b) = \int_a^b f(x)\, dx.
$$

If we have two random variables X and Y that have the **joint density function** $f(x, y)$, the probability that X has values between a and b, Y has values between c and d is

$$
P(a \le X \le b, c \le Y \le d) = \int_{x=a}^{x=b}\int_{y=c}^{y=d} f(x, y)\, dx dy.
$$

More generally, if D is a region in the xy-plane, the probability that the pair (X, Y) has values in D is

$$
P((X, Y) \in D) = \int\int_D f(x, y)\, dx dy.
$$

We have $f(x, y) \ge 0$ for all (x, y) and

$$
\int\int_{\mathbb{R}^2} f(x, y)\, dx dy = 1
$$

since the probability that (X, Y) attains some value is 1.

The random variables are said to be **independent random variables** if

$$
f(x, y) = f_1(x) f_2(y).
$$

Example 3 Assume that the independent random variables X and Y have distribution functions

$$
f_1(x) = \begin{cases} 0 & \text{if } x < 0, \\ \frac{1}{8}e^{-x/8} & \text{if } x \le 0 \end{cases}, \quad \text{and } f_2(y) = \begin{cases} 0 & \text{if } x < 0, \\ \frac{1}{4}e^{-y/4} & \text{if } x \le 0 \end{cases},
$$

respectively. Determine the probability that $X + Y < 16$.

Solution

Since X and Y are independent, their joint density function is

$$
f(x, y) = f_1(x) f_2(y) = \begin{cases} \frac{1}{32}e^{-x/8}e^{-y/4} & \text{if } x \ge 0 \text{ and } y \ge 0, \\ 0 & \text{otherwise.} \end{cases}
$$

Therefore, the probability that $X + Y < 16$ is

$$\int\int_D \frac{1}{32} e^{-x/8} e^{-y/4} dx dy,$$

where D is the triangular region in the first quadrant below the line $x + y = 16$. We have

$$
\begin{aligned}
\int\int_D \frac{1}{32} e^{-x/8} e^{-y/4} dx dy &= \int_{x=0}^{x=16} \int_{y=0}^{16-x} \frac{1}{32} e^{-x/8} e^{-y/4} dy dx \\
&= \frac{1}{32} \int_{x=0}^{x=16} e^{-x/8} \left(\int_{y=0}^{16-x} e^{-y/4} dy \right) dx \\
&= \frac{1}{32} \int_{x=0}^{x=16} e^{-x/8} \left(-4 e^{-y/4} \Big|_0^{16-x} \right) dx \\
&= \frac{1}{8} \int_{x=0}^{x=16} e^{-x/8} \left(1 - e^{-(4-x/4)} \right) dx \\
&= 1 + \frac{1}{e^4} - \frac{2}{e^2} \cong 0.747\,645.
\end{aligned}
$$

□

Recall that a random variable is **normally distributed** if it has a density function of the form

$$\frac{1}{\sigma\sqrt{2\pi}} e^{-(x-\mu)^2/(2\sigma^2)},$$

where μ is the mean and σ is the standard deviation.

Example 4 Assume that X and Y are independent normally distributed random variables with means 5 and 6, and standard deviations 0.2 and 0.1, respectively. Calculate the probability that $4.5 < X < 5.5$ and $5.5 < Y < 6.5$.

Solution

The density functions of X and Y are

$$f_1(x) = \frac{1}{0.2\sqrt{2\pi}} e^{-(x-5)^2/0.08} \quad \text{and} \quad f_2(y) = \frac{1}{0.1\sqrt{2\pi}} e^{-(y-6)^2/0.02},$$

respectively. Therefore,

$$
\begin{aligned}
&P\left(4.5 < X < 5.5 \text{ and } 5.5 < Y < 6.5\right) \\
&= \int_{y=5.5}^{y=6.5} \int_{x=4.5}^{x=5.5} f_1(x) f_2(y) dx dy \\
&= \left(\int_{y=5.5}^{y=6.5} f_2(y) dy \right) \left(\int_{x=4.5}^{x=5.5} f_1(x) dx \right) \\
&= \left(\int_{y=5.5}^{y=6.5} \frac{1}{0.1\sqrt{2\pi}} e^{-(y-6)^2/0.02} dy \right) \left(\int_{x=4.5}^{x=5.5} \frac{1}{0.2\sqrt{2\pi}} e^{-(x-5)^2/0.08} dx \right) \\
&\cong (0.999\,999)(0.987\,581) = 0.987\,58
\end{aligned}
$$

Problems

In problems 1 and 2 determine the mass and center of mass of a thin plate that occupies the region D and has mass density $\rho(x, y)$:

1.
$$D = \{(x,y) : 0 \le x \le 4, \ 0 \le y \le 2\}, \ \rho(x,y) = 1 + x + y$$

2. D is the half-disk
$$D = \{(x,y) : x^2 + y^2 \le 9, \ y \ge 0\}$$

and
$$\rho(x,y) = 81 - \sqrt{x^2 + y^2}$$

Assume that a thin plate occupies the region D in the xy-plane and has mass density $\rho(x,y)$. The **moment of inertia** of the plate **about the x-axis** is

$$I_x = \int\int_D y^2 \rho(x,y) \, dx dy$$

and the **moment of inertia** of the plate **about the y-axis** is

$$I_y = \int\int_D x^2 \rho(x,y) \, dx dy.$$

The **moment of inertia** of the plate **about the origin (the polar moment of inertia)** is

$$I_0 = I_y + I_x = \int\int_D (x^2 + y^2) \rho(x,y) \, dx dy.$$

In problems 3 and 4 calculate the I_x, I_y and I_0 of a thin plate that occupies the region D and has mass density $\rho(x,y)$:

3. D is the interior of the circle that is centered at the origin and has radius r_0. The density has tha constant value ρ_0.

4. D is the half-disk
$$D = \{(x,y) : x^2 + y^2 \le 9, \ y \ge 0\}$$

and
$$\rho(x,y) = 4 - \sqrt{x^2 + y^2}$$

5. Assume that X and Y are independent random variables that have distribution functions

$$f_1(x) = \begin{cases} 0 & \text{if} \quad x < 0, \\ \frac{1}{6}e^{-x/6} & \text{if} \quad x \le 0 \end{cases}, \quad \text{and } f_2(y) = \begin{cases} 0 & \text{if} \quad x < 0, \\ \frac{1}{2}e^{-y/2} & \text{if} \quad x \le 0 \end{cases},$$

respectively. Determine the probability that $X + Y < 8$.

6 [C] Assume that X and Y are independent normally distributed random variables with means 3 and 5, and standard deviations 0.2 and 0.1, respectively. Make use of your computational utility in order to calculate the probability that $2.5 < X < 3.5$ and $4.5 < Y < 5.5$ approximately.

13.5 Triple Integrals

As in the case of double integrals, let's begin by considering a case that leads to a triple integral. Assume that D is a rectangular box $[a,b] \times [c,d] \times [e,f]$ in $\mathbb{R}^3$, and $f(x,y,z)$ is the continuous mass density (say, in kilograms per cubic meter) of some material that occupies D. We will approximate the mass of the material by partitioning D via small rectangular boxes. Let

$$\begin{aligned} a &= x_0 < x_1 < \cdots < x_{j-1} < x_j < \cdots < x_{m-1} < x_m = b, \\ c &= y_0 < y_1 < \cdots < y_{k-1} < y_k < \cdots < y_{n-1} < y_n = d, \\ e &= z_0 < z_1 < \cdots < z_{l-1} < z_l < \cdots < z_{p-1} < z_p = f \end{aligned}$$

be partitions of the intervals, $[a, b]$, $[c, d]$ and $[e, f]$, respectively. We set

$$\Delta x_j = x_j - x_{j-1}, \Delta y_k = y_k - y_{k-1} \text{ and } \Delta z_l = z_l - z_{l-1}.$$

The volume of the box

$$R_{jkl} = [x_j - x_{j-1}] \times [y_k - y_{k-1}] \times [z_l - z_{l-1}]$$

is

$$\Delta V_{jkl} = \Delta x_j \Delta y_k \Delta z_l.$$

We can select a point

$$\left(x_j^*, y_k^*, z_p^*\right) \in [x_j - x_{j-1}] \times [y_k - y_{k-1}] \times [z_l - z_{l-1}]$$

arbitrarily in R_{jkl} and approximate the mass of the material in R_{jkl} by

$$f\left(x_j^*, y_k^*, z_p^*\right) \Delta V_{jkl} = f\left(x_j^*, y_k^*, z_p^*\right) \Delta x_j \Delta y_k \Delta z_l.$$

The total mass of the material is approximated by "a Riemann sum"

$$\sum_{j=1}^{m} \sum_{k=1}^{n} \sum_{l=1}^{p} f\left(x_j^*, y_k^*, z_p^*\right) \Delta V_{jkl} = \sum_{j=1}^{m} \sum_{k=1}^{n} \sum_{l=1}^{p} f\left(x_j^*, y_k^*, z_p^*\right) \Delta x_j \Delta y_k \Delta z_l.$$

It is known that such Riemann sums approximate a number that depends only on f and D with desired accuracy provided that f is continuous and the maximum of the ΔV_{jkl}'s is small enough. That number is called the triple integral of f on D and denoted as

$$\iiint_D f(x, y, z) \, dV = \iiint_D f(x, y, z) \, dx dy dz.$$

Thus

$$\left| \iiint_D f(x, y, z) \, dV - \sum_{j=1}^{m} \sum_{k=1}^{n} \sum_{l=1}^{p} f\left(x_j^*, y_k^*, z_p^*\right) \Delta V_{jkl} \right|$$

$$= \left| \iiint_D f(x, y, z) \, dx dy dz - \sum_{j=1}^{m} \sum_{k=1}^{n} \sum_{l=1}^{p} f\left(x_j^*, y_k^*, z_p^*\right) \Delta x_j \Delta y_k \Delta z_l \right|$$

is as small as desired if the maximum of $\Delta x_j \Delta y_k \Delta z_l$ is sufficiently small. The triple integral corresponds to the mass of the material that occupies D if f is the density of the material. The triple integral is defined for an arbitrary function that is continuous on D. We will see other applications of triple integrals in Chapter 14.

As in the case of double integral, we can evaluate the triple integral as an iterated integral in any order. For example,

$$\iiint_D f(x, y, z) \, dV = \int_{x=a}^{x=b} \int_{y=c}^{y=d} \int_{z=e}^{z=f} f(x, y, z) \, dz dy dx.$$

Example 1 Evaluate

$$\iiint_D f(x, y, z) \, dV,$$

where $f(x, y, z) = yz + xz + xy$ and D is the rectangular box $[0, 1] \times [0, 2] \times [0, 3]$.

Solution

$$\int\int\int_D f\,(x,y,z)\,dV = \int_{z=0}^{3}\int_{y=0}^{2}\int_{x=0}^{x=1}(yz+xz+xy)\,dxdydz.$$

We have

$$\int_{x=0}^{x=1}(yz+xz+xy)\,dx = yzx + \frac{1}{2}x^2z + \frac{1}{2}x^2y\Big|_{x=0}^{1} = yz + \frac{1}{2}z + \frac{1}{2}y.$$

Therefore,

$$\int_{y=0}^{2}\int_{x=0}^{x=1}(yz+xz+xy)\,dxdy = \int_{y=0}^{2}\left(yz + \frac{1}{2}z + \frac{1}{2}y\right)dy$$

$$= \frac{1}{2}y^2z + \frac{1}{2}zy + \frac{1}{4}y^2\Big|_{y=0}^{2}$$

$$= \frac{1}{2}(4)z + \frac{1}{2}(2)z + \frac{1}{4}(2^2)$$

$$= 2z + z + 1 = 3z + 1.$$

Thus,

$$\int_{z=0}^{3}\int_{y=0}^{2}\int_{x=0}^{x=1}(yz+xz+xy)\,dxdydz = \int_{z=0}^{3}(3z+1)\,dz$$

$$= \frac{3}{2}z^2 + z\Big|_{z=0}^{3}$$

$$= \frac{3}{2}(9) + 3 = \frac{33}{2}.$$

$\square$

As in the case of double integrals, the concept of the triple integral is generalized to regions D in $\mathbb{R}^3$ other than rectangular boxes by considering Riemann sums

$$\sum_{j=1}^{m}\sum_{k=1}^{n}\sum_{l=1}^{p}f\,(x_j^*,y_k^*,z_p^*)\,\Delta x_j\Delta y_k\Delta z_l$$

where the sum is over those rectangular boxes with nonempty intersection with the given region. The triple integral of f on D is the number

$$\int\int\int_D f\,(x,y,z)\,dV = \int\int\int_D f\,(x,y,z)\,dxdydz$$

that can be approximated by such sums with desired accuracy provided that the maximum of $\Delta x_j\Delta y_k\Delta z_l$'s is small enough. A triple integral can be evaluated as an iterated integral whereby the given integral is reduced to a double integral. For example, Let D be a region in $\mathbb{R}^3$ that consists of points (x,y,z) where (x,y) varies in some region D_2 in the xy-plane, and

$$g_1\,(x,y) \le z \le g_2\,(x,y) \text{ for each } (x,y) \in D_2.$$

If f is continuous on D

$$\int\int\int_D f\,(x,y,z)\,dV = \int\int_{D_2}\left(\int_{z=g_1(x,y)}^{z=g_2(x,y)} f\,(x,y,z)\,dz\right)dxdy$$

The roles of x, y and z may be interchanged
Note that

$$\iint\int_D dV$$

is simply the volume of D.

Example 2 Evaluate

$$\iint\int_D x^2 dV,$$

where D is the tetrahedron bounded by the coordinate planes and the part of the plane $x+y+z = 1$ in the first octant.

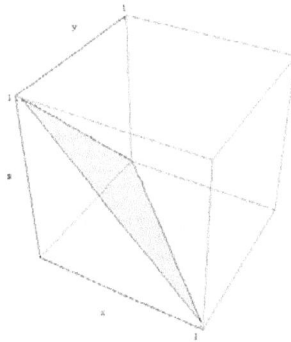

Figure 1

Solution

The region D is of type I. The projection of D onto the xy-plane is the triangle Δ that is bounded by $x + y = 1$ and the coordinate axes.

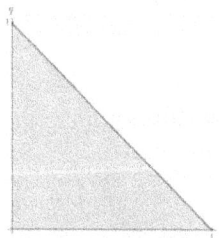

Figure 2

For each $(x, y) \in \Delta$, we have $(x, y, z) \in D$ iff $0 \leq z \leq 1 - x - y$. Therefore,

$$\iint\int_D x^2 dV = \iint_\Delta \left(\int_{z=0}^{z=1-x-y} x^2 dz\right) dxdy = \int_{x=0}^1 \int_{y=0}^{y=1-x} \left(\int_{z=0}^{z=1-x-y} x^2 dz\right) dydx.$$

We have

$$\int_{z=0}^{z=1-x-y} x^2 dz = x^2 z\Big|_{z=0}^{z=1-x-y} = x^2 \left(1 - x - y\right) = x^2 - x^3 - x^2 y.$$

Therefore,

$$\int_{y=0}^{y=1-x} \left(\int_{z=0}^{z=1-x-y} x^2 dz \right) dy = \int_{y=0}^{y=1-x} \left(x^2 - x^3 - x^2 y \right) dy$$

$$= x^2 y - x^3 y - \frac{1}{2} x^2 y^2 \Big|_{y=0}^{y=1-x}$$

$$= x^2 (1-x) - x^3 (1-x) - \frac{1}{2} x^2 (1-x)^2$$

$$= \frac{1}{2} x^2 - x^3 + \frac{1}{2} x^4$$

Thus.

$$\int_{x=0}^{1} \int_{y=0}^{y=1-x} \left(\int_{z=0}^{z=1-x-y} x^2 dz \right) dy dx. = \int_{x=0}^{x=1} \left(\frac{1}{2} x^2 - x^3 + \frac{1}{2} x^4 \right) dx$$

$$= \frac{1}{6} x^3 - \frac{1}{4} x^4 + \frac{1}{10} x^5 \Big|_{0}^{1}$$

$$= \frac{1}{6} - \frac{1}{4} + \frac{1}{10} = \frac{1}{60}$$

$\square$

Example 3 Evaluate

$$\int \int \int_{D} x^2 y^2 dV,$$

where D is the region that is bounded by the surfaces $z = 1 - x^2$ and $z = x^2 - 1$, and the planes $y = -2$ and $y = 2$.

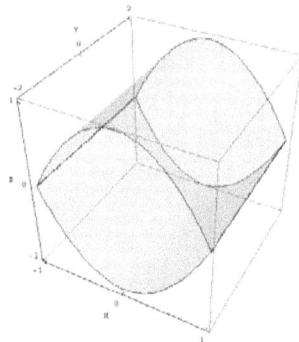

Figure 3

Solution
The surfaces $z = 1 - x^2$ and $z = x^2 - 1$ intersect along the lines that are determined by the solution of the system

$$z = 1 - x^2,$$
$$z = -1 + x^2.$$

Thus,

$$0 = 2 - 2x^2 \Rightarrow x^2 = 1 \Rightarrow x = \pm 1.$$

Therefore, $z = 1 - 1 = 0$. The lines of intersection are the lines $x = \pm 1$ in the xy-plane. Since D is between the planes $y = -2$ and $y = 2$, the projection of D onto the xy-plane is the rectangle $[-1, 1] \times [-2, 2]$. The region D is type I. For any $(x, y) \in [-1, 1] \times [-2, 2]$, the point $(x, y, z) \in D$ iff $-1 + x^2 \le z \le 1 - x^2$. We can express the triple integral as an iterated integral:

$$\iiint_D x^2 y^2 dV = \int_{x=-1}^{x=1} \int_{y=-2}^{y=2} \left(\int_{z=-1+x^2}^{z=1-x^2} x^2 y^2 dz \right) dy dx.$$

We have

$$\int_{z=-1+x^2}^{z=1-x^2} x^2 y^2 dz = x^2 y^2 \int_{z=-1+x^2}^{z=1-x^2} dz = x^2 y^2 \left(z \big|_{-1+x^2}^{1-x^2} \right)$$
$$= x^2 y^2 \left(1 - x^2 + 1 - x^2 \right) = 2x^2 y^2 \left(1 - x^2 \right).$$

Therefore,

$$\iiint_D x^2 y^2 dV = \int_{x=-1}^{x=1} \int_{y=-2}^{y=2} 2x^2 y^2 \left(1 - x^2 \right) dy dx = 2 \left(\int_{x=-1}^{x=1} \left(x^2 - x^4 \right) dx \right) \left(\int_{y=-2}^{y=2} y^2 dy \right)$$
$$= 2 \left(\frac{1}{3} x^3 - \frac{1}{5} x^5 \Big|_{-1}^{1} \right) \left(\frac{1}{3} y^3 \Big|_{-2}^{2} \right)$$
$$= 2 \left(\frac{1}{3} - \frac{1}{5} + \frac{1}{3} - \frac{1}{5} \right) \left(\frac{16}{3} \right)$$
$$= 2 \left(\frac{4}{15} \right) \left(\frac{16}{3} \right) = \frac{128}{45}$$

$\square$

Example 4 Evaluate

$$\iiint_D x dV,$$

where D is the region in the first octant that is bounded by the surface $z = x^2 + y^2$ and the plane $z = 4$.

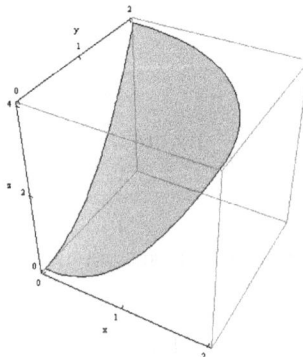

Figure 4

Solution

The projection of D onto the xy-plane is part of the disk $x^2 + y^2 \leq 4$ of radius 2 in the first quadrant. If D_2 denotes that quarter disk, and $(x, y) \in D_2$, then $(x, y, z) \in D$ iff

$$x^2 + y^2 \leq z \leq 4.$$

Therefore,

$$\iiint_D x \, dV = \iint_{D_2} x \left(\int_{z=x^2+y^2}^4 dz \right) dA,$$

The inner integral is

$$\int_{z=x^2+y^2}^4 dz = z\big|_{x^2+y^2}^4 = 4 - \left(x^2 + y^2 \right).$$

Therefore,

$$\iiint_D x \, dV = \iint_{D_2} x \left(4 - (x^2 + y^2) \right) dA.$$

Let's convert to polar coordinates:

$$\iint_{D_2} x \left(4 - (x^2 + y^2) \right) dA = \int_{\theta=0}^{\theta=\pi/2} \int_{r=0}^{r=2} r\cos(\theta) \left(4 - r^2 \right) r \, dr \, d\theta$$

$$= \left(\int_{\theta=0}^{\theta=\pi/2} \cos(\theta) \, d\theta \right) \left(\int_{r=0}^2 \left(4r^2 - r^4 \right) dr \right)$$

$$= \left(\sin(\theta)\big|_0^{\pi/2} \right) \left(\frac{4}{3}r^3 - \frac{1}{5}r^5 \Big|_0^2 \right)$$

$$= \frac{4}{3}2^3 - \frac{1}{5}2^5 = \frac{64}{15}$$

$\square$

Example 5 Assume that D is the region in $\mathbb{R}^3$ that is bounded by the xy-plane, the cylinder $x^2 + y^2 = 4$ and the graph of $z = x^2 + 4$, as shown in the picture:

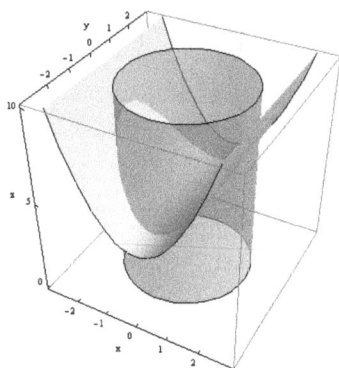

Evaluate

$$\iiint_D \frac{1}{1 + x^2 + y^2} dV.$$

Solution

The projection of the intersection of $x^2 + y^2 = 4$ and $z = x^2 + 4$ onto the xy-plane is the circle $x^2 + y^2 = 4$. This is the circle of radius 2 centered at the origin. Set

$$D_2 = \left\{ (x, y) : x^2 + y^2 \leq 4 \right\}.$$

We have

$$\int \int \int_D dV = \int \int_{D_2} \left(\int_{y=0}^{z=x^2+4} \frac{1}{1 + x^2 + y^2} dz \right) dx dy$$

$$= \int \int_{D_2} \frac{1}{1 + x^2 + y^2} \left(x^2 + 4 \right) dx dy.$$

Let's use polar coordinates in the xy-plane, so that

$$x = r \cos(\theta), \; y = r \sin(\theta).$$

Then,

$$\int \int_{D_2} \frac{x^2 + 4}{1 + x^2 + y^2} dx dy = \int_{\theta=0}^{\theta=2\pi} \int_{r=0}^{2} \frac{4 + r^2 \cos^2(\theta)}{1 + r^2} r dr d\theta$$

$$= \int_{\theta=0}^{\theta=2\pi} \int_{r=0}^{2} \frac{4}{1 + r^2} r dr d\theta + \int_{\theta=0}^{\theta=2\pi} \int_{r=0}^{2} \frac{r^3 \cos^2(\theta)}{1 + r^2} dr d\theta.$$

We have

$$\int_{\theta=0}^{\theta=2\pi} \int_{r=0}^{2} \frac{4}{1 + r^2} r dr d\theta = 2\pi \int_{r=0}^{2} \frac{4}{1 + r^2} r dr$$

$$= 8\pi \left(\frac{1}{2} \ln(1 + r^2) \Big|_0^2 \right) = 4\pi \ln(5).$$

As for the second integral,

$$\int_{\theta=0}^{\theta=2\pi} \int_{r=0}^{2} \frac{r^3 \cos^2(\theta)}{1 + r^2} dr d\theta = \left(\int_{\theta=0}^{\theta=2\pi} \cos^2(\theta) d\theta \right) \left(\int_{r=0}^{2} \frac{r^3}{1 + r^2} dr \right)$$

$$= \left(\int_{\theta=0}^{\theta=2\pi} \left(\frac{1 + \cos(2\theta)}{2} \right) d\theta \right) \left(\int_{r=0}^{2} \frac{r^3}{1 + r^2} dr \right)$$

$$= \left(\frac{1}{2}\theta + \frac{1}{4} \sin(2\theta) \Big|_0^{2\pi} \right) \int_{r=0}^{2} \frac{r^3}{1 + r^2} dr$$

$$= \pi \int_{r=0}^{2} \frac{r^3}{1 + r^2} dr.$$

Finally,

$$\frac{r^3}{1 + r^2} = r - \frac{r}{r^2 + 1},$$

by division, so that

$$\pi \int_{r=0}^{2} \frac{r^3}{1 + r^2} dr = \pi \left(\frac{1}{2} r^2 - \frac{1}{2} \ln(r^2 + 1) \Big|_0^2 \right)$$

$$= \pi \left(2 - \frac{1}{2} \ln(5) \right).$$

Therefore,

$$\iint_{D_2} \frac{x^2 + 4}{1 + x^2 + z^2} dx dz = 4\pi \ln(5) + 2\pi - \frac{\pi}{2} \ln(5)$$
$$= 2\pi + \frac{7\pi}{2} \ln(5).$$

□

Problems

1. Find the volume of the region bounded by the surfaces

$$z = x^2 + y^2 \text{ and } z = 18 - x^2 - y^2.$$

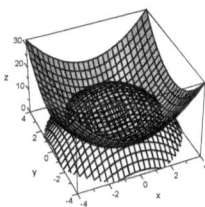

2. Find the volume of the region bounded by the surfaces

$$x^2 + y^2 = 4, \ x + y + z = 1 \text{ and } z = -10.$$

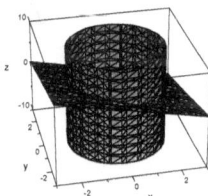

In problems 3-10 evaluate the given integral (Note that the symbol $dxdydz$ stands for dV in Cartesian coordinates and does not specify the order of integration):

3.

$$\iiint_D x^2 dx dy dz,$$

where

$$D = [0, 1] \times [0, 1] \times [0, 1].$$

4.

$$\iiint_D e^{-xy} y \, dx dy dz,$$

where

$$D = [0, 1] \times [0, 1] \times [0, 1]$$

5.

$$\iiint_D ze^{x+y}dxdydz,$$

where

$$D = [0,1] \times [0,1] \times [0,2]$$

6.

$$\iiint_D x^2 \cos(z)\,dxdydz,$$

where D is the region bounded by the planes $z = 0$, $z = \pi/2$, $y = 0$, $y = 1$, $x = 0$ and $x + y = 1$.

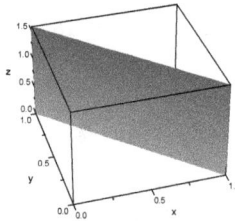

7.

$$\iiint_D zdxdydz,$$

where D is the region bounded by the planes $x = 0$, $y = 0$, $z = 0$, $z = 1$, and the cylinder $x^2 + y^2 = 1$ with $x \geq 0$ and $y \geq 0$.

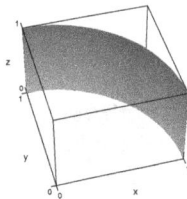

8.

$$\iiint_D x^2dxdydz,$$

where D is the solid tetrahedron with vertices $(0,0,0)$, $(1,0,0)$, $(0,1,0)$, $(0,0,1)$.

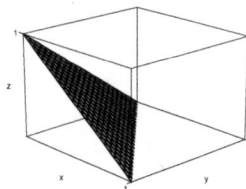

9.

$$\iiint_D xydxdxydz,$$

where D is bounded by the cylinders $y = x^2$, $x = y^2$, the xy-plane and the plane $z = x + y$.

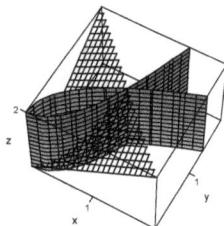

10.

$$\int \int \int_D x^2 dy dx dz,$$

where D is bounded by the cylinder $x^2 + y^2 = 9$, the plane $y + z = 5$ and the plane $z = 1$.

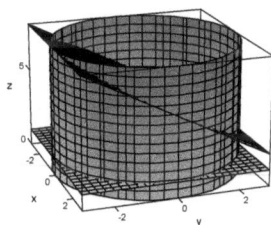

13.6 Triple Integrals in Cylindrical and Spherical Coordinates

Cylindrical Coordinates

We can extend polar coordinates to the three dimensional space $\mathbb{R}^3$ as follows: If the Cartesian coordinates of a point $P \in \mathbb{R}^3$ are x, y and z, then r and θ are polar coordinates for the point (x, y) in the plane. Thus,

$$x = r \cos(\theta) \text{ and } y = r \sin(\theta),$$

so that

$$r = \sqrt{x^2 + y^2}, \ \cos(\theta) = \frac{x}{\sqrt{x^2 + y^2}} \text{ and } \sin(\theta) = \frac{y}{\sqrt{x^2 + y^2}}.$$

The polar angle is determined up to an additive constant that is an integer multiple of 2π. The third coordinate is simply z itself. The coordinates r, θ and z are referred to as the **cylindrical coordinates**.of the point P. We will not allow r to be negative when we utilize cylndrical coordinates. Recall that we can always a value of $\theta \in [-\pi, \pi]$ by setting

$$\theta = \begin{cases} \arccos\left(\frac{x}{\sqrt{x^2+y^2}}\right) & \text{if } y \geq 0, \\ -\arccos\left(\frac{x}{\sqrt{x^2+y^2}}\right) & \text{if } y < 0. \end{cases}$$

Example 1 Determine the cylindrical coordinates of P if

a) $P = (1, \sqrt{3}, 2)$, b) $P = (-1, -1, 2)$ in Cartesian coordinates
Solution

a) We have

$$r = \sqrt{1^2 + \left(\sqrt{3}\right)^2} = \sqrt{4} = 2,$$

and

$$\cos(\theta) = \frac{1}{2}, \ \sin(\theta) = \frac{\sqrt{3}}{2}.$$

Thus, we can set $\theta = \pi/3$ or $\theta = \pi/3 + 2n\pi$, where n is an arbitrary integer. Therefore, the cylndrical coordinates of P are $r = 2$, $\theta = \pi/3 + 2n\pi$ and $z = 2$, where n is an arbitrary integer.
b) We have

$$r = \sqrt{1^2 + (-1)^2} = \sqrt{2},$$

and

$$\cos(\theta) = \frac{-1}{\sqrt{2}}, \ \sin(\theta) = \frac{-1}{\sqrt{2}}.$$

We can set

$$\theta = -\arccos\left(\frac{-1}{\sqrt{2}}\right) = -\frac{3\pi}{4}.$$

$\square$

Triple Integrals in Cylindrical Coordinates

Assume that D is a region in $\mathbb{R}^3$ that consists of points whose cylindrical coordinates satisfy inequalities of the form

$$r_1(\theta) \le r \le r_2(\theta), \ \alpha \le \theta \le \beta \text{ and } z_1(r, \theta) \le z \le z_2(r, \theta).$$

Then

$$\iiint_D f(x, y, z)\, dxdydz = \int_{\theta=\theta_1}^{\theta=\theta_2} \int_{r=r_1(\theta)}^{r=r_2(\theta)} \int_{z_1(r,\theta)}^{z_2(r,\theta)} f(r\cos(\theta), r\sin(\theta), z)\, rdzdrd\theta.$$

Indeed, if D_2 denotes the region in the xy-plane consisting of the points whose polar coordinates satisfy the inequalities

$$r_1(\theta) \le r \le r_2(\theta), \ \alpha \le \theta \le \beta,$$

then

$$\iiint_D f(x, y, z)\, dxdydz = \iint_{D_2} \left(\int_{z_1(r,\theta)}^{z_2(r,\theta)} f(x, y, z)\, dz \right) dxdy$$

$$= \int_{\theta=\theta_1}^{\theta=\theta_2} \int_{r=r_1(\theta)}^{r=r_2(\theta)} \left(\int_{z_1(r,\theta)}^{z_2(r,\theta)} f(r\cos(\theta), r\sin(\theta), z)\, dz \right) rdrd\theta$$

$$= \int_{\theta=\theta_1}^{\theta=\theta_2} \int_{r=r_1(\theta)}^{r=r_2(\theta)} \int_{z_1(r,\theta)}^{z_2(r,\theta)} f(r\cos(\theta), r\sin(\theta), z)\, rdzdrd\theta$$

Thus, the use of cylindrical coordinates in triple integrals amounts to transforming to polar coordinates in the xy-plane.

Example 2 Let D be the region that is bounded by the paraboloid $z = x^2 + y^2$ and the plane $z = 9$. Use cylindrical coordinates to evaluate

$$\int \int \int_D x^2 z dV.$$

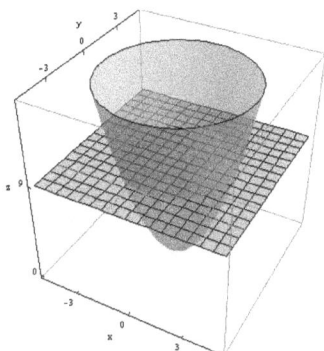

Figure 1

Solution

The equation of the paraboloid in cylindrical coordinates is $z = r^2$. The projection of D to the xy-plane is the disk that is bounded by the circle $x^2 + y^2 = 9$, i.e., the circle of radius 3 centered at the origin. Therefore,

$$
\begin{aligned}
\int \int \int_D x^2 z dV &= \int_{\theta=0}^{\theta=2\pi} \int_{r=0}^{3} \int_{z=r^2}^{z=9} \left(r^2 \cos^2(\theta) \right) zr dz dr d\theta \\
&= \int_{\theta=0}^{2\pi} \int_{r=0}^{3} r^3 \cos^2(\theta) \left(\int_{z=r^2}^{9} z dz \right) dr d\theta.
\end{aligned}
$$

We have

$$\int_{z=r^2}^{9} z dz = \frac{1}{2} z^2 \Big|_{z=r^2}^{z=9} = \frac{81}{2} - \frac{r^4}{2}.$$

Therefore,

$$
\begin{aligned}
\int_{\theta=0}^{2\pi} \int_{r=0}^{3} r^3 \cos^2(\theta) \left(\int_{z=r^2}^{9} z dz \right) dr d\theta &= \int_{\theta=0}^{2\pi} \int_{r=0}^{3} r^3 \cos^2(\theta) \left(\frac{81}{2} - \frac{r^4}{2} \right) dr d\theta \\
&= \left(\int_{\theta=0}^{2\pi} \cos^2(\theta) d\theta \right) \left(\int_{r=0}^{3} r^3 \left(\frac{81}{2} - \frac{r^4}{2} \right) dr \right).
\end{aligned}
$$

We have

$$\int_{\theta=0}^{2\pi} \cos^2(\theta) d\theta = \frac{\theta}{2} + \frac{1}{2} \cos(\theta) \sin(\theta) \Big|_0^{2\pi} = \pi,$$

and

$$
\begin{aligned}
\int_{r=0}^{3} r^3 \left(\frac{81}{2} - \frac{r^4}{2} \right) dr &= \int_0^3 \left(\frac{81}{2} r^3 - \frac{1}{2} r^7 \right) dr \\
&= \frac{81}{8} r^4 - \frac{r^8}{16} \Big|_0^3 = \frac{81}{8} (3^4) - \frac{3^8}{16} = \frac{6561}{16}
\end{aligned}
$$

Therefore,

$$\int \int \int_D x^2 z dV = \frac{6561}{16}\pi.$$

□

Example 3 Let D be the region that is bounded by the cone $z^2 = x^2 + y^2$ and the cylinder $r = 2$. Use cylindrical coordinates to evaluate

$$\int \int \int_D e^{-\sqrt{x^2+y^2}} dV.$$

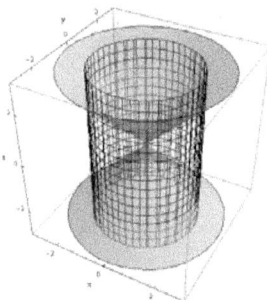

Figure 2

Solution

Since $x^2 + y^2 = r^2$, in cylindrical coordinates the equation of the cone is $z^2 = r^2$. Thus, the cone has two parts, $z = \pm r$. The projection of D onto the xy-plane is the disk of radius 2 centered at the origin. Therefore,

$$\int \int \int_D e^{-\sqrt{x^2+y^2}} dV = \int_{\theta=0}^{\theta=2\pi} \int_{r=0}^{2} \left(\int_{z=-r}^{z=r} e^{-r} r dz \right) dr d\theta = \int_{\theta=0}^{\theta=2\pi} \int_{r=0}^{2} e^{-r} r \left(\int_{z=-r}^{z=r} dz \right) dr d\theta$$

$$= \int_{\theta=0}^{\theta=2\pi} \int_{r=0}^{2} e^{-r} r (2r) \, dr d\theta$$

$$= \left(\int_{\theta=0}^{\theta=2\pi} d\theta \right) \left(2 \int_{r=0}^{2} e^{-r} r^2 dr \right)$$

We have

$$\int_0^2 e^{-r} r^2 dr = 2 - 10e^{-2}$$

(confirm via integration by parts). Therefore,

$$\int \int \int_D e^{-\sqrt{x^2+y^2}} dV = 4\pi \left(2 - 10e^{-2} \right) = 8\pi - \frac{40\pi}{e^2}$$

□

Spherical Coordinates

Let $P = (x, y, z) \neq (0, 0, 0)$. We set ρ to be the distance of P from the origin:

$$\rho = \sqrt{x^2 + y^2 + z^2}.$$

The angle ϕ is the angle between the position vector $\overrightarrow{OP}$ of the point P and the postive z-axis determined so that $0 \leq \phi \leq \pi$. Thus,

$$z = \rho \cos(\phi)$$

and

$$\phi = \arccos\left(\frac{z}{\rho}\right).$$

The distance of the projection Q of P onto the xy-plane from the origin is $\rho \sin(\phi)$. In terms of the polar coordinates of that point, we have

$$x = \rho \sin(\phi) \cos(\theta) \text{ and } y = \rho \sin(\phi) \sin(\theta).$$

Here the polar angle θ is not unique and it is determined up to an integer multiple of 2π.

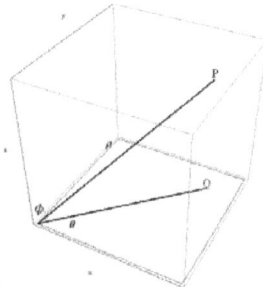

Figure 3

Example 4 Let $P = \left(9, 3\sqrt{3}, 6\right)$. Determine the spherical cooordinates of P.

Solution

We have

$$\rho = \sqrt{9^2 + \left(3\sqrt{3}\right)^2 + 6^2} = 12.$$

Therefore,

$$\cos(\phi) = \frac{z}{\rho} = \frac{6}{12} = \frac{1}{2}.$$

Since $0 \leq \phi \leq \pi$, we have $\phi = \pi/3$. Thus,

$$x = \rho \sin(\phi) \cos(\theta) \text{ and } y = \rho \sin(\phi) \sin(\theta)$$

so that

$$x = (12)\left(\frac{\sqrt{3}}{2}\right) \cos(\theta) = 6\sqrt{3} \cos(\theta) \text{ and } y = 6\sqrt{3} \sin(\theta).$$

Therefore,

$$\cos(\theta) = \frac{9}{6\sqrt{3}} = \frac{3}{2\sqrt{3}} = \frac{\sqrt{3}}{2},$$

and

$$\sin(\theta) = \frac{3\sqrt{3}}{6\sqrt{3}} = \frac{1}{2}.$$

Thus, we can set $\theta = \pi/6$. Therefore,

$$\rho = 12, \ \phi = \pi/3 \text{ and } \theta = \pi/6$$

are spherical coordinates for P. $\square$

Example 5 Let S be the surface that is the graph of the equation

$$z = \sqrt{x^2 + y^2}$$

is Cartesian coordinates. Describe S as the graph of an equation in spherical coordinates and identify the surface geometrically.

Solution

We have

$$
\begin{aligned}
x^2 + y^2 &= \left(\rho \sin\left(\phi\right) \cos\left(\theta\right)\right)^2 + \left(\rho \sin\left(\phi\right) \sin\left(\theta\right)\right)^2 \\
&= \rho^2 \sin^2\left(\phi\right) \cos^2\left(\theta\right) + \rho^2 \sin^2\left(\phi\right) \cos^2\left(\theta\right) \\
&= \rho^2 \sin^2\left(\phi\right).
\end{aligned}
$$

Thus,

$$\sqrt{x^2 + y^2} = \rho \sin\left(\phi\right).$$

Therefore,

$$z = \sqrt{x^2 + y^2} \Rightarrow \rho \cos\left(\phi\right) = \rho \sin\left(\phi\right) \Rightarrow \cos\left(\phi\right) = \sin\left(\phi\right).$$

Since $0 \leq \phi \leq \pi$, we have $\phi = \pi/4$. Thus, a point is on S if the angle ϕ between its position vector and the positive z-axis is $\pi/4$. Since there are no other restriction on the spherical coordinates of the points of S we identify S as a cone that has its vertex at the origin, the z-axis forms its axis, and the opening with the z-axis is $\pi/4$. $\square$

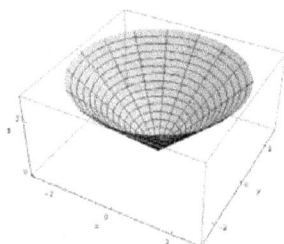

Figure 4: The cone $\phi = \pi/4$

Triple Integrals in Spherical Coordinates

Assume that D is a region in $\mathbb{R}^3$ that consists of points whose spherical coordinates satisfy inequalities of the form

$$\alpha \leq \theta \leq \beta, \; \gamma \leq \phi \leq \delta, \; \rho_1\left(\theta, \phi\right) \leq \rho \leq \rho_2\left(\theta, \phi\right).$$

Then

$$
\begin{aligned}
&\int\int\int_D f\left(x, y, z\right) dx dy dz \\
&= \int_{\theta=\alpha}^{\theta=\beta} \int_{\phi=\gamma}^{\phi=\delta} \int_{\rho=\rho_1(\theta,\phi)}^{\rho=\rho_2(\theta,\phi)} f\left(\rho \sin\left(\phi\right) \cos\left(\theta\right), \rho \sin\left(\phi\right) \sin\left(\theta\right), \rho \cos\left(\phi\right)\right) \rho^2 \sin\left(\phi\right) d\rho d\phi d\theta
\end{aligned}
$$

(the order of θ and ϕ may be interchanged). We are assuming that f and the functions $\rho_1\left(\theta, \phi\right)$ and $\rho_2\left(\theta, \phi\right)$ are continuous. The volume element $dV = dx dy dz$ in Cartesian cooordinates is replaced by the expression

$$\rho^2 \sin\left(\phi\right) d\rho d\phi d\theta.$$

We will not prove the above statement. On the other hand, the expression is plausible: If $\Delta\rho$, $\Delta\phi$ and $\Delta\theta$ small, lets consider a region G whose spherical coordinates are between ρ and $\rho + \Delta\rho$, ϕ and $\phi + \Delta\phi$, θ and $\theta + \Delta\theta$, respectively. The region G is the counterpart of a rectangular box when we deal with spherical coordinates. The volume of G is approximately

$$(\Delta\rho)\,(\rho\Delta\phi)\,(\rho\sin(\phi)\,\Delta\theta) = \rho^2\sin(\phi)\,\Delta\rho\Delta\phi\Delta\theta,$$

as illustrated in Figure 5.

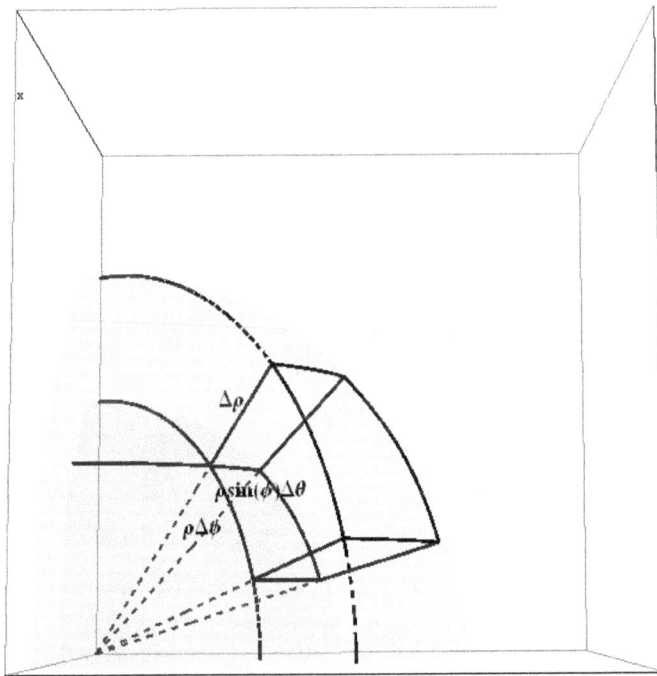

Figure 5

Thus, the integral of f on such a small region is approximately

$$f\left(\rho\sin(\phi)\cos(\theta),\rho\sin(\phi)\sin(\theta),\rho\cos(\phi)\right)\rho^2\sin(\phi)\,\Delta\rho\Delta\phi\Delta\theta.$$

Example 6 Let D be the region that is bounded by the sphere of radius 4 centered at the origin and the xy-plane. Evaluate

$$\iiint_D z^2\,dV.$$

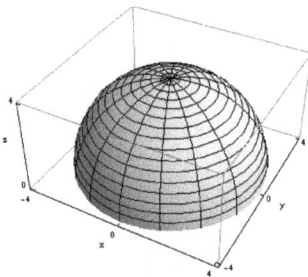

Figure 6

Solution

We will express the integral in spherical coordinates:

$$
\begin{aligned}
\iiint_D z^2 dV &= \int_{\theta=0}^{\theta=2\pi} \int_{\phi=0}^{\pi/2} \left(\int_{\rho=0}^{\rho=4} \rho^2 \cos^2(\phi) \, \rho^2 \sin(\phi) \, d\rho \right) d\phi \, d\theta \\
&= \int_{\theta=0}^{\theta=2\pi} \int_{\phi=0}^{\pi/2} \cos^2(\phi) \sin(\phi) \left(\int_{\rho=0}^{\rho=4} \rho^4 d\rho \right) d\phi \, d\theta \\
&= \int_{\theta=0}^{\theta=2\pi} \int_{\phi=0}^{\pi/2} \cos^2(\phi) \sin(\phi) \left(\frac{4^5}{5} \right) d\phi \, d\theta \\
&= \frac{4^5}{5}(2\pi) \left(\int_{u=1}^{u=0} -u^2 du \right) \\
&= \frac{4^5}{5}(2\pi) \left(\frac{1}{3} \right) = \frac{2048}{15}\pi
\end{aligned}
$$

□

Example 7 Let D be the region inside the cone $\phi = \pi/4$ and the sphere $\rho = 3$. Evaluate

$$
\iiint_D z \, dV.
$$

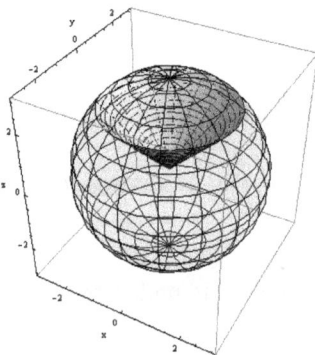

Figure 7

Solution

We express the integral in spherical coordinates:

$$\iiint_D z\,dV = \int_{\theta=0}^{\theta=2\pi} \int_{\phi=0}^{\phi=\pi/4} \int_{\rho=0}^{\rho=3} \rho\cos(\phi)\,\rho^2\sin(\phi)\,d\rho\,d\phi\,d\theta$$

$$= \left(\int_{\theta=0}^{\theta=2\pi} d\theta\right)\left(\int_{\phi=0}^{\phi=\pi/4}\cos(\phi)\sin(\phi)\,d\phi\right)\left(\int_{\rho=0}^{\rho=3}\rho^3\,d\rho\right)$$

$$= 2\pi\left(-\int_{u=1}^{u=\sqrt{2}/2} u\,du\right)\left(\frac{3^4}{4}\right)$$

$$= \frac{3^4\pi}{2}\left(\frac{1}{2}u^2\Big|_{\sqrt{2}/2}^{1}\right)$$

$$= \frac{3^4\pi}{2}\left(\frac{1}{2}-\frac{1}{4}\right) = \frac{81}{8}\pi$$

□

Problems

In problems 1-4 determine the Cartesian coordinates of the point with the given cylindrical coordinates (r,θ,z):

1. $(2,\pi/4,1)$
2. $(4,\pi/6,2)$

3. $(3,2\pi/3,4)$
4. $(2,-\pi/6,4)$

In problems 5-8 determine the cylindrical coordinates (r,θ,z) of the point with the given Cartesian coordinates, such that $-\pi < \theta \le \pi$:

5. $(-1,1,4)$
6. $\left(2,-2\sqrt{3},-1\right)$

7. $(1,-1,2)$
8. $\left(1,\sqrt{3},2\right)$

In problems 9-13 make use of cylindrical coordinates to evaluate the given triple integral (Note that the symbol $dxdydz$ stands for dV in Cartesian coordinates and does not specify the order of integration):

9.

$$\iiint_D \sqrt{x^2+y^2}\,dxdydz,$$

where D is inside the cylinder $x^2+y^2=16$, between the planes $z=-5$ and $z=4$.

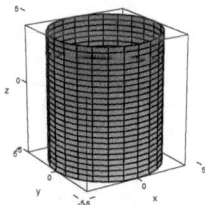

10.

$$\iiint_D \left(x^3+xy^2\right)\,dxdydz,$$

where D is the region in the first octant that lies below the paraboloid $z = 1 - x^2 - y^2$.

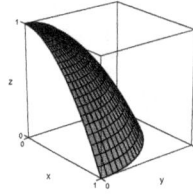

11.

$$\int\int\int_D e^z dx dy dz,$$

where D is bounded by the paraboloid $z = 1 + x^2 + y^2$, the cylinder $x^2 + y^2 = 5$ and the xy-plane.

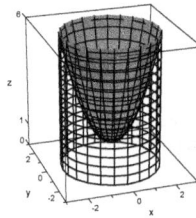

12.

$$\int\int\int_D x^2 dx dy dz,$$

where D is the region that is within the cylinder $x^2 + y^2 = 1$, above the xy-plane, and below the cone $z^2 = 4x^2 + 4y^2$.

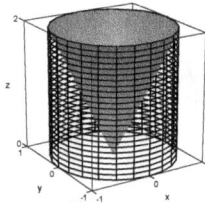

13. Find the volume of the region within both the cylinder $x^2 + y^2 = 1$ and the sphere $x^2 + y^2 + z^2 = 4$.

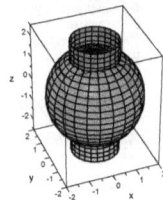

In problems 14-17 determine the Cartesian coordinates of the point with the given spherical coordinates (ρ, θ, ϕ):

14. $(1, 0, 0)$ **16.** $(5, \pi, \pi/2)$
15. $(2, \pi/3, \pi/4)$ **17.** $(4, 3\pi/4, \pi/3)$

In problems 18-21 determine the spherical coordinates (ρ, θ, ϕ) of the point with the given Cartesian coordinates, such that $-\pi < \theta \le \pi$:

18. $\left(1, \sqrt{3}, 2\sqrt{3}\right)$ **20.** $\left(0, \sqrt{3}, 1\right)$
19. $(0, -1, -1)$ **21.** $\left(-1, 1, \sqrt{6}\right)$

In problems 22-29 make use of cylindrical coordinates to evaluate the given triple integral (Note that the symbol $dxdydz$ stands for dV in Cartesian coordinates and does not specify the order of integration):

22.
$$\iiint_D \left(9 - x^2 - y^2\right) dxdydz,$$
where D is the region above the xy-plane that is bounded by the sphere $x^2 + y^2 + z^2 = 9$.

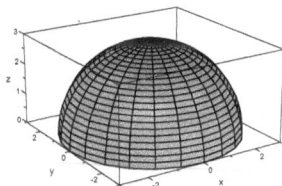

23.
$$\iiint_D z \, dxdydz,$$
where D is in the first octant, between the spheres $\rho = 1$ and $\rho = 2$.

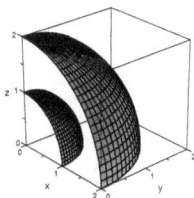

24.
$$\iiint_D e^{\sqrt{x^2 + y^2 + z^2}} \, dxdydz,$$
where D is in the first octant and in the sphere $\rho = 3$.

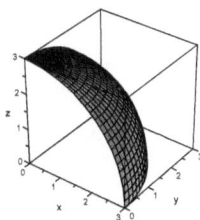

25.

$$\int\int\int_D x^2 dx dy dz,$$

where D is the region that is bounded by the xz-plane and the hemispheres $y = \sqrt{9 - x^2 - z^2}$ and $y = \sqrt{16 - x^2 - z^2}$.

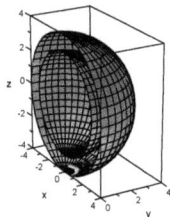

26.

$$\int\int\int_D x^2 z dx dy dz,$$

where D is the region between the spheres $\rho = 2$ and $\rho = 4$ and above the cone $\phi = \pi/3$.

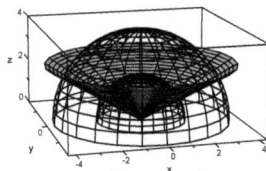

27.

$$\int\int\int_D dx dy dz,$$

where D is the region bounded by the sphere $\rho = 3$ and the cones $\phi = \pi/6$ and $\phi = \pi/3$.

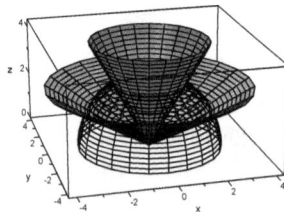

28.

$$\int\int\int_D z dx dy dz,$$

where D is the region that is above the cone $\phi = \pi/3$ and below the sphere $\rho = 4\cos(\phi)$.

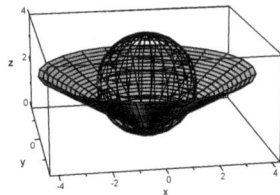

29. Find the volume of the region D within the sphere $x^2 + y^2 + z^2 = 4$, above the xy-plane, and below the cone $z = \sqrt{x^2 + y^2}$.

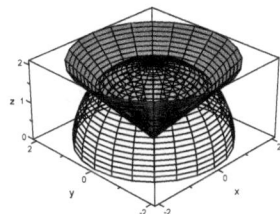

13.7 Change of Variables in Multiple Integrals (the General Case)

Definition 1 Assume that D and D^* are regions in the plane and that T is a function that is defined on D^* and has values in D. We will express this fact by writing

$$T : D^* \to D$$

and refer to T as a **transformation** from D^* into D. If we denote the points in D by (x, y) and the points in D^* by (u, v), so that x, y and u, v are Cartesian coordinates, we will say that D is a region in the xy-plane and D^* is a region in the uv-plane. If

$$T(u, v) = (x(u, v), y(u, v)),$$

we will refer to the functions $x(u, v)$ and $y(u, v)$ as **the coordinate functions** that define the transformation T. The transformation is from D^* **onto** D if each point of D is the value of T at some point in D^*. The transformation is **one-one** if the values of T at distinct points of D^* are distinct points of D.

Example 1 Let θ be a fixed angle and set

$$T(u, v) = (\cos(\theta) u - \sin(\theta) v, \sin(\theta) u + \cos(\theta) v)$$

for each $(u, v) \in \mathbb{R}^2$. Thus, the coordinate functions are

$$x(u, v) = \cos(\theta) u - \sin(\theta) v \text{ and } y(u, v) = \sin(\theta) u + \cos(\theta) v$$

Geometrically, T describes the **rotation** through the angle θ.

For example, if we set $\theta = \pi/3$, we have

$$
\begin{aligned}
T(u,v) &= \left(\cos\left(\frac{\pi}{3}\right) u - \sin\left(\frac{\pi}{3}\right) v, \sin\left(\frac{\pi}{3}\right) u + \cos\left(\frac{\pi}{4}\right) v \right) \\
&= \left(\frac{1}{2}u - \frac{\sqrt{3}}{2}v, \frac{\sqrt{3}}{2}u + \frac{1}{2}v \right).
\end{aligned}
$$

Let D^* be the sector of the unit disk in the uv-plane consisting of points with polar coordinates (r,θ) such that $0 \leq r \leq 1$ and $-\pi/3 \leq \theta \leq 0$. The image of D^* under T is D, the sector of the unit disk in the xy-plane consisting of points with polar coordinates (r,θ) such that $0 \leq r \leq 1$ and $0 \leq \theta \leq \pi/3$. Figure 1 displays D^* and D. $\square$

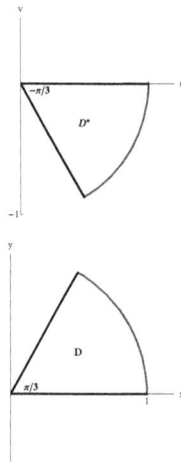

Figure 1

Example 2 We can view the relationship between Cartesian and polar coordinates within the framework of functions from $\mathbb{R}^2$ into $\mathbb{R}^2$.

For given $r > 0$ and $\theta \in \mathbb{R}$, the corresponding Cartesian coordinates are

$$ x = r\cos(\theta) \text{ and } y = r\sin(\theta). $$

Let's consider the Cartesian θr-plane. Thus, we have the θ and r as the orthogonal axes, and any point (θ, r) in that plane is assigned the point

$$ T(r,\theta) = (r\cos(\theta), r\sin(\theta)) $$

in the Cartesian xy-plane. This assignment defines a **transformation** from $\mathbb{R}^2$ into $\mathbb{R}^2$ (from the θr-plane into the xy-plane).

For example, let D^* is the rectangle consisting of the points (θ, r) such that $\pi/4 \leq \theta \leq \pi/2$ and $1 \leq r \leq 2$ in the the θr-plane, and let D be the image of D^* under T. D is bounded by the rays in the xy-plane that correspond to $\theta = \pi/4$ and $\theta = \pi/2$, and the arcs of the circles with radius 1 and 2. Figure 2 displays D^* and D. $\square$

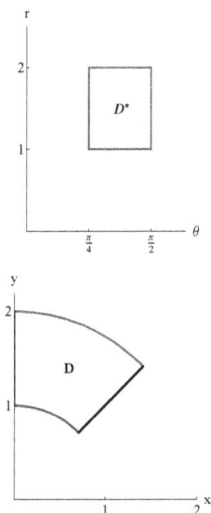

Figure 2

Similarly, we can define transformations in three variables: If $x\left(u,v,w\right)$, $y\left(u,v,w\right)$, $z\left(u,v,w\right)$ define functions of u,v and w, the transformation

$$T\left(u,u,w\right) = \left(x\left(u,v,w\right),y\left(u,v,w\right),z\left(u,v,w\right)\right)$$

is a function that maps a region in the uvw-space onto a region in the xyz-space.

Example 3 Let ρ, θ and ϕ be spherical coordinates in $\mathbb{R}^3$ so that

$$x = \rho\sin\left(\phi\right)\cos\left(\theta\right),\ \ y = \rho\sin\left(\phi\right)\sin\left(\theta\right)\ \text{and}\ z = \rho\cos\left(\phi\right).$$

Here ρ is the distance of the point $\left(x,y,z\right)$ from the origin, i.e.,

$$\rho = \sqrt{x^2+y^2+z^2},$$

ϕ is the angle between the $x\mathbf{i}+y\mathbf{j}+z\mathbf{k}$ and $\mathbf{k}$ $\left(0\le\phi\le\pi\right)$, and θ is a polar angle of the point $\left(x,y,0\right)$ (we can assume that θ is between 0 and 2π).

We can define the transformation from $\mathbb{R}^3$ (the $\rho\phi\theta$-space) into $\mathbb{R}^3$ (the xyz-space) by the rule

$$T\left(\rho,\phi,\theta\right) = \left(\rho\sin\left(\phi\right)\cos\left(\theta\right),\rho\sin\left(\phi\right)\sin\left(\theta\right),\rho\cos\left(\phi\right)\right).$$

If D^* is the box in the $\rho\phi\theta$-space consisting of points $\left(\rho,\phi,\theta\right)$ such that $\rho_1\le\rho\le\rho_2$, $\phi_1\le\phi\le\phi_2$, $\theta_1\le\theta\le\theta_2$, the image of D^* under T in the xyz-space is bounded by the spheres $\rho=\rho_1$, $\rho=\rho_2$, the cones $\phi=\phi_1$, $\phi=\phi_2$ and the planes $\theta=\theta_1$, $\theta=\theta_2$. $\square$

Definition 2 If $T\left(u,v\right) = \left(x\left(u,v\right),y\left(u,v\right)\right)$ is a transformation from D^* into D the **Jacobian** of x and y with respect to u and v is

$$\frac{\partial\left(x,y\right)}{\partial\left(u,v\right)} = \begin{vmatrix} \dfrac{\partial x}{\partial u} & \dfrac{\partial x}{\partial v} \\ \dfrac{\partial y}{\partial u} & \dfrac{\partial y}{\partial v} \end{vmatrix} = \frac{\partial x}{\partial u}\frac{\partial y}{\partial v} - \frac{\partial x}{\partial v}\frac{\partial y}{\partial u}$$

Theorem 1 Assume that $T(u,v) = (x(u,v), y(u,v))$ is a transformation that maps D^* onto D in a one-one manner (i.e., the images of distinct points are distinct), and its coordinate functions are differentiable. Then

$$\int\int_D f(x,y)\,dxdy = \int\int_{D^*} f(x(u,v), y(u,v)) \left|\frac{\partial(x,y)}{\partial(u,v)}\right| dudv$$

for any function f that is continuous on D.

A plausibility Argument for the Theorem

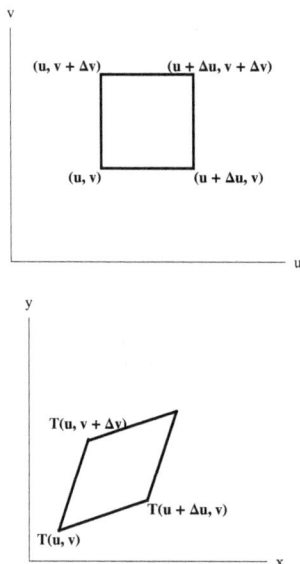

Figure 3

D can be covered by the images of rectangles in the uv-plane. Consider such a rectangle with vertices (u,v), $(u+\Delta u, v)$, $(u, v+\Delta v)$, $(u+\Delta u, v+\Delta v)$. The image of this rectangle can be approximated by a parallelogram with vertices $T(u,v)$, $T(u+\Delta u, v)$ and $T(u, v+\Delta v)$, as illustrated in Figure 3. This parallelogram is spanned by the vectors

$$T(u+\Delta u, v) - T(u,v) \text{ and } T(u, v+\Delta v) - T(u,v).$$

Assuming that Δu and Δv are small,

$$\begin{aligned}
T(u+\Delta u, v) - T(u,v) &= (x(u+\Delta u, v) - x(u,v), y(u+\Delta u, v) - y(u,v)) \\
&\cong \left(\frac{\partial x}{\partial u}(u,v)\Delta u, \frac{\partial y}{\partial u}\Delta u\right) \\
&= \Delta u\left(\frac{\partial x}{\partial u}(u,v), \frac{\partial y}{\partial u}(u,v)\right).
\end{aligned}$$

Similarly,

$$T(u, v+\Delta v) - T(u,v) \cong \Delta v\left(\frac{\partial x}{\partial v}(u,v), \frac{\partial y}{\partial v}(u,v)\right).$$

Thus, the area of the parallelogram that is spanned by the vectors $T(u+\Delta u, v) - T(u,v)$ and

$T(u, v + \Delta v) - T(u, v)$ can be approximated by

$$\left\| \Delta u \left(\frac{\partial x}{\partial u}(u, v), \frac{\partial y}{\partial u}(u, v), 0 \right) \times \Delta v \left(\frac{\partial x}{\partial v}(u, v), \frac{\partial y}{\partial v}(u, v), 0 \right) \right\|$$

$$= \Delta u \Delta v \left\| \begin{matrix} \mathbf{i} & \mathbf{j} & \mathbf{k} \\ \frac{\partial x}{\partial u}(u, v) & \frac{\partial y}{\partial u}(u, v) & 0 \\ \frac{\partial x}{\partial v}(u, v) & \frac{\partial y}{\partial v}(u, v) & 0 \end{matrix} \right\|$$

$$= \Delta u \Delta v \left| \frac{\partial x}{\partial u}(u, v) \frac{\partial y}{\partial v}(u, v) - \frac{\partial y}{\partial u}(u, v) \frac{\partial x}{\partial v}(u, v) \right|$$

$$= \Delta u \Delta v \left| \frac{\partial x}{\partial u}(u, v) \frac{\partial y}{\partial v}(u, v) - \frac{\partial x}{\partial v}(u, v) \frac{\partial y}{\partial u}(u, v) \right|$$

$$= \Delta u \Delta v \left\| \begin{matrix} \frac{\partial x}{\partial u} & \frac{\partial x}{\partial v} \\ \frac{\partial y}{\partial u} & \frac{\partial y}{\partial v} \end{matrix} \right\| = \Delta u \Delta v \left| \frac{\partial(x, y)}{\partial(u, v)} \right|.$$

Therefore, the integral of f on such a parallelogram can be approximated by

$$f(x(u, v), y(u, v)) \left| \frac{\partial(x, y)}{\partial(u, v)} \right| \Delta u \Delta v$$

Since we can approximate

$$\int\int_D f(x, y) \, dx dy$$

by the sum of such integrals, it is indeed plausible that

$$\int\int_D f(x, y) \, dx dy = \int\int_{D^*} f(x(u, v), y(u, v)) \left| \frac{\partial(x, y)}{\partial(u, v)} \right| \, du dv$$

∎

Example 4 Evaluate

$$\int\int_D e^{x^2 + 4y^2} \, dx dy,$$

where D is the region that is bounded by the ellipse $x^2 + 4y^2 = 1$.

Solution

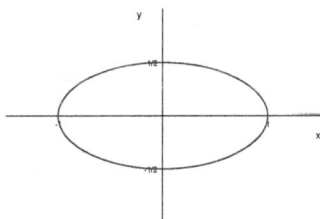

Figure 4

Let's set $u = x$ and $v = 2y$ so that $x = u$ and $y = v/2$. Thus,

$$x^2 + 4y^2 = 1 \Leftrightarrow u^2 + v^2 = 1.$$

The transformation $T(u, v) = (u, v/2)$ maps the disk D^* that is bounded by the unit circle $u^2 + v^2 = 1$ in a one-one manner onto D.

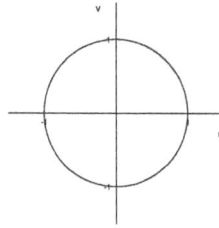

Figure 5

We have

$$\frac{\partial x}{\partial u} = 1, \ \frac{\partial x}{\partial v} = 0, \ \frac{\partial y}{\partial u} = 0, \ \frac{\partial y}{\partial v} = \frac{1}{2}.$$

Therefore,

$$\frac{\partial (x, y)}{\partial (u, v)} = \frac{\partial x}{\partial u} \frac{\partial y}{\partial v} - \frac{\partial x}{\partial v} \frac{\partial y}{\partial u} = \frac{1}{2}.$$

Thus

$$\int \int_D e^{x^2 + 4y^2} dx dy = \int \int_{D^*} e^{u^2 + v^2} \left| \frac{\partial (x, y)}{\partial (u, v)} \right| du dv = \int \int_{D^*} e^{u^2 + v^2} \left(\frac{1}{2} \right) du dv.$$

Let's transform to polar coordinates in the uv-plane so that $u = r \cos(\theta)$ and $v = r \sin(\theta)$. Therefore,

$$\int \int_D e^{x^2 + 4y^2} dx dy = \frac{1}{2} \int \int_{D^*} e^{u^2 + v^2} du dv \ = \ \frac{1}{2} \int_{\theta=0}^{2\pi} \int_{r=0}^{r=1} e^{r^2} 2 dr d\theta$$

$$= \ 2\pi \int_0^1 e^{r^2} r dr.$$

We set $w = r^2$ so that $dw = 2rdr$:

$$2\pi \int_0^1 e^{r^2} r dr = \pi \int_0^1 e^w dw = \pi (e - 1).$$

Therefore,

$$\int \int_D e^{x^2 + 4y^2} dx dy = \pi (e - 1).$$

□

Example 5 We can derive the expression for double integrals in polar coordinates from the general theorem.

We have $x = r \cos(\theta)$ and $y = r \sin(\theta)$. Therefore,

$$\frac{\partial x}{\partial r} = \cos(\theta), \ \frac{\partial x}{\partial \theta} = -r \sin(\theta), \ \frac{\partial y}{\partial r} = \sin(\theta), \ \frac{\partial y}{\partial \theta} = r \cos(\theta).$$

Thus,

$$\frac{\partial (x, y)}{\partial (r, \theta)} \ = \ \frac{\partial x}{\partial r} \frac{\partial y}{\partial \theta} - \frac{\partial x}{\partial \theta} \frac{\partial y}{\partial r}$$

$$= \ \cos(\theta) r \cos(\theta) - (-r \sin(\theta)) \sin(\theta)$$

$$= \ r \cos^2(\theta) + r \sin^2(\theta) = r.$$

Therefore,

$$\int_D f(x,y)\,dxdy \;=\; \int\int_{D^*} f\left(r\cos\left(\theta\right),r\sin\left(\theta\right)\right)\left|\frac{\partial\left(x,y\right)}{\partial\left(r,\theta\right)}\right|\,drd\theta$$

$$=\; \int\int_{D^*} f\left(r\cos\left(\theta\right),r\sin\left(\theta\right)\right)r\,drd\theta.$$

$\square$

There is a counterpart of the theorem on the change of the variables in double integrals for the change of variables in triple integrals.

Definition 3 If $T(u,v,w) = (x(u,v,w),y(u,v,w),z(u,v,w))$ is a transformation from D^* into D the Jacobian of x,y and z with respect to u,v and w is $x(u,v,w)$

$$\frac{\partial\left(x,y,z\right)}{\partial\left(u,v,w\right)} = \begin{vmatrix} \dfrac{\partial x}{\partial u} & \dfrac{\partial x}{\partial v} & \dfrac{\partial x}{\partial w} \\[2mm] \dfrac{\partial y}{\partial u} & \dfrac{\partial y}{\partial v} & \dfrac{\partial y}{\partial w} \\[2mm] \dfrac{\partial x}{\partial u} & \dfrac{\partial z}{\partial v} & \dfrac{\partial z}{\partial w} \end{vmatrix}$$

Theorem 2 Assume that $T(u,v,w) = (x(u,v,w),y(u,v,w),z(u,v,w))$ is a transformation that maps D^* onto D in a one-one manner (i.e., the images of distinct points are distinct), and its coordinate functions are differentiable. Then

$$\int\int_D f(x,y,z)\,dxdy = \int\int_{D^*} f\left(x(u,v,w),y(u,v,w),z(u,v,w)\right)\left|\frac{\partial\left(x,y,z\right)}{\partial\left(u,v,w\right)}\right|\,dudvdw$$

for any function f that is continuous on D.

Just as in the case of two variables, the above theorem is plausible: The volume of the image of a rectangular box with vertices at (u,v,w), $(u+\Delta u,v,w)$, $(u,v+\Delta v,w)$ and $(u,v,w+\Delta w)$ can be approximated by the volume of a parallelepiped with vertices at $T(u,v,w)$, $T(u+\Delta u,v,w)$, $T(u,v+\Delta v,w)$ and $T(u,v,w+\Delta w)$. That parallelepiped is spanned by the vectors

$$T(u+\Delta u,v,w)-T(u,v,w),\;\; T(u,v+\Delta v,w)-T(u,v,w)\;\; \text{and}\;\; T(u,v,w+\Delta w)-T(u,v,w).$$

If Δu, Δv and Δw are small, these vectors can be approximated by

$$\Delta u\left(\frac{\partial x}{\partial u}(u,v,w),\frac{\partial y}{\partial u}(u,v,w),\frac{\partial z}{\partial u}(u,v,w)\right),\;\; \Delta v\left(\frac{\partial x}{\partial v}(u,v),\frac{\partial y}{\partial v}(u,v),\frac{\partial z}{\partial v}(u,v)\right)$$

and

$$\Delta w\left(\frac{\partial x}{\partial w}(u,v),\frac{\partial y}{\partial w}(u,v),\frac{\partial y}{\partial w}(u,v)\right).$$

Thus, we can approximate the volume of the image of a rectangular box with vertices at (u,v,w), $(u+\Delta u,v,w)$, $(u,v+\Delta v,w)$ by the scalar triple product of these vectors:

$$\begin{vmatrix} \dfrac{\partial x}{\partial u} & \dfrac{\partial y}{\partial u} & \dfrac{\partial z}{\partial u} \\[2mm] \dfrac{\partial x}{\partial v} & \dfrac{\partial y}{\partial v} & \dfrac{\partial z}{\partial v} \\[2mm] \dfrac{\partial x}{\partial w} & \dfrac{\partial y}{\partial w} & \dfrac{\partial z}{\partial w} \end{vmatrix}\Delta u\;\Delta v\Delta w = \frac{\partial\left(x,y,z\right)}{\partial\left(u,v,w\right)}\Delta u\;\Delta v\Delta w.$$

Example 6 We can derive the expression for triple integrals in spherical coordinates from the above theorem:

We have

$$x = \rho \sin(\phi) \cos(\theta), \; y = \rho \sin(\phi) \sin(\theta) \text{ and } z = \rho \cos(\phi).$$

Thus,

$$\frac{\partial x}{\partial \rho} = \sin(\phi)\cos(\theta), \; \frac{\partial x}{\partial \phi} = \rho \cos(\phi)\cos(\theta), \; \frac{\partial x}{\partial \theta} = -\rho \sin(\phi)\sin(\theta),$$

$$\frac{\partial y}{\partial \rho} = \sin(\phi)\sin(\theta), \; \frac{\partial y}{\partial \phi} = \rho \cos(\phi)\sin(\theta), \; \frac{\partial y}{\partial \theta} = \rho \sin(\phi)\cos(\theta),$$

$$\frac{\partial z}{\partial \rho} = \cos(\phi), \; \frac{\partial z}{\partial \phi} = -\rho \sin(\phi), \; \frac{\partial z}{\partial \theta} = 0.$$

Therefore,

$$
\begin{aligned}
\frac{\partial(x,y,z)}{\partial(\rho,\phi,\theta)} &= \begin{vmatrix} \sin(\phi)\cos(\theta) & \rho\cos(\phi)\cos(\theta) & -\rho\sin(\phi)\sin(\theta) \\ \sin(\phi)\sin(\theta) & \rho\cos(\phi)\sin(\theta) & \rho\sin(\phi)\cos(\theta) \\ \cos(\phi) & -\rho\sin(\phi) & 0 \end{vmatrix} \\
&= \cos(\phi) \begin{vmatrix} \rho\cos(\phi)\cos(\theta) & -\rho\sin(\phi)\sin(\theta) \\ \rho\cos(\phi)\sin(\theta) & \rho\sin(\phi)\cos(\theta) \end{vmatrix} \\
&\quad + \rho\sin(\phi) \begin{vmatrix} \sin(\phi)\cos(\theta) & -\rho\sin(\phi)\sin(\theta) \\ \sin(\phi)\sin(\theta) & \rho\sin(\phi)\cos(\theta) \end{vmatrix} \\
&= \cos(\phi) \left(\rho^2 \sin(\phi)\cos(\phi)\cos^2(\theta) + \rho^2 \sin(\phi)\cos(\phi)\sin^2(\theta) \right) \\
&\quad + \rho\sin(\phi) \left(\sin^2(\phi)\cos^2(\theta) + \rho\sin^2(\phi)\sin^2(\theta) \right) \\
&= \rho^2 \cos^2(\phi)\sin(\phi) + \rho^2 \sin^3(\phi) \\
&= \rho^2 \sin(\phi) \left(\cos^2(\phi) + \sin^2(\phi) \right) = \rho^2 \sin(\phi).
\end{aligned}
$$

Thus,

$$
\begin{aligned}
\iint_D f(x,y,z)\,dxdy &= \iint_{D^*} f(x(\rho,\phi,\theta), y(\rho,\phi,\theta), z(\rho,\phi,\theta)) \left| \frac{\partial(x,y,z)}{\partial(\rho,\phi,\theta)} \right| d\rho d\phi d\theta \\
&= \iint_{D^*} f(x(\rho,\phi,\theta), y(\rho,\phi,\theta), z(\rho,\phi,\theta)) \rho^2 \sin(\phi)\, d\rho d\phi d\theta.
\end{aligned}
$$

$\square$

Chapter 14

Vector Analysis

In this chapter we will discuss **line integrals** and **surface integrals** that correspond to important topics from Physics, such as the **work** done by a force field on a particle as it moves along a trajectory and the **flux** across a surface of a vector field such as the velocity field of fluid flow or an electrostatic field. The chapter will conclude with three major theorems that have many applications: The theorems of **Green**, **Gauss** and **Stokes**.

14.1 Vector Fields, Divergence and Curl

Vector Fields

A two-dimensional vector field is a function that assigns to each point (x, y) in a subset D of the plane a two-dimensional vector $\mathbf{F}(x, y)$. Thus,

$$\mathbf{F}(x, y) = M(x, y)\,\mathbf{i} + N(x, y)\mathbf{j},$$

where M and N are real-valued functions defined on D. We can visualize such a vector field by attaching the vector $\mathbf{F}(x, y)$ to the point (x, y). Computer algebra systems have special routines for such visualization.

A three-dimensional vector field is a function that assigns to each point (x, y, z) in a subset D of $\mathbb{R}^3$ a three-dimensional vector $\mathbf{F}(x, y, z)$. Thus,

$$\mathbf{F}(x, y, z) = M(x, y, z)\,\mathbf{i} + N(x, y, z)\mathbf{j} + P(x, y, z)\,\mathbf{k},$$

where M, N and P are real-valued functions defined on D. We can visualize $\mathbf{F}(x, y, z)$ as a vector attached to (x, y, z)

Example 1 Let

$$\mathbf{F}(x, y) = \mathbf{r}(x, y) = x\mathbf{i} + y\mathbf{j}.$$

Figure 1 shows the vectors $\mathbf{F}(x, y)$ for a selection of the points (x, y). $\square$

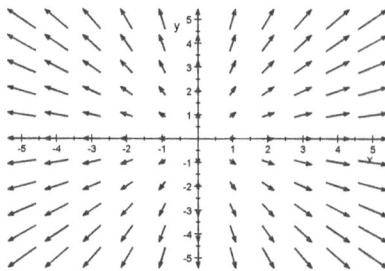

Figure 1: The vector field $\mathbf{F}(x, y) = \mathbf{r}(x, y) = x\mathbf{i} + y\mathbf{j}$

Example 2 Let

$$\mathbf{F}(x, y) = \mathbf{u}_r(x, y) = \frac{\mathbf{r}(x, y)}{||\mathbf{r}(x, y)||} = \frac{1}{\sqrt{x^2 + y^2}}(x\mathbf{i} + y\mathbf{j})$$

for each $(x, y) \neq (0, 0)$.

Thus, F assigns to each point other than the origin the unit vector in the direction of its position vector. We can set $r(x, y) = ||\mathbf{r}(x, y)||$ so that

$$\mathbf{F}(x, y) = \mathbf{u}_r(x, y) = \frac{\mathbf{r}}{r}(x, y).$$

□

Example 3 Let

$$\mathbf{S}(x, y) = -y\mathbf{i} + x\mathbf{j}$$

Note that

$$\mathbf{S}(x, y) \cdot (x\mathbf{i} + y\mathbf{j}) = -yx + xy = 0,$$

so that $\mathbf{S}(x, y)$ is orthogonal to the position vector $\mathbf{r}$ of (x, y). We also have

$$||\mathbf{S}(x, y)|| = \sqrt{y^2 + x^2} = ||x\mathbf{i} + y\mathbf{j}||,$$

so that the length of $\mathbf{S}(x, y)$ is constant on each circle centered at the origin. Figure 2 shows some of the vectors $\mathbf{F}(x, y)$. We may refer to $\mathbf{S}$ as a **spin field**. □

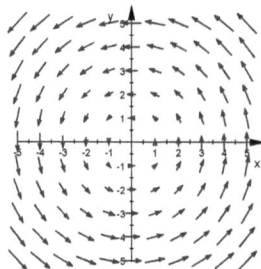

Figure 2: The spin field $\mathbf{S}$

Example 4 (A gravitational field) Newton's inverse-square law says that the magnitude of gravitational attraction between two objects of mass M and m is

$$\frac{GMm}{d^2},$$

where G is a universal constant and d is the distance between the objects. Thus, the force on an object of unit mass at the point (x, y, z) due to the mass M at the origin is

$$\mathbf{F}(x, y, z) = -\frac{GM}{||\mathbf{r}(x, y, z)||^2}\mathbf{u}_r(x, y, z),$$

where $\mathbf{r}(x, y, z) = x\mathbf{i} + y\mathbf{j} + z\mathbf{k} \neq \mathbf{0}$ and

$$\mathbf{u}_r(x, y, z) = \frac{\mathbf{r}}{||\mathbf{r}||}(x, y, z)$$

is the unit vector in the direction of $\mathbf{r}$. This is **the gravitational field due to the mass** M. We can express $\mathbf{F}(x, y, z)$ as

$$\mathbf{F}(x, y, z) = -\frac{GM}{||\mathbf{r}(x, y, z)||^2}\mathbf{u}_r(x, y, z) = -\frac{GM}{||\mathbf{r}(x, y, z)||^2}\left(\frac{\mathbf{r}}{||\mathbf{r}||}(x, y, z)\right) = -\frac{GM}{||\mathbf{r}(x, y, z)||^3}\mathbf{r}(x, y, z).$$

If we set $r(x, y, z) = ||\mathbf{r}(x, y, z)||$,

$$\mathbf{F}(x, y, z) = -\frac{GM}{r^2(x, y, z)}\mathbf{u}_r(x, y, z) = -\frac{GM}{r^3(x, y, z)}\mathbf{r}(x, y, z) = -\frac{GM}{(x^2 + y^2 + z^2)^{3/2}}(x\mathbf{i} + y\mathbf{j} + z\mathbf{k})$$

□

Example 5 (An electrostatic field) By **Coulomb's law,** the force per unit charge at (x, y, z) due to the charge q at the origin is

$$\mathbf{E}(x, y, z) = \frac{q}{4\pi\varepsilon r^2}\mathbf{u}_r,$$

where $\mathbf{u}_r$ is the unit vector in the radial direction. $\mathbf{E}(x, y, z)$ points towards the origin if $q < 0$ and points away from the origin if $q > 0$.□

The fields of examples 1, 2, 11 and 4 are **radial fields**: The value of such a field at (x, y) or (x, y, z) is of the form

$$\frac{C}{r^n}\mathbf{r},$$

where $\mathbf{r}$ is the position vector of (x, y) or (x, y, z), $r = ||\mathbf{r}||$, C is a constant and n is a positive integer.

Recall that **the gradient** of a scalar function $f(x, y)$ is defined by the expression

$$\nabla f(x, y) = \frac{\partial f}{\partial x}\mathbf{i} + \frac{\partial f}{\partial y}\mathbf{j}.$$

Thus, ∇f is a vector field in the plane. We noted that $\nabla f(x, y)$ is orthogonal to the level curve of f that passes through (x, y). Similarly, if f is a scalar function of x, y and z,

$$\nabla f(x, y, z) = \frac{\partial f}{\partial x}\mathbf{i} + \frac{\partial f}{\partial y}\mathbf{j} + \frac{\partial f}{\partial z}\mathbf{k}$$

defines a vector field in three dimensions. The vector $\nabla f(x, y, z)$ is orthogonal to the level surface of f that passes through (x, y, z).

Examples 1, 2, 11 and 4 are examples of gradient fields:

Example 6

a) Let $F(x, y) = r$, where $r = xi + yj$, as in Example 1. Show that

$$\mathbf{F}(x, y) = \nabla\left(\frac{r^2}{2}\right) = \nabla\left(\frac{1}{2}\left(x^2 + y^2\right)\right).$$

b) Let

$$\mathbf{F}(x.y) = \mathbf{u}_r = \frac{\mathbf{r}}{r},$$

as in Example 2. Show that

$$\mathbf{F}(x, y) = \nabla r$$

c) Let

$$\mathbf{F}(x, y, z) = \frac{C}{r^3}\mathbf{r},$$

where C is a constant, as in examples 11 and 4. Show that

$$\mathbf{F}(x, y, z) = \nabla\left(-\frac{C}{r}\right).$$

Solution

a)

$$\nabla\left(\frac{1}{2}r^2\right) = \frac{1}{2}\frac{\partial}{\partial x}\left(x^2 + y^2\right)\mathbf{i} + \frac{\partial}{\partial y}\left(x^2 + y^2\right)\mathbf{j} = \frac{1}{2}\left(2x\right)\mathbf{i} + \frac{1}{2}\left(2y\right)\mathbf{j} = x\mathbf{i} + y\mathbf{j} = \mathbf{r}$$

b)

$$\nabla r = \frac{\partial r}{\partial x}\mathbf{i} + \frac{\partial r}{\partial y}\mathbf{j} = \frac{\partial}{\partial x}\left(x^2 + y^2\right)^{1'2}\mathbf{i} + \frac{\partial}{\partial y}\left(x^2 + y^2\right)^{1/2}\mathbf{j}$$

$$= \frac{1}{2}\left(x^2 + y^2\right)^{-1/2}\left(2x\right)\mathbf{i} + \frac{1}{2}\left(x^2 + y^2\right)^{-1/2}\left(2y\right)\mathbf{j} = \frac{x}{r}\mathbf{i} + \frac{y}{r}\mathbf{j} = \frac{\mathbf{r}}{r} = \mathbf{u}_r$$

c) Since

$$\nabla\left(-\frac{C}{r}\right) = -C\nabla\left(\frac{1}{r}\right),$$

it is sufficient to show that

$$\nabla\left(\frac{1}{r}\right) = -\frac{\mathbf{r}}{r^3}$$

Indeed,

$$\nabla\left(\frac{1}{r}\right) = \frac{\partial}{\partial x}r^{-1}\mathbf{i} + \frac{\partial}{\partial y}r^{-1}\mathbf{j}$$

$$= \left(\frac{\partial}{\partial x}\left(x^2 + y^2 + z^2\right)^{-1/2}\right)\mathbf{i} + \left(\frac{\partial}{\partial y}\left(x^2 + y^2 + z^2\right)^{-1/2}\right)\mathbf{j} + \left(\frac{\partial}{\partial z}\left(x^2 + y^2 + z^2\right)^{-1/2}\right)\mathbf{k}$$

$$= \left(-\frac{1}{2}\left(x^2 + y^2 + z^2\right)^{-3/2}\left(2x\right)\right)\mathbf{i} + \left(-\frac{1}{2}\left(x^2 + y^2 + z^2\right)^{-3/2}\left(2y\right)\right)\mathbf{j}$$

$$+ \left(-\frac{1}{2}\left(x^2 + y^2 + z^2\right)^{-3/2}\left(2z\right)\right)\mathbf{k}$$

$$= -\frac{x}{\left(x^2 + y^2 + z^2\right)^{3/2}}\mathbf{i} + -\frac{y}{\left(x^2 + y^2 + z^2\right)^{3/2}}\mathbf{j} + -\frac{z}{\left(x^2 + y^2 + z^2\right)^{3/2}}\mathbf{k}$$

$$= -\frac{1}{\left(x^2 + y^2 + z^2\right)^{3/2}}\left(x\mathbf{i} + y\mathbf{j} + z\mathbf{k}\right) = -\frac{1}{r^3}\mathbf{r}.$$

□

Definition 1 A field is **conservative** if it is the gradient of a scalar function. Thus, the field **F** is conservative if there exists a scalar function f such that $\mathbf{F} = \nabla f$. Such an f is referred to as a **potential** for **F**.

In Example 6 we showed that gravitational fields and electronic fields are conservative fields.

Remark 1 If f is a potential for **F**, then $f + C$ is also a potential for **F** if C is an arbitrary constant. Indeed,

$$\nabla\left(f + C\right) = \nabla \mathbf{F},$$

since $\nabla C = \mathbf{0}$. ◊

The Divergence and Curl of a Vector Field

Definition 2 The divergence of the vector field

$$\mathbf{F}\left(x, y\right) = M\left(x, y\right)\mathbf{i} + N(x, y)\mathbf{j}$$

is the scalar function

$$\frac{\partial M}{\partial x}\left(x, y\right) + \frac{\partial N}{\partial y}\left(x, y\right).$$

We will denote the divergence of **F** as div **F** so that

$$\operatorname{div} \mathbf{F}\left(x, y\right) = \frac{\partial M}{\partial x}\left(x, y\right) + \frac{\partial N}{\partial y}\left(x, y\right).$$

Let

$$\nabla = \frac{\partial}{\partial x}\mathbf{i} + \frac{\partial}{\partial y}\mathbf{j}$$

be the *symbolic vector* that is used in denoting the gradient of a scalar function. Usually, we will denote the divergence of a vector field by the *symbolic dot product*

$$\nabla \cdot \mathbf{F}\left(x, y\right) = \left(\frac{\partial}{\partial x}\mathbf{i} + \frac{\partial}{\partial y}\mathbf{j}\right) \cdot \left(M\left(x, y\right)\mathbf{i} + N(x, y)\mathbf{j}\right) = \frac{\partial M}{\partial x}\left(x, y\right) + \frac{\partial N}{\partial y}\left(x, y\right).$$

Similarly, if $\mathbf{F}(x, y, z) = M\left(x, y, z\right)\mathbf{i} + N(x, y, z)\mathbf{j} + P\left(x, y, z\right)\mathbf{k}$ is a vector field in $\mathbf{R}^3$, we define the divergence of **F** by setting

$$\operatorname{div} \mathbf{F}\left(x, y, z\right) = \frac{\partial M}{\partial x} + \frac{\partial N}{\partial y} + \frac{\partial P}{\partial z}.$$

In terms of the symbolic vector

$$\nabla = \frac{\partial}{\partial x}\mathbf{i} + \frac{\partial}{\partial y}\mathbf{j} + \frac{\partial}{\partial z}\mathbf{k},$$

$$\operatorname{div} \mathbf{F}\left(x, y, z\right) = \nabla \cdot \mathbf{F}\left(x, y, z\right) = \left(\frac{\partial}{\partial x}\mathbf{i} + \frac{\partial}{\partial y}\mathbf{j} + \frac{\partial}{\partial z}\mathbf{k}\right) \cdot \left(M\left(x, y, z\right)\mathbf{i} + N(x, y, z)\mathbf{j} + P\left(x, y, z\right)\mathbf{k}\right)$$
$$= \frac{\partial M}{\partial x} + \frac{\partial N}{\partial y} + \frac{\partial P}{\partial z}.$$

Example 7

a) Let $\mathbf{F}(x, y) = \mathbf{r} = x\mathbf{i} + y\mathbf{j}$, as in Example 1. Then

$$\operatorname{div} \mathbf{F}(x, y) = \boldsymbol{\nabla} \cdot \mathbf{F}(x, y) = \frac{\partial}{\partial x}(x) + \frac{\partial}{\partial y}(y) = 1 + 1 = 2.$$

b) $\mathbf{S}(x, y) = -y\mathbf{i} + x\mathbf{j}$ be the spin field of Example 3. Then

$$\boldsymbol{\nabla} \cdot \mathbf{S}(x, y) = \frac{\partial}{\partial x}(-y) + \frac{\partial}{\partial y}(x) = 0.$$

Later in this chapter we will see that the divergence of a vector field will be a measure of the pointwise expansion or contraction of a vector field. The vector field of part a) "expands" with respect to the origin. The vector field of part b) does not contract or expand.□

Definition 3 The curl of a three-dimensional vector field $\mathbf{F}(x, y, z)$ is the vector field that is defined by the symbolic cross product

$$\boldsymbol{\nabla} \times \mathbf{F}(x, y, z) = \left(\frac{\partial}{\partial x}\mathbf{i} + \frac{\partial}{\partial y}\mathbf{j} + \frac{\partial}{\partial z}\mathbf{k} \right) \times (M(x, y, z)\,\mathbf{i} + N(x, y, z)\mathbf{j} + P(x, y, z)\,\mathbf{k})$$

$$= \begin{vmatrix} \mathbf{i} & \mathbf{j} & \mathbf{k} \\ \dfrac{\partial}{\partial x} & \dfrac{\partial}{\partial y} & \dfrac{\partial}{\partial z} \\ M & N & P \end{vmatrix} = \left(\frac{\partial P}{\partial y} - \frac{\partial N}{\partial z} \right)\mathbf{i} - \left(\frac{\partial P}{\partial x} - \frac{\partial M}{\partial z} \right)\mathbf{j} + \left(\frac{\partial N}{\partial x} - \frac{\partial M}{\partial y} \right)\mathbf{k}$$

If $\mathbf{F}(x, y) = M(x, y)\,\mathbf{i} + N(x, y)\mathbf{j}$ is a two-dimensional vector field, we can consider $\mathbf{F}$ as a vector field in three dimensions by setting

$$\mathbf{F}(x, y, z) = M(x, y)\,\mathbf{i} + N(x, y)\mathbf{j} + 0\mathbf{k}.$$

With this understanding,

$$\boldsymbol{\nabla} \times \mathbf{F}(x, y) = \begin{vmatrix} \mathbf{i} & \mathbf{j} & \mathbf{k} \\ \dfrac{\partial}{\partial x} & \dfrac{\partial}{\partial y} & \dfrac{\partial}{\partial z} \\ M(x, y) & N(x, y) & 0 \end{vmatrix} = -\frac{\partial N}{\partial z}\mathbf{i} + \frac{\partial M}{\partial z}\mathbf{j} + \left(\frac{\partial N}{\partial x} - \frac{\partial M}{\partial y} \right)\mathbf{k}$$

$$= \left(\frac{\partial N}{\partial x} - \frac{\partial M}{\partial y} \right)\mathbf{k}$$

Thus, the curl of a vector field in the xy-plane is always orthogonal to the xy-plane.

Example 8 Let $\omega \mathbf{S}(x, y) = -\omega y\mathbf{i} + \omega x\mathbf{j}$, where $\mathbf{S}$ is the spin field of Example 3. Calculate $\boldsymbol{\nabla} \times (\omega \mathbf{S})$.

Solution

We have

$$\boldsymbol{\nabla} \times (\omega \mathbf{S})(x, y) = \left(\frac{\partial}{\partial x}(\omega x) - \frac{\partial}{\partial y}(-\omega y) \right)\mathbf{k} = 2\omega\mathbf{k}$$

Thus, the curl of $\omega \mathbf{S}$ at any point is a vector of magnitude $|2\omega|$ that is orthogonal to the xy-plane. We can imagine that $\omega \mathbf{S}$ is the velocity field of a fictitious fluid that rotates about the origin, and we may refer to 2ω as the angular velocity. Later in this chapter we will see that the curl of a vector field at a point is a measure of the rotation of the vector field about that point. □

Definition 4 A vector field $\mathbf{F}$ such that $\boldsymbol{\nabla} \times \mathbf{F} = \mathbf{0}$ is said to be irrotational.

Proposition 1 A gradient field is irrotational:

$$\nabla \times (\nabla f) = \mathbf{0}$$

Proof

$$\nabla \times (\nabla f) = \begin{vmatrix} \mathbf{i} & \mathbf{j} & \mathbf{k} \\ \frac{\partial}{\partial x} & \frac{\partial}{\partial y} & \frac{\partial}{\partial z} \\ \frac{\partial f}{\partial x} & \frac{\partial f}{\partial y} & \frac{\partial f}{\partial z} \end{vmatrix}$$

$$= \left(\frac{\partial}{\partial y} \left(\frac{\partial f}{\partial z} \right) - \frac{\partial}{\partial z} \left(\frac{\partial f}{\partial y} \right) \right) \mathbf{i} - \left(\frac{\partial}{\partial x} \left(\frac{\partial f}{\partial z} \right) - \frac{\partial}{\partial z} \left(\frac{\partial f}{\partial x} \right) \right) \mathbf{j}$$

$$+ \left(\frac{\partial}{\partial x} \left(\frac{\partial f}{\partial y} \right) - \frac{\partial}{\partial y} \left(\frac{\partial f}{\partial x} \right) \right) \mathbf{k} = \mathbf{0}$$

∎

Example 9 Let

$$\mathbf{F}(x, y, z) = \frac{\mathbf{r}}{r^3},$$

where $\mathbf{r} = x\mathbf{i} + y\mathbf{j} + z\mathbf{k}$ and $r = \sqrt{x^2 + y^2 + z^2}$.

In Example 6 we showed that $\mathbf{F}$ is a gradient field. Therefore, $\mathbf{F}$ is irrotational. In particular, gravitational fields and electrostatic fields are irrotational. Later in this chapter we will see that any irrotational vector field $\mathbf{F}$ is a gradient field if $\mathbf{F}$ satisfies certain smoothness conditions. □

Problems

In problems 1-6 determine
a) The divergence of $\mathbf{F}$,
b) The curl of $\mathbf{F}$.

1.

$$\mathbf{F}(x, y, z) = xyz\mathbf{i} - x^2 y\mathbf{k}$$

2.

$$\mathbf{F}(x, y, z) = e^{-x} xy^2 \mathbf{i} + e^{-x} x^2 y\mathbf{j}$$

3.

$$\mathbf{F}(x, y, z) = \cos(xz)\mathbf{j} - \sin(xy)\mathbf{k}$$

4.

$$\mathbf{F}(x, y, z) = \arctan\left(\frac{y}{x}\right)\mathbf{i} + \arctan\left(\frac{x}{y}\right)\mathbf{j}$$

5.

$$\mathbf{F}(x, y, z) = 2xy\mathbf{i} + \left(x^2 + 2yz\right)\mathbf{j} + y^2\mathbf{k}$$

6.

$$\mathbf{F}(x, y, z) = \ln(x)\,i + \ln(xy)\,\mathbf{j} + \ln(xyz)\,\mathbf{k}$$

14.2   Line Integrals

The Integral of a Scalar Function with respect to arc length

Let C be a curve in the plane that is parametrized by the path $\boldsymbol{\sigma} : [a, b] \to \mathbb{R}^2$. Assume that f is a real-valued function of two variables. You may imagine that C is a wire, and $f(\boldsymbol{\sigma}(t)) = f(x(t), y(t))$ is the lineal density of the wire. How can we calculate the mass of the wire? It is reasonable to approximate mass by sums of the form

$$\sum_{k=1}^{n} f(\boldsymbol{\sigma}(t_k)) \Delta s_k = \sum_{k=1}^{n} f(x(t_k), y(t_k)) \Delta s_k,$$

where $\{t_0, t_1, \ldots, t_{n-1}, t_n\}$ is a partition of $[a, b]$ and Δs_k is the length of the part of C corresponding to the interval $[t_{k-1}, t_k]$. Thus

$$\Delta s_k = \int_{t_{k-1}}^{t_k} \|\boldsymbol{\sigma}'(t)\| \, dt.$$

By the Mean Value Theorem for integrals,

$$\int_{t_{k-1}}^{t_k} \|\boldsymbol{\sigma}'(t)\| \, dt = \boldsymbol{\sigma}'(t_k^*)(t_k - t_{k-1}) = \boldsymbol{\sigma}'(t_k^*) \Delta t_k,$$

where t_k^* is some point between t_{k-1} and t_k. Therefore,

$$\sum_{k=1}^{n} f(\boldsymbol{\sigma}(t_k)) \Delta s_k = \sum_{k=1}^{n} f(\boldsymbol{\sigma}(t_k)) \|\boldsymbol{\sigma}'(t_k^*)\| \Delta t_k$$

This is almost a Riemann sum for the integral

$$\int_{a}^{b} f(\boldsymbol{\sigma}(t)) \|\boldsymbol{\sigma}'(t)\| \, dt$$

and approximates the integral with desired accuracy provided that f and $\sigma'(t)$ are continuous. This leads to the following definition:

Definition 1 The integral of the scalar function f with respect to arc length on the curve C that is parametrized by the function $\boldsymbol{\sigma} : [a, b] \to \mathbb{R}^2$ is defined as

$$\int_{a}^{b} f(\boldsymbol{\sigma}(t)) \|\boldsymbol{\sigma}'(t)\| \, dt$$

Symbolically,

$$ds = \frac{ds}{dt} dt = \left\| \frac{d\boldsymbol{\sigma}}{dt} \right\| dt,$$

so that we can express the integral as

$$\int_{C} f \, ds.$$

In terms of the coordinate functions $x(t)$ and $y(t)$ of $\sigma(t)$,

$$\int_{C} f \, ds = \int_{a}^{b} f(x(t), y(t)) \sqrt{\left(\frac{dx}{dt}\right)^2 + \left(\frac{dy}{dt}\right)^2} \, dt.$$

Example 1 Evaluate

$$\int_C x^2 y^2 ds,$$

where C is the semicircle $x^2 + y^2 = 4$ that is parametrized so that C is traversed from $(2, 0)$ to $(-2, 0)$.

Solution

We can set

$$\sigma(t) = (2\cos(t), 2\sin(t)), \ 0 \le t \le \pi.$$

Then σ parametrizes C as required.

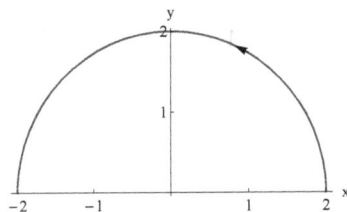

Figure 1

Therefore,

$$\sigma'(t) = -2\sin(t)\,\mathbf{i} + 2\cos(t)\,\mathbf{j} \Rightarrow ||\sigma'(t)|| = \sqrt{4\sin^2(t) + 4\cos^2(t)} = 2.$$

Thus,

$$\int_C x^2 y^2 ds = \int_0^\pi \left(4\cos^2(t)\right)\left(4\sin^2(t)\right)||\sigma'(t)||\,dt = 16\int_0^\pi \cos^2(t)\sin^2(t)\,(2)\,dt$$

$$= 32\int_0^\pi \cos^2(t)\sin^2(t)\,dt.$$

As we saw in Section 7.3,

$$\int \cos^2(t)\sin^2(t)\,dt = \frac{1}{8}x - \frac{1}{8}\sin(x)\cos^3(x) + \frac{1}{8}\sin^3(x)\cos(x).$$

Therefore,

$$32\int_0^\pi \cos^2(t)\sin^2(t)\,dt = 4x - 4\sin(x)\cos^3(x) + 4\sin^3(x)\cos(x).\Big|_0^\pi = 4\pi$$

$\sqcup$

A curve C can be parametrized by different functions and may be traversed from point A to point B or vice versa. Thus, there is a notion of "the orientation" of a curve. Assume that C is a curve in the plane that is parametrized initially by $\sigma : [a, b] \to \mathbb{R}^2$ and $\sigma(t)$ traces C from "**the initial point**" P_1 to "**the terminal point**" P_2 as t increases from a to b. We will say that $\tilde{\sigma} : [\alpha, \beta] \to \mathbb{R}^2$ is an **orientation preserving parametrization** of C if $\tilde{\sigma}(\tau)$ traces C from the initial point P_1 to the terminal point P_2 as the new parameter τ increases from α to β. We will say that $\tilde{\sigma} : [\alpha, \beta] \to \mathbb{R}^2$ is an **orientation reversing parametrization** of C if $\tilde{\sigma}(\tau)$ traces C from the terminal point P_2 to the initial point P_1 as the new parameter τ increases from α to β. **The positive orientation** of C is the orientation provided by the initial parametrization σ and **the negative orientation** of C is the orientation provided by any orientation reversing parametrization.You can find the precise defintions at the end of this section. The line integral of a scalar function is does not change if the curve is parametrized by any orientation preserving or orientation reversing parametrization:

Proposition 1 If C is parametrized by $\sigma : [a,b] \to \mathbb{R}^2$ and $\tilde{\sigma} : [\alpha,\beta] \to \mathbb{R}^2$ is an orientation preserving or orientation reversing parametrization of C, then

$$\int_C f ds = \int_{t=a}^{t=b} f\left(\sigma\left(t\right)\right) \left\|\frac{d\sigma}{dt}\right\| dt = \int_{\tau=\alpha}^{\tau=\beta} f\left(\tilde{\sigma}\left(\tau\right)\right) \left\|\frac{d\tilde{\sigma}}{d\tau}\right\| d\tau$$

You can find the proof of Proposition 1 at the end of this section.

Example 2 Let $P_1 = (1,2)$ and $P_2 = (2,4)$, and let C be the directed line segment $P_1 P_2$. We can parametrize C by

$$\sigma\left(t\right) = OP_1 + tP_1P_2 = (1,2) + t(1,2) = (1+t, 2+2t),\ 0 \le t \le 1.$$

If we set

$$\tilde{\sigma}\left(\tau\right) = OP_1 + 2\tau P_1P_2 = (1,2) + 2\tau(1,2) = (1+2\tau, 2+4\tau),\ 0 \le \tau \le \frac{1}{2},$$

then σ is an orientation preserving parametrization of C.
If we set

$$\sigma^*\left(\mu\right) = OP_2 + \mu\tau P_2P_1 = (2,4) + \mu(-1,-2) = (2-\mu, 4-2\mu),\ 0 \le \mu \le 1,$$

then σ^* is an orientation reversing parametrization of C. Confirm that the statements of Proposition 1 are valid in the special case

$$\int_C \sin\left(\frac{\pi}{2}(x-y)\right) ds$$

Solution

We have

$$\frac{d\sigma}{dt} = (1,2)\ \text{ so that }\ \left\|\frac{d\sigma}{dt}\right\| = \sqrt{1+4} = \sqrt{5}.$$

Therefore

$$\begin{aligned}
\int_C \sin\left(\frac{\pi}{2}(x-y)\right) ds &= \int_0^1 \sin\left(\frac{\pi}{2}\left((1+t)-(2+2t)\right)\right) \left\|\frac{d\sigma}{dt}\right\| dt \\
&= \int_0^1 \sin\left(\frac{\pi}{2}(-t-1)\right) \sqrt{5} dt \\
&= \sqrt{5} \int_0^1 -\sin\left(\frac{\pi}{2}(t+1)\right) dt.
\end{aligned}$$

We set

$$u = \frac{\pi}{2}(t+1) \Rightarrow du = \frac{\pi}{2} dt.$$

Thus,

$$\sqrt{5}\int_0^1 -\sin\left(\frac{\pi}{2}(t+1)\right) dt = \frac{2\sqrt{5}}{\pi}\int_{\pi/2}^\pi -\sin\left(u\right) du = \frac{2\sqrt{5}}{\pi}\left(\cos\left(u\right)\big|_{u=\pi/2}^{u=\pi}\right) = -\frac{2\sqrt{5}}{\pi}.$$

Now let's use the orientation preserving parametrization σ. We have

$$\frac{d\sigma}{d\tau} = (2,4) \text{ so that } \left\|\frac{d\sigma}{dt}\right\| = \sqrt{4+16} = \sqrt{20} = 2\sqrt{5}.$$

Therefore

$$
\begin{aligned}
\int_C \sin\left(\frac{\pi}{2}(x-y)\right)ds &= \int_0^{1/2} \sin\left(\frac{\pi}{2}((1+2\tau)-(2+4\tau))\right)\left\|\frac{d\sigma}{d\tau}\right\|d\tau \\
&= \int_0^{1/2} \sin\left(\frac{\pi}{2}(-2\tau-1)\right) 2\sqrt{5}d\tau \\
&= -2\sqrt{5}\int_0^{1/2} \sin\left(\frac{\pi}{2}(2\tau+1)\right)d\tau.
\end{aligned}
$$

We set

$$u = \frac{\pi}{2}(2\tau+1) \Rightarrow du = \pi d\tau.$$

Thus

$$2\sqrt{5}\int_0^{1/2} -\sin\left(\frac{\pi}{2}(2\tau+1)\right)d\tau = \frac{2\sqrt{5}}{\pi}\int_{\pi/2}^{\pi} -\sin(u)\,du = -\frac{2\sqrt{5}}{\pi},$$

as in the case of the parametrization by σ.

$$\sigma^*(\mu) = OP_2 + \mu\tau P_2 P_1 = (2,4) + \mu(-1,-2) = (2-\mu, 4-2\mu),\ 0 \le \mu \le 1,$$

Now let's use the orientation reversing parametrization σ^*. We have

$$\frac{d\sigma^*}{d\mu} = (-1,-2) \text{ so that } \left\|\frac{d\sigma^*}{d\mu}\right\| = \sqrt{5}.$$

Therefore

$$
\begin{aligned}
\int_0^1 \sin\left(\frac{\pi}{2}((2-\mu)-(4-2\mu))\right)\left\|\frac{d\sigma^*}{d\mu}\right\|d\mu &= \int_0^1 \sin\left(\frac{\pi}{2}(\mu-2)\right)\sqrt{5}d\mu \\
&= \sqrt{5}\int_0^1 -\sin\left(\frac{\pi}{2}(2-\mu)\right)d\mu.
\end{aligned}
$$

We set

$$u = \frac{\pi}{2}(2-\mu) \Rightarrow du = -\frac{\pi}{2}d\mu.$$

Thus

$$\sqrt{5}\int_0^1 -\sin\left(\frac{\pi}{2}(2-\mu)\right)d\mu = \frac{2\sqrt{5}}{\pi}\int_{\pi}^{\pi/2} \sin(u)\,du = \frac{2\sqrt{5}}{\pi}.$$

Therefore

$$\int_0^1 \sin(\sigma^*(\mu))\left\|\frac{d\sigma^*}{d\mu}\right\|d\mu = -\int_C fds,$$

as predicted by Proposition 1

The Line Integral of a Vector Field in the Plane

Assume that $\mathbf{F}(x,y) = M(x,y)\mathbf{i} + N(x,y)\mathbf{j}$ is a two-dimensional force field, $\boldsymbol{\sigma} : [a,b] \to \mathbb{R}^2$, and a particle is at $\boldsymbol{\sigma}(t) = (x(t), y(t))$ at time t. Let C be the curve that is parametrized by $\boldsymbol{\sigma}$. How should we calculate the **work** done by the force $\mathbf{F}$ as the particle moves from $(x(a), y(a))$ to $(x(b), y(b))$ along C? Our starting point is the definition of the work done by a constant force $\mathbf{F}$ on a particle that moves along a line. If the displacement of the particle is represented by the vector $\mathbf{w}$, the work done is

$$\mathbf{F} \cdot \mathbf{w} = \|\mathbf{F}\| \, \|\mathbf{w}\| \cos(\theta),$$

where θ is the angle between $\mathbf{F}$ and $\mathbf{w}$. Let us subdivide the interval $[a,b]$ into subintervals by a partition

$$P = \{a = t_0 < t_1 < \cdots < t_{k-1} < t_k < \cdots < t_{n-1} < t_n = b\}.$$

As usual we set $\Delta t_k = t_k - t_{k-1}$. Let $\Delta\boldsymbol{\sigma}_k = \boldsymbol{\sigma}(t_k) - \boldsymbol{\sigma}(t_{k-1})$. If the norm of the partition P is small, we can approximate the work done by $\boldsymbol{F}$ as the particle moves from $\boldsymbol{\sigma}(t_{k-1})$ to $\boldsymbol{\sigma}(t_k)$ by

$$\mathbf{F}(\boldsymbol{\sigma}(t_k)) \cdot \Delta\boldsymbol{\sigma}_k = \mathbf{F}(\sigma(t_k)) \cdot (\boldsymbol{\sigma}(t_k) - \boldsymbol{\sigma}(t_{k-1})) \cong \mathbf{F}(\boldsymbol{\sigma}(t_k)) \cdot \frac{\boldsymbol{\sigma}(t_k) - \boldsymbol{\sigma}(t_{k-1})}{\Delta t_k} \Delta t_k$$

$$\cong \mathbf{F}(\boldsymbol{\sigma}(t_k)) \cdot \frac{d\boldsymbol{\sigma}}{dt}(t_k) \Delta t_k$$

Therefore, the total work done is approximately

$$\sum_{k=1}^{n} \mathbf{F}(\boldsymbol{\sigma}(t_k)) \cdot \frac{d\boldsymbol{\sigma}}{dt}(t_k) \Delta t_k$$

This is a Riemann sum for the integral

$$\int_a^b \mathbf{F}(\boldsymbol{\sigma}(t)) \cdot \frac{d\boldsymbol{\sigma}}{dt} dt.$$

We will define **the work done by the force $\mathbf{F}$ on the particle as it moves along the curve** C that is parametrized by $\boldsymbol{\sigma}$ by this integral. In more general terms, this is a line integral:

Definition 2 Assume that the curve C in the plane is parametrized by $\boldsymbol{\sigma} : [a,b] \to \mathbb{R}^2$ and $\mathbf{F}$ is a continuous vector field in the plane. **The line integral of $\mathbf{F}$ on C is**

$$\int_C \mathbf{F} \cdot d\boldsymbol{\sigma} = \int_a^b \mathbf{F}(\sigma(t)) \cdot \frac{d\boldsymbol{\sigma}}{dt} dt.$$

The notation $\int_C \mathbf{F} \cdot d\boldsymbol{\sigma}$ is appropriate since the line integral is approximated by sums of the form

$$\sum_{k=1}^{n} \mathbf{F}(\sigma(t_k)) \cdot \Delta\boldsymbol{\sigma}_k.$$

Symbolically,

$$d\boldsymbol{\sigma} = \frac{d\boldsymbol{\sigma}}{dt} dt.$$

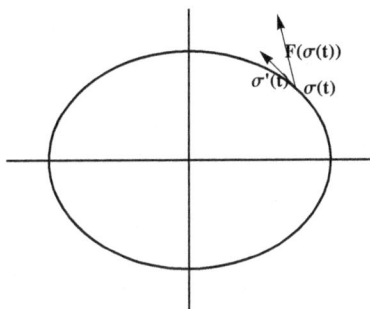

Figure 2

Example 3 Let $\mathbf{F}(x, y) = 2x\mathbf{i} + y\mathbf{j}$ and assume that C is parametrized by $\boldsymbol{\sigma}(t) = (\cosh(t), \sinh(t))$, where $0 \leq t \leq 3$. Calculate

$$\int_C \mathbf{F} \cdot d\boldsymbol{\sigma}$$

Solution

Figure 3 shows the curve C.

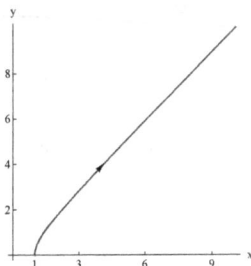

Figure 3

We have

$$\frac{d\boldsymbol{\sigma}}{dt} = \sinh(t)\mathbf{i} + \cosh(t)\mathbf{j}.$$

Therefore,

$$\int_C \mathbf{F} \cdot d\boldsymbol{\sigma} = \int_0^3 (2\cosh(t)\mathbf{i} + \sinh(t)\mathbf{j}) \cdot (\sinh(t)\mathbf{i} + \cosh(t)\mathbf{j})\, dt$$

$$= \int_0^3 (2\cosh(t)\sinh(t) + \sinh(t)\cosh(t))\, dt$$

$$= \int_0^3 3\cosh(t)\sinh(t)\, dt.$$

If we set $u = \cosh(t)$ then $du = \sinh(t)\, dt$, so that

$$\int_0^3 3\cosh(t)\sinh(t)\, dt. = 3\int_{\cosh(0)}^{\cosh(3)} u\, du = 3\left(\frac{u^2}{2}\bigg|_1^{\cosh(3)}\right) = \frac{3}{2}\left(\cosh^2(3) - 1\right)$$

□

We stated that the integral of a scalar function with respect to arc length does not depend on an orientation preserving or orientation reversing parametrization of C. In the case of the line integral of a vector field, an orientation reversing reparametrization introduces a minus sign:

Proposition 2 Assume that the curve C is parametrized by $\boldsymbol{\sigma} : [a, b] \to \mathbb{R}^2$. If $\tilde{\boldsymbol{\sigma}} : [\alpha, \beta] \to \mathbb{R}^2$ is an orientation preserving parametrization of C, then

$$\int_\alpha^\beta \mathbf{F}\left(\tilde{\boldsymbol{\sigma}}\left(\tau\right)\right) \cdot \frac{d\tilde{\boldsymbol{\sigma}}\left(\tau\right)}{d\tau} d\tau = \int_a^b \mathbf{F}\left(\boldsymbol{\sigma}\left(t\right)\right) \cdot \frac{d\boldsymbol{\sigma}}{dt} dt.$$

If $\tilde{\boldsymbol{\sigma}} : [\alpha, \beta] \to \mathbb{R}^2$ is an orientation reversing parametrization of C, then

$$\int_\alpha^\beta \mathbf{F}\left(\tilde{\boldsymbol{\sigma}}\left(\tau\right)\right) \cdot \frac{d\tilde{\boldsymbol{\sigma}}\left(\tau\right)}{d\tau} d\tau = - \int_a^b \mathbf{F}\left(\boldsymbol{\sigma}\left(t\right)\right) \cdot \frac{d\boldsymbol{\sigma}}{dt} dt$$

You can find the proof of Proposition 2 at the end of this section.

Remark   Assume that the curve C is parametrized by $\boldsymbol{\sigma} : [a, b] \to \mathbb{R}^2$. and that $\tilde{\boldsymbol{\sigma}} : [\alpha, \beta] \to \mathbb{R}^2$ is an orientation reversing parametrization of C. We will set

$$\int_{-C} \mathbf{F} \cdot d\boldsymbol{\sigma} = \int_\alpha^\beta \mathbf{F}\left(\tilde{\boldsymbol{\sigma}}\left(\tau\right)\right) \cdot \frac{d\tilde{\boldsymbol{\sigma}}\left(\tau\right)}{d\tau} d\tau.$$

By Proposition 2,

$$\int_{-C} \mathbf{F} \cdot d\boldsymbol{\sigma} = - \int_a^b \mathbf{F}\left(\boldsymbol{\sigma}\left(t\right)\right) \cdot \frac{d\boldsymbol{\sigma}}{dt} dt = - \int_C \mathbf{F} \cdot d\boldsymbol{\sigma}.$$

If C is a piecewise smooth curve that can be expressed as a sequence of smooth curves $C_1, C_2, \ldots, C_m$, we will write

$$C = C_1 + C_2 + \cdots + C_m.$$

With this understanding, we set

$$\int_C \mathbf{F} \cdot d\boldsymbol{\sigma} = \int_{C_1 + C_2 + \cdots + C_m} \mathbf{F} \cdot d\boldsymbol{\sigma} = \int_{C_1} \mathbf{F} \cdot d\boldsymbol{\sigma} + \int_{C_2} \mathbf{F} \cdot d\boldsymbol{\sigma} + \cdots + \int_{C_m} \mathbf{F} \cdot d\boldsymbol{\sigma}$$

$\Diamond$

Example 4 Let

$$\mathbf{F}\left(x, y\right) = -y\mathbf{i} + x\mathbf{j}.$$

Evaluate

$$\int_{C_1 + C_2} \mathbf{F} \cdot d\boldsymbol{\sigma},$$

where C_1 is the line segment from $(-1, 0)$ to $(1, 0)$ on the x-axis, and C_2 is the semicircle of radius 1 centered at the origin traversed in the counterclockwise direction

Solution

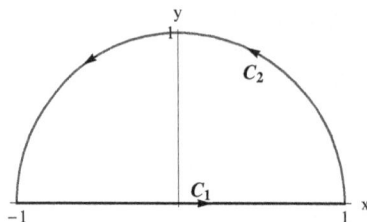

Figure 4

We can parametrize the line segment by $\sigma_1(t) = (t, 0)$, where $t \in [-1, 1]$. The semicircle can be parametrized by $\sigma_2(t) = (\cos(t), \sin(t))$, where $t \in [0, \pi]$. Therefore,

$$\int_{C_1} \mathbf{F} \cdot d\boldsymbol{\sigma} = \int_{-1}^{1} \mathbf{F}(\boldsymbol{\sigma}(t)) \cdot \frac{d\boldsymbol{\sigma}}{dt} dt = \int_{-1}^{1} t\mathbf{j} \cdot \mathbf{i} dt = 0,$$

and

$$\int_{C_2} \mathbf{F} \cdot d\boldsymbol{\sigma} = \int_{0}^{\pi} \mathbf{F}(\boldsymbol{\sigma}(t)) \cdot \frac{d\boldsymbol{\sigma}}{dt} dt = \int_{0}^{\pi} (-\sin(t)\mathbf{i} + \cos(t)\mathbf{j}) \cdot (-\sin(t)\mathbf{i} + \cos(t)\mathbf{j}) dt$$

$$= \int_{0}^{\pi} (\sin^2(t) + \cos^2(t)) dt = \int_{0}^{\pi} 1 dt = \pi.$$

Thus,

$$\int_{C_1 + C_2} \mathbf{F} \cdot d\boldsymbol{\sigma} = \int_{C_1} \mathbf{F} \cdot d\boldsymbol{\sigma} + \int_{C_2} \mathbf{F} \cdot d\boldsymbol{\sigma} = \pi.$$

☐

The Line Integral as an Integral with respect to Arc Length

We can express the line integral of $\mathbf{F}$ on the curve C as **the integral of the tangential component of F with respect to arc length** (Recall that $\mathbf{T}(t)$ denotes the unit tangent to C at $\boldsymbol{\sigma}(t)$):

$$\int_{C} \mathbf{F} \cdot d\boldsymbol{\sigma} = \int_{a}^{b} \mathbf{F}(\boldsymbol{\sigma}(t)) \cdot \frac{d\boldsymbol{\sigma}}{dt} dt = \int_{a}^{b} \mathbf{F}(\boldsymbol{\sigma}(t)) \cdot \frac{\frac{d\boldsymbol{\sigma}}{dt}}{\left\| \frac{d\boldsymbol{\sigma}}{dt} \right\|} \left\| \frac{d\boldsymbol{\sigma}}{dt} \right\| dt = \int_{a}^{b} \mathbf{F}(\boldsymbol{\sigma}(t)) \cdot \mathbf{T}(t) \left\| \frac{d\boldsymbol{\sigma}}{dt} \right\| dt$$

$$= \int_{a}^{b} \mathbf{F}(\boldsymbol{\sigma}(t)) \cdot \mathbf{T}(t) \frac{ds}{dt} dt$$

$$= \int_{C} \mathbf{F} \cdot \mathbf{T} ds.$$

Thus, we have the following fact:

Proposition 3
$$\int_{C} \mathbf{F} \cdot d\boldsymbol{\sigma} = \int_{C} \mathbf{F} \cdot \mathbf{T} ds.$$

Example 5 Let
$$\mathbf{F}(x, y) = -y\mathbf{i} + x\mathbf{j}$$

and let C be the circle of radius 2 that is centered at the origin and is traversed counterclockwise. Evaluate the line integral of $\mathbf{F}$ on C as an integral with respect to arc length.

Solution

We can parametrize C via the function $\boldsymbol{\sigma}(t) = (2\cos(t), 2\sin(t))$, where $0 \leq t \leq 2\pi$. We have

$$\boldsymbol{\sigma}'(t) = -2\sin(t)\mathbf{i} + 2\cos(t)\mathbf{j},$$

so that

$$\mathbf{F}(\boldsymbol{\sigma}(t)) = \mathbf{F}(2\cos(t), 2\sin(t)) = -2\sin(t)\mathbf{i} + 2\cos(t)\mathbf{j},$$

$$ds = \|\boldsymbol{\sigma}'(t)\| \, dt = \sqrt{4\sin^2(t) + 4\cos^2(t)} \, dt = 2dt,$$

and

$$\mathbf{T}(t) = \frac{\boldsymbol{\sigma}'(t)}{\|\boldsymbol{\sigma}'(t)\|} = -\sin(t)\mathbf{i} + \cos(t)\mathbf{j}.$$

Therefore,

$$\begin{aligned}
\int_C \mathbf{F} \cdot d\boldsymbol{\sigma} = \int_C \mathbf{F} \cdot \mathbf{T} ds &= \int_0^{2\pi} \mathbf{F}(\boldsymbol{\sigma}(t)) \cdot \mathbf{T}(t)\, ds \\
&= \int_0^{2\pi} (-2\sin(t)\mathbf{i} + 2\cos(t)\mathbf{j}) \cdot (-\sin(t)\mathbf{i} + \cos(t)\mathbf{j}) \, 2dt \\
&= \int_0^{2\pi} \left(4\sin^2(t) + 4\cos^2(t)\right) dt \\
&= \int_0^{2\pi} 4dt = 8\pi.
\end{aligned}$$

$\square$

The Differential Form Notation

There is still another useful way to express a line integral. If $\mathbf{F}(x,y) = M(x,y)\mathbf{i} + N(x,y)\mathbf{j}$ and the curve C is parametrized by $\boldsymbol{\sigma}(t) = (x(t), y(t))$, where $a \leq t \leq b$,

$$\begin{aligned}
\int_C \mathbf{F} \cdot d\boldsymbol{\sigma} &= \int_a^b (M(x(t), y(t))\mathbf{i} + N(x(t), y(t))\mathbf{j}) \cdot \left(\frac{dx}{dt}\mathbf{i} + \frac{dy}{dt}\mathbf{j}\right) dt \\
&= \int_a^b \left(M(x(t), y(t))\frac{dx}{dt} + N(x(t), y(t))\frac{dy}{dt}\right) dt \\
&= \int_a^b M(x(t), y(t))\frac{dx}{dt} dt + \int_a^b N(x(t), y(t))\frac{dy}{dt} dt.
\end{aligned}$$

Using the formalism,

$$dx = \frac{dx}{dt}dt \text{ and } dy = \frac{dy}{dt}dt,$$

we can express the integrals as

$$\int_a^b M(x(t), y(t))\, dx + \int_a^b N(x(t), y(t))\, dy.$$

This motivates the notation

$$\int_C Mdx + Ndy$$

for the line integral of $\boldsymbol{F} = M\mathbf{i} + N\mathbf{j}$ on the curve C.

Definition 3 If M and N are functions of x and y, and C is a smooth curve in the plane that is parametrized by $\boldsymbol{\sigma} : [a,b] \to \mathbb{R}^2$ we set

$$\int_C Mdx + Ndy = \int_a^b M(x(t), y(t))\frac{dx}{dt}dt + \int_a^b N(x(t), y(t))\frac{dy}{dt}dt.$$

Thus,

$$\int_C \mathbf{F} \cdot d\boldsymbol{\sigma} = \int_C Mdx + Ndy$$

if $\mathbf{F} = M\mathbf{i} + N\mathbf{j}$. The expression $Mdx + Ndy$ is referred to as **a differential form**, and we will refer to the above expression for a line integral as **the integral of a differential form on the curve** C.

Example 6 Evaluate the line integral of $\mathbf{F}(x, y) = y^3\mathbf{i} + x^2\mathbf{j}$ on the ellipse that is parametrized by

$$\boldsymbol{\sigma}(t) = (4\cos(t), \sin(t)), \ 0 \le t \le 2\pi,$$

by expressing the line integral as the integral of a differential form.

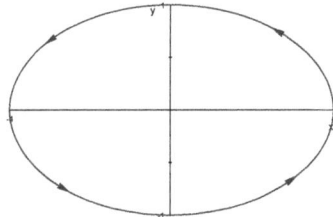

Figure 5

Solution

$$\int_C \mathbf{F} \cdot d\boldsymbol{\sigma} = \int_C y^3 dx + x^2 dy = \int_0^{2\pi} \left(\sin^3(t) \frac{dx}{dt} + 16\cos^2(t) \frac{dy}{dt} \right) dt$$

$$= \int_0^{2\pi} \left(\sin^3(t)(-4\sin(x)) + 16\cos^2(t)(\cos(t)) \right) dt$$

$$= \int_0^{2\pi} \left(-4\sin^4(t) + 16\cos^3(t) \right) dt = -3\pi$$

(Check: You will have to refresh your memory with respect to the integrals of powers of sines and cosines as we discussed in Section 7.3). $\square$

Example 7 Let C be the boundary of the square $[-1, 1] \times [-1, 1]$ that is traversed in the counterclockwise direction. Calculate

$$\int_C -ydx + xdy.$$

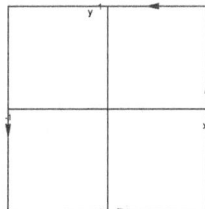

Figure 6

Solution

We can express C as $C_1 + C_2 + C_3 + C_4$, as indicated in Figure 4.
We can parametrize C_1 by setting $x(t) = t$ and $y(t) = -1$, where $-1 \le t \le 1$. Then,

$$\int_{C_1} -ydx + xdy = \int_{-1}^{1} \left(\frac{dx}{dt} + t\frac{dy}{dt} \right) dt = \int_{-1}^{1} dt = 2$$

We can express C_3 as $-C_3$, where $-C_3$ is parametrized by setting $x(t) = t$ and $y(t) = 1$, where $-1 \le t \le 1$. Then,

$$\int_{C_3} -ydx + xdy = -\int_{-C_3} -ydx + xdy = -\int_{-1}^{1} \left(-\frac{dx}{dt} + t\frac{dy}{dt} \right) dt = \int_{-1}^{1} dt = 2$$

We can parametrize C_2 by setting $x(t) = 1$ and $y(t) = t$, where $-1 \le t \le 1$. Then,

$$\int_{C_2} -ydx + xdy = \int_{-1}^{1} \left(-t\frac{dx}{dt} + \frac{dy}{dt} \right) dt = \int_{-1}^{1} dt = 2$$

Similarly, $C_4 = -C_2$, where $-C_2$ is parametrized by $x(t) = -1$ and $y(t) = t$, $-1 \le t \le 1$. Then,

$$\int_{C_4} -ydx + xdy = -\int_{-C_2} -ydx + xdy = -\int_{-1}^{1} \left(-\frac{dy}{dt} \right) dt = \int_{-1}^{1} dt = 2$$

Therefore,

$$\int_{C} -ydx + xdy. = \sum_{k=1}^{4} \int_{C_k} -ydx + xdy = 2 + 2 + 2 + 2 = 8.$$

$\square$

Curves in $\mathbb{R}^3$

Our discussion of integrals with respect to arc length and line integrals extends to curves in three dimensions. Thus, assume that $\boldsymbol{\sigma} : [a, b] \to \mathbb{R}^3$ is a smooth function and parametrizes the curve C. If f is a continuous scalar function of three variables, we define the integral of f with respect to arc length on C as

$$\int_{C} fds = \int_{a}^{b} f(\boldsymbol{\sigma}(t)) \|\boldsymbol{\sigma}'(t)\| dt.$$

If $\boldsymbol{\sigma}(t) = (x(t), y(t), z(t))$,

$$\int_{C} fds = \int_{a}^{b} f(x(t), y(t), z(t)) \sqrt{\left(\frac{dx}{dt}\right)^2 + \left(\frac{dy}{dt}\right)^2 + \left(\frac{dz}{dt}\right)^2} dt.$$

As in the case of curves in the plane, the above integral is the same for orientation preserving and orientation reversing parametrizations of C.
If F is a vector field in $\mathbb{R}^3$ so that

$$\mathbf{F}(x, y, z) = M(x, y, z)\mathbf{i} + N(x, y, z)\mathbf{j} + P(x, y, z)\mathbf{k},$$

the line integral of $\mathbf{F}$ on the curve C is

$$\int_{C} \mathbf{F} \cdot d\boldsymbol{\sigma} = \int_{a}^{b} \mathbf{F}(\boldsymbol{\sigma}(t)) \cdot \frac{d\boldsymbol{\sigma}}{dt} dt.$$

In the differential form notation,

$$\int_C \mathbf{F} \cdot d\boldsymbol{\sigma} = \int_C M\,dx + N\,dy + P\,dz$$
$$= \int_a^b \left(M\left(x\left(t\right)\right), y\left(t\right), z\left(t\right) \frac{dx}{dt} + N\left(x\left(t\right)\right), y\left(t\right), z\left(t\right) \frac{dy}{dt} + P\left(x\left(t\right)\right), y\left(t\right), z\left(t\right) \frac{dz}{dt} \right) dt.$$

The line integral can be expressed as an integral with respect to arc length:

$$\int_C \mathbf{F} \cdot d\boldsymbol{\sigma} = \int_C \mathbf{F} \cdot \mathbf{T}\,ds,$$

where $\mathbf{T}$ denotes the unit tangent to the curve C. As in the case of two-dimensional vector fields,

$$\int_C \mathbf{F} \cdot d\boldsymbol{\sigma}$$

is unchanged by an orientation preserving parametrization of C and multiplied by -1 if the parametrization reverses the orientation.

Example 8 Let C be the **helix** that is parametrized by

$$\sigma\left(t\right) = \left(\cos\left(t\right), \sin\left(t\right), t\right), \ 0 \le t \le 4\pi,$$

and let $\mathbf{F}\left(x, y, z\right) = -y\mathbf{i} + x\mathbf{j} + z\mathbf{k}$. Evaluate

a)
$$\int_C z\,ds,$$

b)
$$\int_C \mathbf{F} \cdot d\boldsymbol{\sigma}$$

Solution

The picture shows the helix.

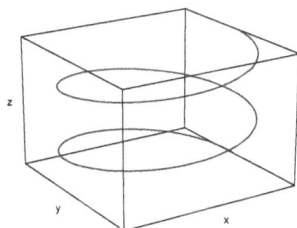

Figure 7

a) We have
$$\sigma'\left(t\right) = -\sin\left(t\right)\mathbf{i} + \cos\left(t\right)\mathbf{j} + \mathbf{k}$$

Therefore,

$$ds = ||\sigma'\left(t\right)|| = \sqrt{\sin^2\left(t\right) + \cos^2\left(t\right) + 1} = \sqrt{1+1} = \sqrt{2}.$$

Thus,

$$\int_C z\,ds = \int_0^{4\pi} t\,\|\boldsymbol{\sigma}'(t)\|\,dt = \int_0^{4\pi} t\sqrt{2}\,dt = \sqrt{2}\left(\frac{t^2}{2}\bigg|_0^{4\pi}\right) = \sqrt{2}\left(\frac{16\pi^2}{2}\right) = 8\sqrt{2}\pi^2$$

b)

$$\int_C \mathbf{F}\cdot d\boldsymbol{\sigma} = \int_0^{4\pi} \left(-\sin(t)\,\mathbf{i} + \cos(t)\,\mathbf{j} + t\mathbf{k}\right)\cdot\left(-\sin(t)\,\mathbf{i} + \cos(t)\,\mathbf{j} + \mathbf{k}\right)dt$$

$$= \int_0^{4\pi} \left(\sin^2(t) + \cos^2(t) + t\right)dt$$

$$= \int_0^{4\pi} (1+t)\,dt = t + \frac{t^2}{2}\bigg|_0^{4\pi} = 4\pi + \frac{16\pi^2}{2} = 4\pi + 8\pi^2.$$

$\square$

Precise Definitions and Proofs

Let's begin by stating the precise definitions of orientation preserving and orientation reversing parametrizations of a curve:

Definition 4 Assume that C is a curve in the plane that is parametrized by $\boldsymbol{\sigma} : [a, b] \to \mathbb{R}^2$. We say that $\tilde{\boldsymbol{\sigma}} : [\alpha, \beta] \to \mathbb{R}^2$ is an **orientation preserving parametrization of** C if there exists a differentiable increasing function $h : [\alpha, \beta] \to [a, b]$ such that $\tilde{\boldsymbol{\sigma}}(\tau) = \boldsymbol{\sigma}(h(\tau))$ for each $\tau \in [\alpha, \beta]$. The function $\tilde{\boldsymbol{\sigma}}$ is an **orientation reversing parametrization of** C if there exists a differentiable decreasing function $h : [\alpha, \beta] \to [a, b]$ such that $\tilde{\boldsymbol{\sigma}}(\tau) = \boldsymbol{\sigma}(h(\tau))$ for each $\tau \in [\alpha, \beta]$.

Now let's prove Proposition 1:

Assume that C is a curve in the plane that is parametrized by $\boldsymbol{\sigma} : [a, b] \to \mathbb{R}^2$ and that $\tilde{\boldsymbol{\sigma}} : [\alpha, \beta] \to \mathbb{R}^2$ is an orientation preserving or reversing parametrization of C. Then

$$\int_a^b f(\boldsymbol{\sigma}(t))\left\|\frac{d\boldsymbol{\sigma}}{dt}\right\|dt = \int_\alpha^\beta f(\tilde{\boldsymbol{\sigma}}(\tau))\left\|\frac{d\tilde{\boldsymbol{\sigma}}}{dt}\right\|d\tau.$$

Proof

Assume that $\tilde{\boldsymbol{\sigma}}(\tau) = \boldsymbol{\sigma}(h(\tau))$, where h is an increasing or decreasing and differentiable function from $[\alpha; \beta]$ onto $[a, b]$. By the chain rule,

$$\frac{d\boldsymbol{\sigma}}{d\tau} = \frac{dh}{d\tau}\frac{d\boldsymbol{\sigma}}{dt}(h(\tau)).$$

Therefore,

$$\left\|\frac{d\boldsymbol{\sigma}}{d\tau}\right\| = \left|\frac{dh}{d\tau}\right|\left\|\frac{d\boldsymbol{\sigma}}{dt}(h(\tau))\right\|$$

Thus,

$$\int_\alpha^\beta f(\sigma(h(\tau)))\left\|\frac{d\boldsymbol{\sigma}}{d\tau}\right\|d\tau = \int_\alpha^\beta f(\sigma(h(\tau)))\left|\frac{dh}{d\tau}\right|\left\|\frac{d\boldsymbol{\sigma}}{dt}(h(t))\right\|d\tau$$

If $dh/d\tau > 0$,

$$\int_\alpha^\beta f(\sigma(h(\tau)))\left|\frac{dh}{d\tau}\right|\left\|\frac{d\boldsymbol{\sigma}}{dt}\right\|d\tau = \int_\alpha^\beta f(\sigma(h(\tau)))\frac{dh}{d\tau}\left\|\frac{d\boldsymbol{\sigma}}{dt}\right\|d\tau.$$

By the substitution rule,

$$\int_{\alpha}^{\beta} f\left(\sigma\left(h\left(\tau\right)\right)\right) \frac{dh}{d\tau} \left\|\frac{d\sigma}{dt}\right\| d\tau = \int_{\alpha}^{\beta} f\left(\sigma\left(h\left(\tau\right)\right)\right) \left\|\frac{d\sigma}{dt}\right\| \frac{dh}{d\tau} d\tau = \int_{h(\alpha)}^{h(\beta)} f\left(t\right) \left\|\frac{d\sigma}{dt}\right\| dt$$

$$= \int_{a}^{b} f\left(t\right) \left\|\frac{d\sigma}{dt}\right\| dt.$$

Thus,

$$\int_{\alpha}^{\beta} f\left(\sigma\left(h\left(\tau\right)\right)\right) \left\|\frac{d\sigma}{d\tau}\right\| d\tau = \int_{a}^{b} f\left(t\right) \left\|\frac{d\sigma}{dt}\right\| dt,$$

as claimed.

Similarly, if $dh/d\tau < 0$,

$$\int_{\alpha}^{\beta} f\left(\sigma\left(h\left(\tau\right)\right)\right) \left\|\frac{d\sigma}{d\tau}\right\| d\tau = \int_{\alpha}^{\beta} f\left(\sigma\left(h\left(\tau\right)\right)\right) \left|\frac{dh}{d\tau}\right| \left\|\frac{d\sigma}{dt}\right\| d\tau$$

$$= -\int_{\alpha}^{\beta} f\left(\sigma\left(h\left(\tau\right)\right)\right) \frac{dh}{d\tau} \left\|\frac{d\sigma}{dt}\right\| d\tau$$

$$= \int_{\beta}^{\alpha} f\left(\sigma\left(h\left(\tau\right)\right)\right) \frac{dh}{d\tau} \left\|\frac{d\sigma}{dt}\right\| d\tau$$

$$= \int_{h(\beta)}^{h(\alpha)} f\left(\sigma\left(t\right)\right) \left\|\frac{d\sigma}{dt}\right\| dt = \int_{a}^{b} f\left(\sigma\left(t\right)\right) \left\|\frac{d\sigma}{dt}\right\| dt.$$

■

Now we will prove Proposition 2 with regard to the effect of reparametrization on the line integral of a vector field. Recall the statement:

Assume that the curve C is parametrized by $\boldsymbol{\sigma} : [a, b] \to \mathbb{R}^2$. If $\tilde{\boldsymbol{\sigma}} : [\alpha, \beta] \to \mathbb{R}^2$ is an orientation preserving parametrization of C, then

$$\int_{\alpha}^{\beta} \mathbf{F}\left(\tilde{\boldsymbol{\sigma}}\left(\tau\right)\right) \cdot \frac{d\tilde{\boldsymbol{\sigma}}\left(\tau\right)}{d\tau} d\tau = \int_{a}^{b} \mathbf{F}\left(\boldsymbol{\sigma}\left(t\right)\right) \cdot \frac{d\boldsymbol{\sigma}}{dt} dt.$$

If $\tilde{\boldsymbol{\sigma}} : [\alpha, \beta] \to \mathbb{R}^2$ is an orientation reversing parametrization of C, then

$$\int_{\alpha}^{\beta} \mathbf{F}\left(\tilde{\boldsymbol{\sigma}}\left(\tau\right)\right) \cdot \frac{d\tilde{\boldsymbol{\sigma}}\left(\tau\right)}{d\tau} d\tau = -\int_{a}^{b} \mathbf{F}\left(\boldsymbol{\sigma}\left(t\right)\right) \cdot \frac{d\boldsymbol{\sigma}}{dt} dt$$

Proof

a) By the chain rule,

$$\frac{d\boldsymbol{\sigma}\left(h\left(\tau\right)\right)}{d\tau} = \frac{dh}{d\tau} \frac{d\boldsymbol{\sigma}}{dt}\left(h\left(\tau\right)\right).$$

Therefore,

$$\int_\alpha^\beta \mathbf{F}\left(\tilde{\boldsymbol{\sigma}}\left(\tau\right)\right) \cdot \frac{d\tilde{\boldsymbol{\sigma}}\left(\tau\right)}{d\tau} d\tau = \int_\alpha^\beta \mathbf{F}\left(\boldsymbol{\sigma}\left(h\left(\tau\right)\right)\right) \cdot \frac{d\boldsymbol{\sigma}\left(h\left(\tau\right)\right)}{d\tau} d\tau$$

$$= \int_\alpha^\beta \mathbf{F}\left(\boldsymbol{\sigma}\left(h\left(\tau\right)\right)\right) \cdot \left(\frac{dh}{d\tau}\frac{d\boldsymbol{\sigma}}{dt}\left(h\left(\tau\right)\right)\right) d\tau$$

$$= \int_\alpha^\beta \mathbf{F}\left(\boldsymbol{\sigma}\left(h\left(\tau\right)\right)\right) \cdot \frac{d\boldsymbol{\sigma}}{dt}\left(h\left(\tau\right)\right) \frac{dh}{d\tau} d\tau$$

$$= \int_{h(\alpha)}^{h(\beta)} \mathbf{F}\left(\boldsymbol{\sigma}\left(t\right)\right) \cdot \frac{d\boldsymbol{\sigma}}{dt} dt$$

$$= \int_a^b \mathbf{F}\left(\boldsymbol{\sigma}\left(t\right)\right) \cdot \frac{d\boldsymbol{\sigma}}{dt} dt,$$

by the substitution rule.

b) As in part a)

$$\int_\alpha^\beta \mathbf{F}\left(\tilde{\boldsymbol{\sigma}}\left(\tau\right)\right) \cdot \frac{d\tilde{\boldsymbol{\sigma}}\left(\tau\right)}{d\tau} d\tau = \int_\alpha^\beta \mathbf{F}\left(\boldsymbol{\sigma}\left(h\left(\tau\right)\right)\right) \cdot \frac{d\boldsymbol{\sigma}}{dt}\left(h\left(\tau\right)\right) \frac{dh}{d\tau} d\tau$$

$$= \int_{h(\alpha)}^{h(\beta)} \mathbf{F}\left(\boldsymbol{\sigma}\left(t\right)\right) \cdot \frac{d\boldsymbol{\sigma}}{dt} dt.$$

Since $h\left(\tau\right)$ is a decreasing function, we have $h\left(\alpha\right) = b$ and $h\left(\beta\right) = a$. Therefore,

$$\int_{h(\alpha)}^{h(\beta)} \mathbf{F}\left(\boldsymbol{\sigma}\left(t\right)\right) \cdot \frac{d\boldsymbol{\sigma}}{dt} dt = \int_b^a \mathbf{F}\left(\boldsymbol{\sigma}\left(t\right)\right) \cdot \frac{d\boldsymbol{\sigma}}{dt} dt = -\int_a^b \mathbf{F}\left(\boldsymbol{\sigma}\left(t\right)\right) \cdot \frac{d\boldsymbol{\sigma}}{dt} dt,$$

as claimed. ∎

Problems

In problems 1-5 evaluate the given integral of a scalar function with respect to arc length:

1.

$$\int_C y^3 ds,$$

where C is parametrized by

$$\sigma\left(t\right) = \left(t^3, t\right), \ 0 \leq t \leq 2.$$

2.

$$\int_C xy^2 ds,$$

where C is parametrized by

$$\sigma\left(t\right) = \left(2\cos\left(t\right), 2\sin\left(t\right)\right), \ -\frac{\pi}{2} \leq t \leq \frac{\pi}{2}.$$

3.

$$\int_C xe^y ds,$$

where C is the line segment traversed from $\left(2, 1\right)$ to $\left(4, 5\right).$

4.

$$\int_C \frac{y}{x^2 + y^2} ds,$$

where C is the semicircle of radius 2 that is centered at $(4, 3)$ and traversed from $(6, 3)$ to $(2, 3)$ in the counterclockwise direction.

5.

$$\int_C (x - 2) \, e^{(y-3)(z-4)} ds,$$

where C is the line segment from $(2, 3, 4)$ to $(3, 5, 7)$.

In problems 6-10 evaluate the line integral

$$\int_C \mathbf{F} \cdot d\boldsymbol{\sigma}$$

6.

$$\mathbf{F}(x, y) = x^2 \mathbf{i} + y^2 \mathbf{j}$$

and C is the line segment from $(1, 2)$ to $(3, 4)$.

7.

$$\mathbf{F}(x, y) = y\mathbf{i} - x\mathbf{j}$$

and C is the part of the unit circle traversed from $(0, 1)$ to $(-1, 0)$ in the counterclockwise direction.

8.

$$\mathbf{F}(x, y) = \ln(y)\, \mathbf{i} - e^x \mathbf{j}$$

and C is parametrized by

$$\sigma(t) = \left(\ln(t), t^3 \right), \ 1 \le t \le e.$$

9.

$$\mathbf{F}(x, y, z) = \cos(y)\, \mathbf{i} + \sin(z)\, \mathbf{j} + x\mathbf{k}$$

and C is parametrized by

$$\sigma(t) = (\cos(t), t, t), \ \frac{\pi}{4} \le t \le \frac{\pi}{2}$$

10.

$$F(x, y) = xy\mathbf{i} + (x - y)\,\mathbf{j}$$

and $C = C_1 + C_2$, where C_1 is the line segment from $(0, 0)$ to $(2, 0)$ and C_2 is the line segment from $(2, 0)$ to $(3, 2)$.

In problems 11 and 12 evaluate

$$\int_C \mathbf{F} \cdot \mathbf{T} ds$$

11.

$$\mathbf{F}(x, y) = -2xy\mathbf{i} + (y + 1)\,\mathbf{j}$$

and C is the part of the circle of radius 2 centered at the origin traversed from $(-2, 0)$ to $(0, 2)$ in the clockwise direction.

12.

$$\mathbf{F}(x, y, z) = e^{x+y}\mathbf{i} + xz\mathbf{j} + y\mathbf{k}$$

and C is the line segment from $(1, 2, 3)$ to $(-1, -2, -3)$.

In problems 13 and 14

a) Express the line integral

$$\int_C \mathbf{F} \cdot d\boldsymbol{\sigma}$$

in the differential form notation,

b) Evaluate the expression that you obtained in part a).

13.

$$\mathbf{F}(x,y) = -xy\mathbf{i} + \frac{1}{x^2+1}\mathbf{j}$$

and C is parametrized by

$$\sigma(t) = \left(t, t^2\right), \quad -4 \leq t \leq -1$$

14.

$$F(x,y) = y\mathbf{i} - x\mathbf{j}$$

and C is the part of the unit circle that is traversed from $(0,-1)$ to $(0,1)$ in the counterclockwise direction.

In problems 15-18 evaluate the given line integral:

15.

$$\int_C 3x^2 dx - 2y^3 dy$$

where C is the part of the unit circle traversed from $(1,0)$ to $(0,1)$.

16.

$$\int_C e^x dx + e^y dy,$$

where C is the part of the ellipse $x^2 + 4y^2 = 4$ traversed from $(0,1)$ to $(2,0)$ in the clockwise direction.

Hint: Parametrize C by a function of the form $(a\cos(\theta), b\sin(\theta))$.

17.

$$\int_C -\sin(x)\,dx + \cos(x)\,dy,$$

where C is the part of the parabola $y = x^2$ traversed from $(0,0)$ to $\left(\pi, \pi^2\right)$.

18.

$$\int_C x^3 dx + y^2 dy + z\,dz,$$

where C is the line segment from the origin to $(2,3,4)$.

14.3   Line Integrals of Conservative Vector Fields

In this section we will see that the line integral of the gradient of a scalar function on a curve is simply the difference between the values of the function at the endpoints of the curve. Thus, such a line integral does not depend on the particular path that connects two given points. We will discuss conditions under which a vector field is the gradient of a scalar function.

The Fundamental Theorem for Line Integrals

Theorem 1 Assume that the initial point of a curve C in the plane is (x_1, y_1) and its terminal point is (x_2, y_2). Then

$$\int_C \nabla f(x, y) \cdot d\sigma = f(x_2, y_2) - f(x_1, y_1).$$

Proof
Assume that $\sigma(t) = (x(t), y(t))$, where $a \leq t \leq b$, parametrizes C. Then

$$\int_C \nabla f \cdot d\sigma = \int_a^b \nabla f(\sigma(t)) \cdot \frac{d\sigma}{dt} dt$$

$$= \int_a^b \left(\frac{\partial f}{\partial x}(x(t), y(t)) \mathbf{i} + \frac{\partial f}{\partial y}(x(t), y(t)) \mathbf{j} \right) \cdot \left(\frac{dx}{dt} \mathbf{i} + \frac{dy}{dt} \mathbf{j} \right) dt$$

$$= \int_a^b \left(\frac{\partial f}{\partial x}(x(t), y(t)) \frac{dx}{dt} + \frac{\partial f}{\partial y}(x(t), y(t)) \frac{dy}{dt} \right) dt.$$

By the chain rule,

$$\frac{\partial f}{\partial x}(x(t), y(t)) \frac{dx}{dt} + \frac{\partial f}{\partial y}(x(t), y(t)) \frac{dy}{dt} = \frac{d}{dt} f(x(t), y(t)).$$

By the Fundamental Theorem of Calculus,

$$\int_a^b \frac{d}{dt} f(x(t), y(t)) \, dt = f(x(b), y(b)) - f(x(a), y(a)) = f(x_2, y_2) - f(x_1, y_1).$$

Therefore,

$$\int_C \nabla f \cdot d\sigma = f(x_2, y_2) - f(x_1, y_1),$$

as claimed. ∎

Remark 1 Theorem 1can be expressed in the differential form notation: We have

$$\int_C \nabla f \cdot d\sigma = \int_a^b \left(\frac{\partial f}{\partial x}(x(t), y(t)) \frac{dx}{dt} + \frac{\partial f}{\partial y}(x(t), y(t)) \frac{dy}{dt} \right) dt$$

$$= \int_C \frac{\partial f}{\partial x} dx + \frac{\partial f}{\partial y} dy,$$

so that

$$\int_C \frac{\partial f}{\partial x} dx + \frac{\partial f}{\partial y} dy = f(x_2, y_2) - f(x_1, y_1).$$

Since

$$df = \frac{\partial f}{\partial x} dx + \frac{\partial f}{\partial y} dy,$$

we can also write

$$\int_C df = f(x_2, y_2) - f(x_1, y_1).$$

Thus the line integral of the differential of a scalar function f on C is the difference between the values of f at the endpoints of C. A similar statement is valid for gradient fields in $\mathbb{R}^3$: Assume that the initial point of a curve C in R^3 is (x_1, y_1, z_1) and its terminal point is (x_2, y_2, z_2). Then

$$\int_C \nabla f \cdot d\sigma = \int_C df = f(x_2, y_2, z_2) - f(x_1, y_1, z_1).$$

◊

Definition 1 We will say that the line integral

$$\int_C F \cdot d\sigma$$

is **independent of path** in the region D if its value is the same for all curves in D that have the same initial point and terminal point.

Thus, **the line integral of a gradient field is independent of path.**

Example 1 Let

$$r(x,y) = x\mathbf{i} + y\mathbf{j}$$

be the position vector of the point (x,y), $r(x,y) = \sqrt{x^2 + y^2} = ||r(x,y)||$ and

$$u_r(x,y) = \frac{r}{r}(x,y) = \frac{x}{\sqrt{x^2+y^2}}\mathbf{i} + \frac{y}{\sqrt{x^2+y^2}}\mathbf{j}.$$

Thus, u_r is the unit vector in the radial direction. Show that

$$\int_C u_r(x,y) \cdot d\sigma$$

is independent of path in any region that does not contain the origin. Determine such a line integral if C joins the point (x_1, y_1) to (x_2, y_2) and does not pass through the origin.

Solution

In Example 6 of Section 14.1 we showed that

$$\nabla r(x,y) = u_r(x,y).$$

By Theorem 1,

$$\int_C u_r(x,y) \cdot d\sigma = \int_C \nabla r(x,y) \cdot d\sigma \quad = \quad r(x_2, y_2) - r(x_1, y_1)$$
$$= \quad \sqrt{x_2^2 + y_2^2} - \sqrt{x_1^2 + y_1^2}.$$

□

Example 2 Let

$$r = r(x,y,z) = x\mathbf{i} + y\mathbf{j} + z\mathbf{k} \text{ and } r = r(x,y,z) = ||r|| = \sqrt{x^2 + y^2 + z^2},$$

and

$$\mathbf{F}(x,y,z) = -\frac{GM}{r^3}r$$

be the gravitational field due to the mass M at the origin, as we discussed in Section 15.1. Show that

$$\int_{Cr} \mathbf{F}(x,y,z) \cdot d\sigma$$

is independent of path in any region that does not contain the origin. Determine such a line integral if C joins the point (x_1, y_1, z_1) to (x_2, y_2, z_1) and does not pass through the origin.

Solution

As we showed in Example 6 of Section 14.1,

$$\nabla\left(\frac{1}{r}\right) = -\frac{\boldsymbol{r}}{r^3}.$$

Thus,

$$\boldsymbol{F}(x,y,z) = -\frac{GM}{r^3}\boldsymbol{r} = \nabla\left(\frac{GM}{r}\right).$$

Therefore,

$$\int_C -\frac{GM}{r^3}\boldsymbol{r}\cdot d\boldsymbol{\sigma} = \int_C \nabla\left(\frac{GM}{r}\right)\cdot d\boldsymbol{\sigma} = \frac{GM}{\sqrt{x_2^2+y_2^2+z_2^2}} - \frac{GM}{\sqrt{x_1^2+y_1^2+z_1^2}}.$$

Thus, the work done by the gravitational field in moving a unit mass from one point to another depends only on the distances of the points from the mass M, irrespective of the path that is followed. □

Definition 2 A vector field $\boldsymbol{F}$ is said to be **conservative** in a region D if $\boldsymbol{F}$ is the gradient of a scalar function f in D. Such a scalar function is referred to as a **potential** for **F**.

Thus, the field of Example 1 and the gravitational field that was discussed in Example 2 are conservative fields in a region that does not contain the origin.
By Theorem 1, the line integral of a field **F** on a curve C depends only on the endpoints of C if **F** is conservative in a region that contains C. In particular, if C is closed,

$$\int_C \mathbf{F}\cdot d\boldsymbol{\sigma} = \int_C \nabla f\cdot d\boldsymbol{\sigma} = 0.$$

Remark 2 (Conservation of Energy) In Physics and some fields of engineering, a function ϕ is referred to as a potential for **F** if $\mathbf{F} = -\nabla\phi$. The adjective "conservative" has its origin in mechanics. Assume that $\mathbf{F} = -\nabla\phi$ is a force field in $\mathbb{R}^3$ and $\boldsymbol{\sigma}(t)$ is the position of an object at time t. The work done by **F** on the object as it moves from the point P_1 to P_2 along the curve C that is parametrized by $\boldsymbol{\sigma}(t)$ is

$$\int_C \mathbf{F}\cdot d\boldsymbol{\sigma} = \int_C -\nabla\phi\cdot d\boldsymbol{\sigma} = -(\phi(P_2)-\phi(P_1)) = \phi(P_1)-\phi(P_2).$$

By **Newton's second law of motion,** if the mass of the object is m, $\boldsymbol{a}(t)$ is its acceleration and $\boldsymbol{v}(t)$ is its velocity at time t, then

$$\mathbf{F} = m\boldsymbol{a}(t) = m\frac{d\boldsymbol{v}}{dt}.$$

If $P_1 = \boldsymbol{\sigma}(t_1)$ and $P_2 = \boldsymbol{\sigma}(t_1)$,

$$\int_C \mathbf{F}\cdot d\boldsymbol{\sigma} = \int_{t_1}^{t_2} m\frac{d\boldsymbol{v}}{dt}\cdot\frac{d\boldsymbol{\sigma}}{dt}dt = m\int_{t_1}^{t_2}\frac{d\boldsymbol{v}}{dt}\cdot\boldsymbol{v}dt$$

$$= m\int_{t_1}^{t_2}\frac{1}{2}\frac{d}{dt}(\boldsymbol{v}\cdot\boldsymbol{v})\,dt$$

$$= m\int_{t_1}^{t_2}\frac{1}{2}\frac{d}{dt}\|\boldsymbol{v}\|^2\,dt$$

$$= \frac{1}{2}m\|\boldsymbol{v}(t_2)\|^2 - \frac{1}{2}m\|\boldsymbol{v}(t_1)\|^2.$$

Thus,

$$\phi\left(P_1\right) - \phi\left(P_2\right) = \frac{1}{2}m\left\|\boldsymbol{v}\left(t_2\right)\right\|^2 - \frac{1}{2}m\left\|\boldsymbol{v}\left(t_1\right)\right\|^2,$$

so that

$$\phi\left(P_1\right) + \frac{1}{2}m\left\|\boldsymbol{v}\left(t_1\right)\right\|^2 = \phi\left(P_2\right) + \frac{1}{2}m\left\|\boldsymbol{v}\left(t_2\right)\right\|^2.$$

The quantity $\phi\left(P\right)$ is called **the potential energy at** P and the quantity

$$\frac{1}{2}m\left\|\mathbf{v}\left(t\right)\right\|^2$$

is **the kinetic energy at time** t. Their sum is **the total energy.** Therefore, the above equality says that the total energy is conserved as the object moves from a point P_1 to a point P_2 in a force field that has a potential. $\Diamond$

Conditions for a Field to be Conservative

Given a vector field $\mathbf{F}$, we would like to be able to determine whether $\mathbf{F}$ is conservative by examining the components of $\mathbf{F}$. Let's begin with a vector field

$$\mathbf{F}\left(x, y\right) = M\left(x, y\right)\mathbf{i} + N\left(x, y\right)\mathbf{j}$$

in the plane. Assume that $\mathbf{F}\left(x, y\right) = \boldsymbol{\nabla} f\left(x, y\right)$ for each $\left(x, y\right)$ in the region D. Thus,

$$M\left(x, y\right) = \frac{\partial f}{\partial x}\left(x, y\right) \text{ and } N\left(x, y\right) = \frac{\partial f}{\partial y}\left(x, y\right)$$

for each $\left(x, y\right) \in D$. Therefore,

$$\frac{\partial M}{\partial y} = \frac{\partial^2 f}{\partial y \partial x} \text{ and } \frac{\partial N}{\partial x} = \frac{\partial^2 f}{\partial x \partial y}.$$

If we assume that the second partial derivatives of f are continuous in D, we have

$$\frac{\partial^2 f}{\partial y \partial x}\left(x, y\right) = \frac{\partial^2 f}{\partial x \partial y}\left(x, y\right),$$

so that

$$\frac{\partial M}{\partial y}\left(x, y\right) = \frac{\partial N}{\partial x}\left(x, y\right)$$

for each $\left(x, y\right) \in D$.
Let's record this important fact:

Proposition 1 A necessary condition for $\mathbf{F}\left(x, y\right) = M\left(x, y\right)\mathbf{i} + N\left(x, y\right)\mathbf{j}$ to be conservative and to have a potential function with continuous second-order partial derivatives in the region D is that

$$\frac{\partial M}{\partial y}\left(x, y\right) = \frac{\partial N}{\partial x}\left(x, y\right)$$

for each $\left(x, y\right) \in D$.

Remark 3 (Caution) The above condition is necessary but not sufficient for a vector field to be conservative. Indeed, let

$$\mathbf{F}(x, y) = -\frac{y}{x^2 + y^2}\mathbf{i} + \frac{x}{x^2 + y^2}\mathbf{j}.$$

We have

$$\frac{\partial}{\partial y}\left(-\frac{y}{x^2 + y^2}\right) = -\frac{(x^2 + y^2) - y(2y)}{(x^2 + y^2)^2} = \frac{-x^2 + y^2}{(x^2 + y^2)^2}$$

and

$$\frac{\partial}{\partial x}\left(\frac{x}{x^2 + y^2}\right) = \frac{(x^2 + y^2) - x(2x)}{(x^2 + y^2)^2} = \frac{-x^2 + y^2}{(x^2 + y^2)^2}.$$

Thus,

$$\frac{\partial}{\partial y}\left(-\frac{y}{x^2 + y^2}\right) = \frac{\partial}{\partial x}\left(\frac{x}{x^2 + y^2}\right)$$

for each $(x, y) \neq (0, 0)$. If $\mathbf{F}$ were conservative in $D = \{(x, y) \in \mathbb{R}^2 : (x, y) \neq 0\}$, the line integral of $\mathbf{F}$ would have been 0 around any closed curve in D. Let C be the unit circle so that C can be parametrized by $\boldsymbol{\sigma}(t) = (\cos(t), \sin(t))$, where $0 \leq t \leq 2\pi$. We have

$$\begin{aligned}
\mathbf{F}(\cos(t), \sin(t)) &= -\frac{\sin(t)}{\cos^2(t) + \sin^2(t)}\mathbf{i} + \frac{\cos(t)}{\cos^2(t) + \sin^2(t)}\mathbf{j} \\
&= -\sin(t)\mathbf{i} + \cos(t)\mathbf{j}
\end{aligned}$$

and

$$\frac{d\boldsymbol{\sigma}}{dt} = -\sin(t)\mathbf{i} + \cos(t)\mathbf{j}.$$

Therefore,

$$\begin{aligned}
\int_C \mathbf{F} \cdot d\boldsymbol{\sigma} &= \int_0^{2\pi}(-\sin(t)\mathbf{i} + \cos(t)\mathbf{j}) \cdot (-\sin(t)\mathbf{i} + \cos(t)\mathbf{j})\, dt \\
&= \int_0^{2\pi}(\sin^2(t) + \cos^2(t))\, dt = \int_0^{2\pi} dt = 2\pi \neq 0.
\end{aligned}$$

Therefore $\mathbf{F}$ is not conservative. As we will see in the following sections, the problem is that the vector field is not defined at $(0, 0)$ so that the necessary conditions are not satisfied at $(0, 0)$: The origin forms a "hole" in the region D. $\Diamond$

Even though the condition that is stated in Proposition 1is not sufficient for the existence of a potential function, at least we can rule out the existence of such a function if the condition is not satisfied. We can go ahead and try to find a potential function if the condition is satisfied.

Example 3 Let

$$\mathbf{F}(x, y) = x^2 y^2 \mathbf{i} + 2xy\mathbf{j}$$

Show that $\mathbf{F}$ is not conservative.

Solution

We have

$$\frac{\partial}{\partial y}(x^2 y^2) = 2x^2 y \text{ and } \frac{\partial}{\partial x}(2xy) = 2y.$$

Now,

$$2x^2 y = 2y \Leftrightarrow y(x^2 - 1) = 0 \Leftrightarrow y = 0 \text{ or } x = \pm 1.$$

Thus, the necessary condition for the existence of a potential is satisfied only on the x-axis and on the vertical lines $x = \pm 1$. This implies that the necessary condition is not satisfied in any open region D. Therefore $\mathbf{F}$ does not have a potential function in any open region. $\square$

Example 4 Let

$$\mathbf{F}(x, y) = \frac{x}{x^2 + y^2}\mathbf{i} + \frac{y}{x^2 + y^2}\mathbf{j}$$

a) Determine a potential functions for F.

b) Evaluate the line integral of F on any curve that joins $(1, 0)$ to $(0, \sqrt{e})$ and does not pass through the origin.

Solution

a) We have

$$\frac{\partial}{\partial y}\left(\frac{x}{x^2 + y^2}\right) = \frac{-2xy}{(x^2 + y^2)^2},$$

and

$$\frac{\partial}{\partial x}\left(\frac{y}{x^2 + y^2}\right) = \frac{-2xy}{(x^2 + y^2)^2}.$$

Therefore,

$$\frac{\partial}{\partial y}\left(\frac{x}{x^2 + y^2}\right) = \frac{\partial}{\partial x}\left(\frac{y}{x^2 + y^2}\right)$$

if $(x, y) \neq (0, 0)$. Thus the necessary condition for the existence of a potential function is satisfied everywhere except at the origin. We saw that in such a case a potential function need not exist in a region that contains the origin(2). Let's try to find a potential in a region that excludes the origin. We have $\mathbf{F}(x, y) = \nabla f(x, y)$ if and only if

$$\frac{\partial f}{\partial x} = \frac{x}{x^2 + y^2} \text{ and } \frac{\partial f}{\partial y} = \frac{y}{x^2 + y^2}$$

By the first condition

$$f(x, y) = \int \frac{\partial f}{\partial x}dx = \int \frac{x}{x^2 + y^2}dx.$$

If we set $u = x^2 + y^2$.

$$\frac{\partial u}{\partial x} = 2x.$$

We keep y constant and set $du = 2xdx$. Since any function of y is treated as a constant, there exists $g(y)$ such that

$$\int \frac{x}{x^2 + y^2}dx = \frac{1}{2}\int \frac{1}{u}du = \frac{1}{2}\ln(|u|) + g(y) = \frac{1}{2}\ln(x^2 + y^2) + g(y).$$

Thus,

$$f(x, y) = \frac{1}{2}\ln(x^2 + y^2) + g(y)$$

Therefore,

$$\frac{\partial f}{\partial y} = \frac{1}{2(x^2 + y^2)}(2y) + \frac{dg(y)}{dy} = \frac{y}{x^2 + y^2} + \frac{dg(y)}{dy}.$$

Thus,

$$\frac{y}{x^2 + y^2} + \frac{dg(y)}{dy} = \frac{\partial f}{\partial y} = \frac{y}{x^2 + y^2}$$

This implies that

$$\frac{dg(y)}{dy} = 0$$

so that g is a constant K. Therefore,

$$f(x,y) = \frac{1}{2}\ln(x^2 + y^2) + K.$$

b) By part a),the line integral of $\boldsymbol{F}$ on a curve that that joins $(1,0)$ to $(0,\sqrt{e})$ and does not pass through the origin is

$$f(0,\sqrt{e}) - f(1,0) = \frac{1}{2}\ln(e) - \frac{1}{2}\ln(1) = \frac{1}{2}.$$

□

Now let's consider conditions for the existence of a potential function for a vector field in three dimensions. Let

$$\boldsymbol{F}(x,y,z) = M(x,y,z)\mathbf{i} + N(x,y,z)\mathbf{j} + P(x,y,z)\mathbf{k}$$

and assume that $\boldsymbol{F} = \boldsymbol{\nabla}f$. Then

$$M = \frac{\partial f}{\partial x}, \; N = \frac{\partial f}{\partial y} \text{ and } P = \frac{\partial f}{\partial z}.$$

Assuming that f has continuous second-order partial derivatives,

$$\frac{\partial M}{\partial y} = \frac{\partial^2 f}{\partial y \partial x} = \frac{\partial^2 f}{\partial x \partial y} = \frac{\partial N}{\partial x},$$

$$\frac{\partial M}{\partial z} = \frac{\partial^2 f}{\partial z \partial x} = \frac{\partial^2 f}{\partial x \partial z} = \frac{\partial P}{\partial x},$$

$$\frac{\partial N}{\partial z} = \frac{\partial^2 f}{\partial z \partial y} = \frac{\partial^2 f}{\partial y \partial z} = \frac{\partial P}{\partial y}.$$

Let's record these facts:

Proposition 2 The necessary conditions for

$$\boldsymbol{F}(x,y,z) = M(x,y)\mathbf{i} + N(x,y)\mathbf{j} + P(x,y,z)\mathbf{k}$$

to be conservative and to have a potential function with continuous second-order partial derivatives in the region D are the equalities

$$\frac{\partial M}{\partial y} = \frac{\partial N}{\partial x}, \; \frac{\partial M}{\partial z} = \frac{\partial P}{\partial x}, \; \frac{\partial N}{\partial z} = \frac{\partial P}{\partial y}$$

for each $(x,y,z) \in D$.

These equalities are equivalent to the condition that $\boldsymbol{\nabla} \times \mathbf{F} = \mathbf{0}$ since

$$\boldsymbol{\nabla} \times \mathbf{F} = \begin{vmatrix} \mathbf{i} & \mathbf{j} & \mathbf{k} \\ \frac{\partial}{\partial x} & \frac{\partial}{\partial y} & \frac{\partial}{\partial z} \\ M & N & P \end{vmatrix} = \left(\frac{\partial P}{\partial y} - \frac{\partial N}{\partial z}\right)\mathbf{i} + \left(\frac{\partial M}{\partial z} - \frac{\partial P}{\partial x}\right)\mathbf{j} + \left(\frac{\partial N}{\partial x} - \frac{\partial M}{\partial y}\right)\mathbf{k}$$

In the next chapter we will examine the sufficiency of this condition for the existence of a potential function.

Remark 4 If $\mathbf{F}(x,y) = M(x,y)\mathbf{i} + N(x,y)\mathbf{j}$ is a two-dimensional vector field, we define $\boldsymbol{\nabla} \times \mathbf{F}$ as

$$\boldsymbol{\nabla} \times (M(x,y)\mathbf{i} + N(x,y)\mathbf{j} + 0\mathbf{k}) = \left(\frac{\partial N}{\partial x} - \frac{\partial M}{\partial y}\right)\mathbf{k}.$$

Therefore, the condition that is stated in Proposition 1 can be considered to be a special case of Proposition 2. ◊

Example 5 Let

$$\mathbf{F}(x, y, z) = -2xe^{-x^2-y^2}\sin(z)\,\mathbf{i} - 2ye^{-x^2-y^2}\sin(z)\,\mathbf{j} + e^{-x^2-y^2}\cos(z)\,\mathbf{k}.$$

a) Show that F satisfies the necessary condition for the existence
b) Determine a potential function for **F**.
c) Determine the line integral of **F** *on any curve C that connects* $(0, 0, 0)$ *to* $(1, 1, \pi/2)$.

Solution

a)

$$\boldsymbol{\nabla} \times \mathbf{F}(x, y, z) = \begin{vmatrix} \mathbf{i} & \mathbf{j} & \mathbf{k} \\ \dfrac{\partial}{\partial x} & \dfrac{\partial}{\partial y} & \dfrac{\partial}{\partial z} \\ -2xe^{-x^2-y^2}\sin(z) & -2ye^{-x^2-y^2}\sin(z) & e^{-x^2-y^2}\cos(z) \end{vmatrix}$$

$$= \left(\frac{\partial}{\partial y}\left(e^{-x^2-y^2}\cos(z) \right) - \frac{\partial}{\partial z}\left(-2ye^{-x^2-y^2}\sin(z) \right) \right)\mathbf{i}$$

$$- \left(\frac{\partial}{\partial x}\left(e^{-x^2-y^2}\cos(z) \right) - \frac{\partial}{\partial z}\left(-2xe^{-x^2-y^2}\sin(z) \right) \right)\mathbf{j}$$

$$+ \left(\frac{\partial}{\partial x}\left(-2ye^{-x^2-y^2}\sin(z) \right) - \frac{\partial}{\partial y}\left(-2xe^{-x^2-y^2}\sin(z) \right) \right)\mathbf{k}$$

$$= \left(-2ye^{-x^2-y^2}\cos(z) + 2ye^{-x^2-y^2}\cos(z) \right)\mathbf{i}$$

$$- \left(-2xe^{-x^2-y^2}\cos(z) + 2xe^{-x^2-y^2}\cos(z) \right)\mathbf{j}$$

$$+ \left(4xye^{-x^2-y^2}\sin(z) - 4xye^{-x^2-y^2}\sin(z) \right)\mathbf{k}$$

$$= \mathbf{0}.$$

b) We have $\mathbf{F} = \boldsymbol{\nabla} f$ if

$$\frac{\partial f}{\partial x}(x, y, z) = -2xe^{-x^2-y^2}\sin(z),$$

$$\frac{\partial f}{\partial y}(x, y, z) = -2ye^{-x^2-y^2}\sin(z),$$

$$\frac{\partial f}{\partial z}(x, y, z) = e^{-x^2-y^2}\cos(z).$$

By the first equation,

$$f(x, y, z) = \int -2xe^{-x^2-y^2}\sin(z)\,dx$$

$$= e^{-y^2}\sin(z)\int -2xe^{-x^2}\,dx$$

$$= e^{-y^2}\sin(z)\,e^{-x^2} + g(y, z) = e^{-x^2-y^2}\sin(z) + g(y, z),$$

since any function of x and y has to be treated as a constant when we integrate with respect to x.
Therefore,

$$-2ye^{-x^2-y^2}\sin(z) = \frac{\partial f}{\partial y}(x, y, z) = -2ye^{-x^2-y^2}\sin(z) + \frac{\partial g(y, z)}{\partial y},$$

so that

$$\frac{\partial g\left(y,z\right)}{\partial y} = 0 \Rightarrow g\left(y,z\right) = h\left(z\right),$$

for some function h of z. Thus

$$f\left(x,y,z\right) = e^{-x^2-y^2}\sin\left(z\right) + g\left(y,z\right) = e^{-x^2-y^2}\sin\left(z\right) + h\left(z\right).$$

Therefore,

$$e^{-x^2-y^2}\cos\left(z\right) = \frac{\partial f}{\partial z} = e^{-x^2-y^2}\cos\left(z\right) + \frac{dh}{dz}.$$

Thus,

$$\frac{dh}{dz} = 0 \Rightarrow h\left(z\right) = K,$$

where K is a constant. Therefore,

$$f\left(x,y,z\right) = e^{-x^2-y^2}\sin\left(z\right) + K.$$

c)

$$\int_C \mathbf{F}\cdot d\boldsymbol{\sigma} = f\left(1,1,\pi\right) - f\left(0,0,0\right) = e^{-2}\sin\left(\pi/2\right) - \sin\left(0\right) = e^{-2}.$$

$\square$

Problems

In problems 1-4
a) Check whether the conditions that are necessary for the vector field $\mathbf{F}$ to be conservative are satisfied,
b) If your response to a) is in the affirmative, find a a potential f for $\mathbf{F}$.

1.

$$\mathbf{F}\left(x,y\right) = \left(2x-3y\right)\mathbf{i} + \left(-3x+4y-8\right)\mathbf{j}$$

2.

$$\mathbf{F}\left(x,y\right) = e^x\cos\left(y\right)\mathbf{i} + e^x\sin\left(y\right)\mathbf{j}$$

3.

$$\mathbf{F}\left(x,y\right) = e^x\sin\left(y\right)\mathbf{i} + e^x\cos\left(y\right)\mathbf{j}$$

4.

$$\mathbf{F}\left(x,y,z\right) = -\cos\left(z\right)e^{-x-y}\mathbf{i} - \cos\left(z\right)e^{-x-y}\mathbf{j} - \sin\left(z\right)e^{-x-y}\mathbf{k}$$

In problems 5-7,
a) Find a potential f for $\mathbf{F}$,
b) Make use of f to evaluate the line integral

$$\int_C \mathbf{F}\cdot d\boldsymbol{\sigma}$$

5.

$$\mathbf{F}\left(x,y\right) = xy^2\mathbf{i} + x^2y\mathbf{j}$$

and C is an arbitrary curve from $\left(0,1\right)$ to $\left(2,1\right)$.

6.

$$\mathbf{F}\left(x,y\right) = \left(ye^{xy}-1\right)\mathbf{i} + xe^{xy}\mathbf{j}$$

and C is an arbitrary curve from $(0, 1)$ to $(4, \ln(2))$.

7.

$$F(x, y, z) = yz\mathbf{i} + xz\mathbf{j} + (xy + 2z)\mathbf{k}$$

and C is an arbitrary curve from $(1, 0, -2)$ to $(4, 6, 3)$.

In problems 8 and 9 evaluate the line integral

$$\int_C M(x, y)\, dx + N(x, y)\, dy$$

by finding a function $f(x, y)$ such that

$$df = M(x, y)\, dx + N(x, y)\, dy.$$

8.

$$\int_C \frac{y}{x^2 + y^2}\, dx - \frac{x}{x^2 + y^2}\, dy$$

where C is a curve that is in the upper half-plane $(y > 0)$ and joins $(0, 1)$ to $(1, 1)$.

9.

$$\int_C \frac{y^2}{1 + x^2}\, dx + 2y \arctan(x)\, dy$$

where C is an arbitrary curve from $(0, 0)$ to $(1, 2)$.

10. Evaluate

$$\int_C y^2 \cos(z)\, dx + 2xy \cos(z)\, dy - xy^2 \sin(z)\, dz,$$

where C is an arbitrary curve from $(0, 0, 0)$ to $(1, 1, \pi)$ by finding $f(x, y, z)$ such that

$$df = y^2 \cos(z)\, dx + 2xy \cos(z)\, dy - xy^2 \sin(z)\, dz$$

14.4   Parametrized Surfaces and Tangent Planes

In this section we will introduce parametrizes surfaces. Graphs of functions of two variables are special cases. A surface such as a sphere cannot be expressed as the graph of a single function of two variables. In such a case the parametric representation of the surface is very useful. We will be able to calculate normal vectors to such surfaces. That will turn out to be very important in the calculation of surface integrals in the next section.

Parametrized Surfaces

Assume that $\mathbf{\Phi}(u, v) = (x(u, v), y(u, v), z(u, v))$ for each (u, v) in a region $D \subset \mathbb{R}^2$. Thus $\mathbf{\Phi}$ is a function of two variables and has values in the three-dimensional space $\mathbb{R}^3$. We will use the notation $\mathbf{\Phi} : D \subset \mathbb{R}^2 \to \mathbb{R}^3$ to refer to such a function. The symbols u and v are referred to as **parameters**, and D is **the parameter domain**. We will identify the position vector

$$\mathbf{\Phi}(u, v) = x(u, v)\mathbf{i} + y(u, v)\mathbf{j} + z(u, v)\mathbf{k}$$

with the point $\mathbf{\Phi}(u, v)$, so that we may refer to the coordinate functions as **the component functions of $\mathbf{\Phi}$**. The surface M **that is parametrized by the function $\mathbf{\Phi}$** is the image of $\mathbf{\Phi}$, i.e., S is the set of all points (x, y, z) in $\mathbb{R}^3$ such that $x = x(u, v)$, $y = y(u, v)$ and $z = z(u, v)$ where (u, v) varies in D.

Example 1 Let

$$\mathbf{\Phi}\left(u, v\right) = \left(\cosh(u)\cos(v), 2\cosh\left(u\right)\sin\left(v\right), 4\sinh\left(u\right)\right),$$

where $1 \leq u \leq 1$ and $0 \leq v \leq 2\pi$.

Thus $\mathbf{\Phi} : D \to \mathbb{R}^3$, where the parameter domain D is the rectangle $[-1, 1] \times [0, 2\pi]$ in the uv-plane. The parameters are u and v. The coordinate functions of $\mathbf{\Phi}$ are

$$x\left(u, v\right) = \cosh(u)\cos(v), \ y\left(u, v\right) = 2\cosh\left(u\right)\sin\left(v\right), \text{ and } z\left(u, v\right) = 4\sinh\left(u\right).$$

The surface M that is parametrized by $\mathbf{\Phi}$ is shown in Figure 1.

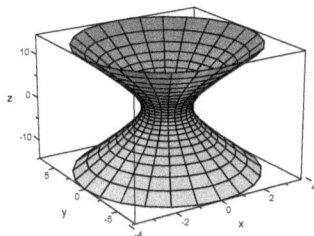

Figure 1

The picture gives the impression that M is a hyperboloid of one sheet. Indeed,

$$16x^2\left(u, v\right) + 8y^2\left(u, v\right) - z^2\left(u, v\right)$$
$$= 16\cosh^2(u)\cos^2(v) + 16\cosh^2\left(u\right)\sin^2\left(v\right) - 16\sinh^2\left(u\right)$$
$$= 16\cosh^2(u)\left(\cos^2\left(v\right) + \sin^2\left(v\right)\right) - 16\sinh^2\left(u\right)$$
$$= 16\cosh^2(u) - 16\sin^2\left(u\right)$$
$$= 16\left(\cosh^2\left(u\right) - \sinh^2\left(u\right)\right) = 16.$$

Thus, the points(x, y, z) on the surface M satisfy the equation

$$16x^2 + 8y^2 - z^2 = 16.$$

The intersection of M with a horizontal plane is an ellipse (or empty) The intersection of M with planes of the form $x = $ constant or $y = $ constant are hyperbolas (or empty). $\square$

The graph of a scalar function of two variables can be considered to be a parametrized surface: If $f\left(x, y\right)$ is defined for each $(x, y) \in D \subset \mathbb{R}^2$, we set

$$\mathbf{\Phi}\left(x, y\right) = \left(x, y, f\left(x, y\right)\right).$$

Thus, x and y are the parameters. The parameter domain coincides with the domain of f.

Example 2 Let $f\left(x, y\right) = x^2 - y^2$.

The graph of f can be considered to be the surface M that is parametrized by the function $\mathbf{\Phi} : \mathbb{R}^2 \to \mathbb{R}^3$, where

$$\mathbf{\Phi}\left(x, y\right) = \left(x, y, x^2 - y^2\right).$$

The associated vector-valued function is

$$\mathbf{\Phi}\left(x, y\right) = x\mathbf{i} + y\mathbf{j} + \left(x^2 - y^2\right)\mathbf{k}.$$

Figure 2 shows S. $\square$

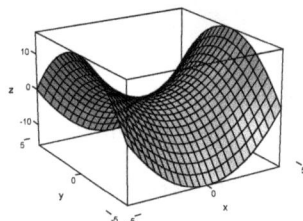

Figure 2

Assume that a surface S is the graph of the equation $\rho = f(\phi, \theta)$, where ρ, ϕ and θ are **spherical coordinates**. Thus,

$$x(\phi, \theta) = \rho \sin(\phi) \cos(\theta) = f(\phi, \theta) \sin(\phi) \cos(\theta),$$
$$y(\phi, \theta) = \rho \sin(\phi) \sin(\theta) = f(\phi, \theta) \sin(\phi) \sin(\theta),$$
$$z(\phi, \theta) = \rho \cos(\phi) = f(\phi, \theta) \cos(\phi),$$

where $0 \le \phi \le \pi$ and $0 \le \theta \le 2\pi$. Therefore, the surface M is parametrized by the function

$$\mathbf{\Omega}(\phi, \theta) = (f(\phi, \theta) \sin(\phi) \cos(\theta), f(\phi, \theta) \sin(\phi) \sin(\theta), f(\phi, \theta) \cos(\phi)).$$

Example 3 Let M be the sphere of radius 2 centered at the origin. In spherical coordinates, M

The sphere can be parametrized by

$$\mathbf{\Omega}(\phi, \theta) = (2 \sin(\phi) \cos(\theta), 2 \sin(\phi) \sin(\theta), 2 \cos(\phi)),$$

where $0 \le \phi \le \pi$ and $0 \le \theta \le 2\pi$. Figure 3 shows S. $\square$

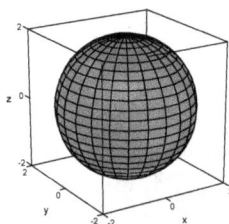

Figure 3

Example 4 Let M be the cone $\phi = \phi_0$. Determine a parametrization of S.

Solution

We set

$$\mathbf{\Omega}(\rho, \theta) = (\rho \sin(\phi_0) \cos(\theta), \rho \sin(\phi_0) \sin(\theta), \rho \cos(\phi_0)),$$

where $\rho \ge 0$ and $0 \le \theta \le 2\pi$. Figure 4 shows the cone $\phi = \pi/6$. $\square$

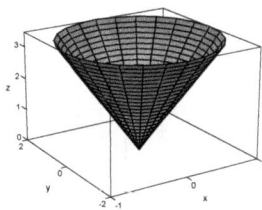

Figure 4

Assume that r, θ and z are **cylindrical coordinates,** so that

$$r = x^2 + y^2, \ \cos(\theta) = \frac{x}{r}, \ \sin(\theta) = \frac{y}{r}.$$

If the surface S is the graph of $z = g(r, \theta)$, we can view S as the surface that is parametrized by

$$\mathbf{\Phi}(r, \theta) = (r \cos(\theta), r \sin(\theta), g(r, \theta)).$$

Example 5 Let M be the surface that is described by the equation $z = r^2$ in cylindrical coordinates. Determine a parametrization of M.

Solution

The surface S can be parametrized by the function

$$\mathbf{\Phi}(r, \theta) = \left(r \cos(\theta), r \sin(\theta), r^2\right),$$

where $r \geq 0$ and $0 \leq \theta \leq 2\pi$. Note that M is a paraboloid. Figure 5 shows the paraboloid. $\square$

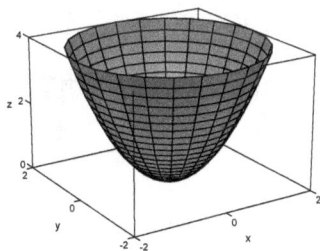

Figure 5

Example 6 Determine a parametrization of the cylinder: $r = 2$.

Solution

The cylinder can be parametrized by

$$\mathbf{\Phi}(\theta, z) = (2 \cos(\theta), 2 \sin(\theta), z),$$

where $0 \leq \theta \leq 2\pi$ and $-\infty < z < +\infty$. Figure 6 shows the cylinder. $\square$

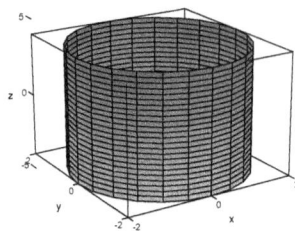

Figure 6

Surfaces of Revolution can be expressed parametrically. Assume that we form a surface S by rotating the graph of $z = f(x)$ on the interval $[a, b]$ about the x-axis. If the point $(x, 0, f(x))$ is rotated by angle θ the resulting point is $(x, f(x) \sin(\theta), f(x) \cos(\theta))$. Thus, S can be parametrized by the function

$$\Omega(x, \theta) = (x, f(x) \sin(\theta), f(x) \cos(\theta)),$$

where $a \leq x \leq b$ and $0 \leq \theta \leq 2\pi$.

Example 7 Let M be the surface generated by revolving the graph of $z = 1 + x^2$ on the interval $[1, 2]$ about the x-axis. Determine a parametrization of M.

Solution

We can set

$$\Omega(x, \theta) = \left(x, \left(1 + x^2\right) \sin(\theta), \left(1 + x^2\right) \cos(\theta)\right),$$

where $1 \leq x \leq 2$ and $0 \leq \theta \leq 2\pi$. Then M is parametrized by the function Ω. Figure 7 shows S. $\square$

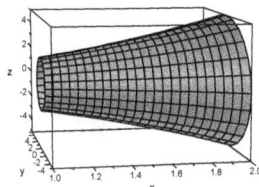

Figure 7

Assume that we rotate $z = f(x)$ about the z-axis. The resulting surface of revolution can be parametrized by the function

$$(x \cos(\theta), x \sin(\theta), f(x)), \ 0 \leq \theta \leq 2\pi.$$

For a given x, we have

$$\sqrt{y^2 + x^2} = x.$$

This is the radius of the circular cross section. The point $(x \cos(\theta), x \sin(\theta), f(x))$ is at an angle θ from the x-direction.

Example 8 Let M be the surface generated by revolving the graph of $z = 1 + x^2$ on the interval $[1, 2]$ about the z-axis. Determine a parametrization of M.

Solution

We can set

$$\Omega\left(x,\theta\right)=\left(x\cos\left(\theta\right),x\sin\left(\theta\right),1+x^{2}\right),$$

where $1\le x\le 2$ and $0\le\theta\le 2\pi$. Then S is parametrized by the function Ω. Figure 8 shows M. $\square$

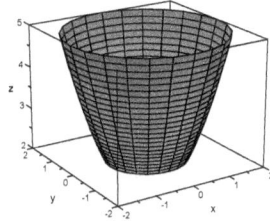

Figure 8

Normal Vectors, Tangent Planes and Orientation

Assume that the surface M is parametrized by the function $\boldsymbol{\Phi}:D\subset\mathbb{R}^{2}\to\mathbb{R}^{3}$, where

$$\boldsymbol{\Phi}\left(u,v\right)=\left(x\left(u,v\right),y\left(u,v\right),z\left(u,v\right)\right)$$

for each $(u,v)\in D$. Given a point $(u_0,v_0)\in D$, the restriction of $\boldsymbol{\Phi}$ to the line $v=v_0$, i.e., the function

$$u\to\boldsymbol{\Phi}\left(u,v_0\right)$$

parametrizes a curve on the surface M. We will refer to this curve as **the u-coordinate curve on M** that passes through $\Phi\left(u_0,v_0\right)$. Similarly, **the v-coordinate curve on M** that passes through $\Phi\left(u_0,v_0\right)$ is parametrized by the restriction of Φ to the line $u=u_0$, i.e., the function

$$v\to\boldsymbol{\Phi}\left(u_0,v\right).$$

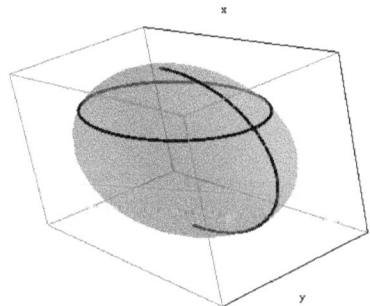

Figure 9: Coordinate Curves on a Surface

We will define **the partial derivatives of** Φ as the vector-valued functions

$$\partial_u\boldsymbol{\Phi}\left(u,v\right)=\frac{\partial\boldsymbol{\Phi}}{\partial u}\left(u,v\right)=\frac{\partial x}{\partial u}\left(u,v\right)\mathbf{i}+\frac{\partial y}{\partial u}\left(u,v\right)\mathbf{j}+\frac{\partial z}{\partial u}\left(u,v\right)\mathbf{k},$$

and

$$\partial_v \mathbf{\Phi} (u,v) = \frac{\partial \mathbf{\Phi}}{\partial v} (u,v) = \frac{\partial x}{\partial v} (u,v) \mathbf{i} + \frac{\partial y}{\partial v} (u,v) \mathbf{j} + \frac{\partial z}{\partial v} (u,v) \mathbf{k}.$$

Thus, the vectors $\partial_u \mathbf{\Phi} (u_0, v_0)$ and $\partial_v \mathbf{\Phi} (u_0, v_0)$ are tangent to the u-coordinate curve and the v-coordinate curve, respectively, that pass through $\mathbf{\Phi} (u_0, v_0)$.

Definition 1 Assume that $\partial_u \mathbf{\Phi} (u_0, v_0)$ and $\partial_v \mathbf{\Phi} (u_0, v_0)$ exist and $\mathbf{N} (u_0, v_0) = \partial_u \mathbf{\Phi} (u_0, v_0) \times \partial_v \mathbf{\Phi} (u_0, v_0) \neq \mathbf{0}$. In this case we will say that the surface M that is parametrized by $\mathbf{\Phi}$ is **smooth** at $\mathbf{\Phi} (u_0, v_0)$. The vector $\mathbf{N} (u_0, v_0)$ is the **normal to the surface M** at $\mathbf{\Phi}(u_0, v_0)$. **The tangent plane** to S at $P_0 = \mathbf{\Phi}(u_0, v_0)$ is the set of points P such that

$$\mathbf{N} (u_0, v_0) \cdot \overrightarrow{P_0 P} = 0.$$

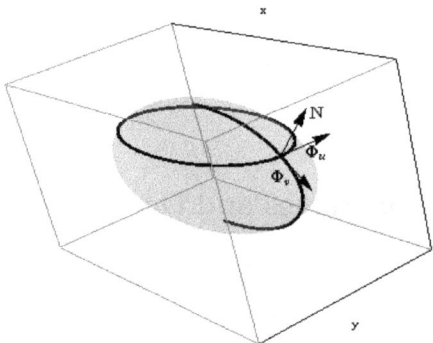

Figure 10: Tangent and normal vectors

Example 9 The sphere M of radius ρ_0 can be parametrized by the function

$$\mathbf{\Phi} (\phi, \theta) = (\rho_0 \sin (\phi) \cos (\theta) , \rho_0 \sin (\phi) \sin (\theta) , \rho_0 \cos (\phi)) ,$$

where $0 \leq \phi \leq \pi$ and $0 \leq \theta \leq 2\pi$ (spherical coordinates).
a) Identify the coordinate curves on M
b) Determine the Norrmal to M at $\Phi (\phi, \theta)$.
c) Set the radius to be 1 and determine the plane that is tangent to M at $\Phi (\pi/4, \pi/6)$

Solution

a) If $\phi = \phi_0$, the corresponding θ-coordinate curve on the sphere is parametrized by the function

$$\theta \to \mathbf{\Phi} (\phi_0, \theta) = (\rho_0 \sin (\phi_0) \cos (\theta) , \rho_0 \sin (\phi_0) \sin (\theta) , \rho_0 \cos (\phi_0)) ,$$

where $0 \leq \theta \leq 2\pi$. Thus, z has the constant value $\rho_0 \cos (\phi_0)$ and we would expect that this curve is a circle on the plane $z = \rho_0 \cos (\phi_0)$ that is centered at $(0, 0, \rho_0 \cos (\phi_0))$. Indeed, the distance of $\mathbf{\Phi} (\phi_0, \theta)$ from $(0, 0, \rho_0 \cos (\phi_0))$ is

$$\sqrt{(\rho_0 \sin (\phi_0) \cos (\theta))^2 + (\rho_0 \sin (\phi_0) \sin (\theta))} = \sqrt{\rho_0^2 \sin^2 (\phi_0) \cos^2 (\theta) + \rho_0^2 \sin^2 (\phi_0) \sin^2 (\theta)}$$

$$= \rho_0 \sin (\phi_0) \sqrt{\cos^2 (\theta) + \sin^2 (\theta)} = \rho_0 \sin (\phi_0) .$$

Therefore, a ϕ-coordinate curve that passes through $\mathbf{\Phi} (\phi_0, \theta_0)$ is a circle of radius $\rho_0 \sin (\phi_0)$. For example, if $\phi_0 = \pi/2$, the circle is "the equator" on the xy-plane, centered at the origin and having radius ρ_0. If $\phi_0 = 0$, the circle degenerates to the "north pole" $(0, 0, \rho_0)$.

As for the ϕ-coordinate curves, if $\theta = \theta_0$, the corresponding coordinate curve on the sphere is parametrized by the function

$$\phi \to \mathbf{\Phi}(\phi, \theta_0) = (\rho_0 \sin(\phi) \cos(\theta_0), \rho_0 \sin(\phi) \sin(\theta_0), \rho_0 \cos(\phi)),$$

where $0 \le \phi \le \pi$. The curve is a semicircle of radius ρ_0 in the plane $\theta = \theta_0$ that is centered at the origin ("a meridian").

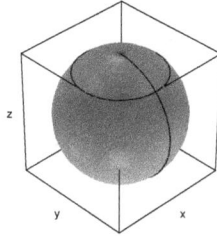

Figure 11: Coordinate curves on a sphere

b) We have

$$\partial_\phi \mathbf{\Phi}(\phi, \theta) = \partial_\phi (\rho_0 \sin(\phi) \cos(\theta)) \mathbf{i} + \partial_\phi (\rho_0 \sin(\phi) \sin(\theta)) \mathbf{j} + \partial_\phi (\rho_0 \cos(\phi)) \mathbf{k}$$
$$= \rho_0 \cos(\phi) \cos(\theta) \mathbf{i} + \rho_0 \cos(\phi) \sin(\theta) \mathbf{j} - \rho_0 \sin(\phi) \mathbf{k},$$

and

$$\partial_\theta \mathbf{\Phi}(\phi, \theta) = \partial_\theta (\rho_0 \sin(\phi) \cos(\theta)) \mathbf{i} + \partial_\theta (\rho_0 \sin(\phi) \sin(\theta)) \mathbf{j} + \partial_\theta (\rho_0 \cos(\phi)) \mathbf{k}$$
$$= -\rho_0 \sin(\phi) \sin(\theta) \mathbf{i} + \rho_0 \sin(\phi) \cos(\theta) \mathbf{j}.$$

Therefore,

$$\mathbf{N}(\phi, \theta) = \partial_\phi \mathbf{\Phi}(\phi, \theta) \times \partial_\theta \mathbf{\Phi}(\phi, \theta)$$

$$= \begin{vmatrix} \mathbf{i} & \mathbf{j} & \mathbf{k} \\ \rho_0 \cos(\phi) \cos(\theta) & \rho_0 \cos(\phi) \sin(\theta) & -\rho_0 \sin(\phi) \\ -\rho_0 \sin(\phi) \sin(\theta) & \rho_0 \sin(\phi) \cos(\theta) & 0 \end{vmatrix}$$

$$= \begin{vmatrix} \rho_0 \cos(\phi) \sin(\theta) & -\rho_0 \sin(\phi) \\ \rho_0 \sin(\phi) \cos(\theta) & 0 \end{vmatrix} \mathbf{i} - \begin{vmatrix} \rho_0 \cos(\phi) \cos(\theta) & -\rho_0 \sin(\phi) \\ -\rho_0 \sin(\phi) \sin(\theta) & 0 \end{vmatrix} \mathbf{j}$$

$$+ \begin{vmatrix} \rho_0 \cos(\phi) \cos(\theta) & \rho_0 \cos(\phi) \sin(\theta) \\ -\rho_0 \sin(\phi) \sin(\theta) & \rho_0 \sin(\phi) \cos(\theta) \end{vmatrix} \mathbf{k}$$

$$= \rho_0^2 \sin^2(\phi) \cos(\theta) \mathbf{i} + \rho_0^2 \sin^2(\phi) \sin(\theta) \mathbf{j}$$
$$+ \left(\rho_0^2 \sin(\phi) \cos(\phi) \cos^2(\theta) + \rho_0^2 \sin(\phi) \cos(\theta) \sin^2(\theta) \right) \mathbf{k}$$

$$= \rho_0^2 \sin^2(\phi) \cos(\theta) \mathbf{i} + \rho_0^2 \sin^2(\phi) \sin(\theta) \mathbf{j} + \rho_0^2 \sin(\phi) \cos(\phi) \mathbf{k}$$

$$= \rho_0 \sin(\phi) \left(\rho_0 \sin(\phi) \cos(\theta) \mathbf{i} + \rho_0 \sin(\phi) \sin(\theta) \mathbf{j} + \rho_0 \cos(\phi) \mathbf{k} \right)$$

$$= \rho_0 \sin(\phi) \mathbf{\Phi}(\phi, \theta).$$

Thus, the normal $\mathbf{N}(\phi, \theta)$ is in the radial direction, as we would have expected.

c) In particular, if $\rho_0 = 1$,

$$\mathbf{N}(\pi/4, \pi/6) = \sin^2(\pi/4)\cos(\pi/6)\mathbf{i} + \sin^2(\pi/4)\sin(\pi/6)\mathbf{j} + \sin(\pi/4)\cos(\pi/4)\mathbf{k}$$

$$= \left(\frac{1}{2}\right)\left(\frac{\sqrt{3}}{2}\right)\mathbf{i} + \left(\frac{1}{2}\right)\left(\frac{1}{2}\right)\mathbf{j} + \left(\frac{\sqrt{2}}{2}\right)\left(\frac{\sqrt{2}}{2}\right)\mathbf{k}$$

$$= \frac{\sqrt{3}}{4}\mathbf{i} + \frac{1}{4}\mathbf{j} + \frac{1}{2}\mathbf{k},$$

and

$$\mathbf{\Phi}(\pi/4, \pi/6) = (\sin(\pi/4)\cos(\pi/6), \sin(\pi/4)\sin(\pi/6), \cos(\pi/4))$$

$$= \left(\left(\frac{\sqrt{2}}{2}\right)\left(\frac{\sqrt{3}}{2}\right), \left(\frac{\sqrt{2}}{2}\right)\left(\frac{1}{2}\right), \frac{\sqrt{2}}{2}\right)$$

$$= \left(\frac{\sqrt{6}}{4}, \frac{\sqrt{2}}{4}, \frac{\sqrt{2}}{2}\right).$$

Therefore, the tangent plane to the sphere at $\mathbf{\Phi}(\pi/4, \pi/6)$ is the graph of the equation

$$\frac{\sqrt{3}}{4}\left(x - \frac{\sqrt{6}}{4}\right) + \frac{1}{4}\left(y - \frac{\sqrt{2}}{4}\right)j + \frac{1}{2}\left(z - \frac{\sqrt{2}}{2}\right) = 0.$$

$\square$

With reference to the above example, the normal $\mathbf{N}(\phi, \theta)$ points towards the exterior of the sphere. if we interchange the order the parameters ϕ and θ we have $\mathbf{N}(\phi, \theta) = -\mathbf{N}(\phi, \theta)$ (confirm), so that $\mathbf{N}(\phi, \theta)$ points towards the interior of the sphere. The sphere is an example of an "**orientable surface**". Intuitively, if a surface M is orientable we can tell the difference between "inside" and "outside" on M. More precisely, **a surface is orientable** provided that $\mathbf{N}(u, v)$ is **continuous** on D if (u, v) varies in "the parameter domain" D. The normal vector may point outward, and we may label that case as "**the positive orientation**" of the surface. Then we label the case when the normal points inward as "**the negative orientation**" of the surface. Thus, in the case of the sphere of the above example, $\mathbf{N}(\phi, \theta) = -\mathbf{N}(\phi, \theta)$ specifies the negative orientation of the surface. In the general case, Assume that M is parametrized initially by $\mathbf{\Phi} : D \to \mathbb{R}^3$. If $\mathbf{N}(u, v)$ is the normal determined by the parametrization $\mathbf{\Phi}$ and determines "**the positive orientation**" of M, we say that $\mathbf{\Omega} : D^* \to \mathbb{R}^2$ is a **positively oriented parametrization of** M if there is a one-one correspondence between the parameter regions D and D^* such that normal $\mathbf{N}^*(u^*, v^*)$ determined by $\mathbf{\Omega}$ points in the same direction as the corresponding $\mathbf{N}(u, v)$. If $\mathbf{\Omega} : D^* \to \mathbb{R}^2$ is a parametrization of M so that $\mathbf{N}^*(u^*, v^*)$ determined by $\mathbf{\Omega}$ points in the direction of $-\mathbf{N}(u, v)$, we say that $\mathbf{\Omega}$ is a parametrization of M with **negative orientation**, The precise definition of the orientation of surface is given at the end of this section. The informal intuitive understanding of orientability and the positive or negative orientation of a surface will be adequate for our purposes. All of the previous examples are orientable surfaces. In any case, let's note that there are nonorientable surfaces:

Example 10 **A Mobius band** is not orientable. This is a surface that you can construct by identifying two opposite sides of a rectangle so that A matches A' and B matches B' is in the picture:

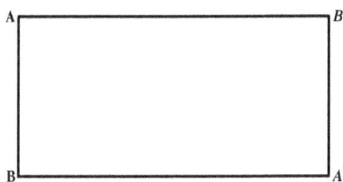

Figure 12

The result is a surface that looks like this:

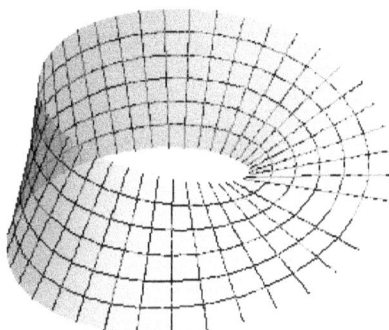

Figure 13

The entire surface does not have two distinct sides that can be identified as "inside" and "outside". If you imagine that a bug that does not fall off the surface starts crawling on one side of a small patch the surface, it will eventually end up on the other side of the same patch. The normal vector $\mathbf{N}$ at the beginning is twisted on the way and comes back as $-\mathbf{N}$. $\square$

It is worth noting the **special case of a surface that is the graph of a function of two variables.** Assume that the surface M is the graph of $z = f(x, y)$, where $(x, y) \in D \subset \mathbb{R}^2$. Then S can be parametrized by the function $\mathbf{\Phi} : D \to \mathbb{R}^3$, where

$$\mathbf{\Phi}(x, y) = (x, y, f(x, y)).$$

In this case, the x-coordinate curve that passes through the point $\mathbf{\Phi}(x_0, y_0) = (x_0, y_0, f(x_0, y_0))$ can be parametrized by the function

$$x \to \mathbf{\Phi}(x, y_0) = (x, y_0, f(x, y_0)),$$

just as we defined in Section 12.5. Similarly, the y-coordinate curve that passes through the point $\mathbf{\Phi}(x_0, y_0)$ can be parametrized by the function

$$y \to \mathbf{\Phi}(x_0, y) = (x_0, y, f(x_0, y)),$$

just as we defined in Section 12.5.
We have

$$\partial_x \mathbf{\Phi}(x_0, y_0) = (1, 0, \partial_x f(x_0, y_0))$$

and

$$\partial_y \mathbf{\Phi}(x_0, y_0) = (0, 1, \partial_y f(x_0, y_0)).$$

Therefore,

$$\mathbf{N}\left(x_0, y_0\right) = \left(\mathbf{i} + \frac{\partial f}{\partial x}\left(x_0, y_0\right)\mathbf{k}\right) \times \left(\mathbf{j} + \frac{\partial f}{\partial y}\left(x_0, y_0\right)\mathbf{k}\right)$$

$$= \begin{vmatrix} \mathbf{i} & \mathbf{j} & \mathbf{k} \\ 1 & 0 & f_x\left(x_0, y_0\right) \\ 0 & 1 & f_y\left(x_0, y_0\right) \end{vmatrix}$$

$$= -\partial_x f\left(x_0, y_0\right)\mathbf{i} - \partial_y f\left(x_0, y_0\right)\mathbf{j} + \mathbf{k},$$

as in Section 12.5 (Remark 1 of Section 12.5). Thus, we can express the tangent plane as the graph of the equation

$$\mathbf{N}(x_0, y_0) \cdot \left(\left(x - x_0\right)\mathbf{i} + \left(y - y_0\right)\mathbf{j} + \left(z - z_0\right)\mathbf{k}\right) = 0,$$

i.e.,

$$-\partial_x f\left(x_0, y_0\right)\left(x - x_0\right) - \partial_y f\left(x_0, y_0\right)\left(y - y_0\right) + \left(z - f\left(x_0, y_0\right)\right) = 0.$$

Thus,

$$z = f\left(x_0, y_0\right) + f_x\left(x_0, y_0\right)\left(x - x_0\right) + f_y\left(x_0, y_0\right)\left(y - y_0\right),$$

as before. $\Diamond$

Example 11 Let $f\left(x, y\right) = x^2 - y^2$ and let M be the graph of f.
a) Identify the coordinate curves on M that pass through $(2, 1, 1)$, and determine vectors that are tangent to these curves at $(2, 1, 1)$.
b) Determine a normal vector to M at $(2, 1, 1)$ and the tangent plane at that point.

Solution

Figure 14 shows the graph of f.

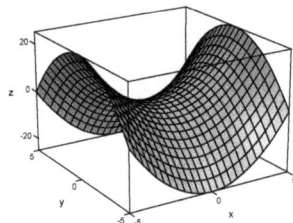

Figure 14

a) Let

$$\boldsymbol{\Phi}\left(x, y\right) = \left(x, y, f\left(x, y\right)\right) = \left(x, y, x^2 - y^2\right).$$

The x-coordinate curve that passes through $(2, 1, 1)$ is parametrized by the function

$$x \to \boldsymbol{\Phi}\left(x, 1, x^2 - 1\right) = \left(x, 1, x^2 - 1\right).$$

Thus, the curve is in the plane $y = 1$ and its projection on the xz-plane is the parabola $z = x^2 - 1$. The y-coordinate curve that passes through $(2, 1, 1)$ is parametrized by the function

$$y \to \boldsymbol{\Phi}\left(2, y, 4 - y^2\right) = \left(2, y, 4 - y^2\right).$$

Thus, the curve is in the plane $x = 2$ and its projection on the yz-plane is the parabola $z = 4 - y^2$.

b) We have

$$\partial_x \Phi (x, y) = (1, 0, 2x) \text{ and } \partial_y \Phi (x, y) = (0, 1, -2y).$$

Therefore,

$$\partial_x \Phi (2, 1) = (1, 0, 4) \text{ and } \partial_y \Phi (2, 1) = (0, 1, -2.)$$

These vectors are tangent to the x- and y-coordinate curves, respectively, at $(2, 1, 1)$. We have

$$\mathbf{N} (x, y) = \partial_x \Phi (2, 1) \times \partial_y \Phi (2, 1) = \begin{vmatrix} \mathbf{i} & \mathbf{j} & \mathbf{k} \\ 1 & 0 & 4 \\ 0 & 1 & -2 \end{vmatrix}$$

$$= \begin{vmatrix} 0 & 4 \\ 1 & -2 \end{vmatrix} \mathbf{i} - \begin{vmatrix} 1 & 4 \\ 0 & -2 \end{vmatrix} \mathbf{j} + \begin{vmatrix} 1 & 0 \\ 0 & 1 \end{vmatrix} \mathbf{k}$$

$$= -4\mathbf{i} + 2\mathbf{j} + \mathbf{k}.$$

Therefore, the tangent plane to the graph of f at $(2, 1, 1)$ is the graph of the equation

$$-4 (x - 2) + 2 (y - 1) + (z - 1) = 0$$

□

Example 12 Assume that the surface M is parametrized by

$$\Phi (\theta, z) = (2 \cos (\theta), 2 \sin (\theta), z),$$

where $0 \le \theta \le 2\pi$ and $-\infty < z < +\infty$ (cylindrical coordinates).
a) Identify the coordinate curves on M.
b) Determine the tangent plane to M at $\Phi (\pi/4, 1).$

Solution

Figure 15 shows the surface M. Note that M is a cylinder whose axis is along the z-axis, since a point (x, y, z) on M satisfies the equation

$$x^2 + y^2 = (2 \cos (\theta))^2 + (2 \sin (\theta))^2 = 4.$$

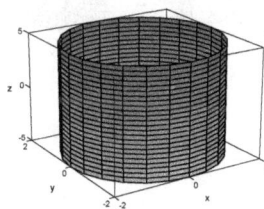

Figure 15

a) If $\theta = \theta_0$, then

$$z \to \Phi (\theta_0, z) = (2 \cos (\theta_0), 2 \sin (\theta_0), z)$$

parametrizes a line that is parallel to the z-axis and passes through the point $(x_0, y_0, 0)$, where $x_0 = 2 \cos (\theta_0)$ and $y_0 = 2 \sin (\theta_0)$.
If $z = z_0,$

$$\theta \to \Phi (\theta, z_0) = (2 \cos (\theta), 2 \sin (\theta), z_0)$$

parametrizes a circle of radius 2 that is in the plane $z = z_0$ and centered at $(0, 0, z_0).$

b) We have

$$\partial_\theta \mathbf{\Phi}(\theta, z) = -2\sin(\theta)\mathbf{i} + 2\cos(\theta)\mathbf{j}, \text{ and } \partial_z \mathbf{\Phi}(\theta, z) = \mathbf{k}.$$

Therefore,

$$\mathbf{N}(\theta, z) = \partial_\theta \mathbf{\Phi}(\theta, z) \times \partial_z \mathbf{\Phi}(\theta, z) = \begin{vmatrix} \mathbf{i} & \mathbf{j} & \mathbf{k} \\ -2\sin(\theta) & 2\cos(\theta) & 0 \\ 0 & 0 & 1 \end{vmatrix}$$

$$= 2\cos(\theta)\mathbf{i} + 2\sin(\theta)\mathbf{j}$$

We have

$$\mathbf{\Phi}(\pi/4, 1) = \left(2\cos\left(\frac{\pi}{4}\right), 2\sin\left(\frac{\pi}{4}\right), 1\right) = \left(\sqrt{2}, \sqrt{2}, 1\right),$$

and

$$\mathbf{N}(\pi/4, 1) = \sqrt{2}\mathbf{i} + \sqrt{2}\mathbf{j}.$$

Therefore, the equation of the plane that is tangent to M at $\left(\sqrt{2}, \sqrt{2}, 1\right)$ is

$$\sqrt{2}\left(x - \sqrt{2}\right) + \sqrt{2}\left(y - \sqrt{2}\right) = 0.$$

□

Example 13 Assume that $a > b > 0$. Let

$$\mathbf{\Phi}(\theta, \phi) = ((a + b\cos(\phi))\cos(\theta), (a + b\cos(\phi))\sin(\theta), b\sin(\phi)),$$

where $0 \le \theta \le 2\pi$ and $0 \le \phi \le 2\pi$. The surface M that is parametrized by $\mathbf{\Phi}$ is a **torus**. S is generated by revolving a disk of radius b that is perpendicular to the xy-plane, about the z-axis, with its center at a distance a from the z-axis. Figure 16 shows M when $a = 4$ and $b = 1$.

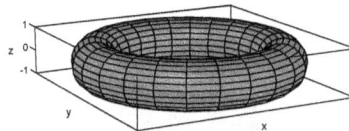

Figure 16

If $\phi = \phi_0$, the corresponding θ-coordinate curve is parametrized by the function

$$\theta \to \mathbf{\Phi}(\theta, \phi_0) = ((a + b\cos(\phi_0))\cos(\theta), (a + b\cos(\phi_0))\sin(\theta), b\sin(\phi_0)),$$

where $0 \le \theta \le 2\pi$. We notice that the z-coordinate has the constant value $b\sin(\phi_0)$, and that the curve is a circle of radius $(a + b\cos(\phi_0))$ centered at $(0, 0, b\sin(\phi_0))$. For example, if $\phi_0 = 0$, the curve is the circle of radius $a + b$ in the xy-plane that is centered at the origin. If $\phi_0 = \pi$, the curve is the circle of radius $a - b$ in the xy-plane that is centered at the origin. If $\theta = \theta_0$, the corresponding ϕ-coordinate curve is parametrized by the function

$$\phi \to \mathbf{\Phi}(\theta_0, \phi) = ((a + b\cos(\phi))\cos(\theta_0), (a + b\cos(\phi))\sin(\theta_0), b\sin(\phi)),$$

where $0 \leq \phi \leq 2\pi$. This is the circle of radius b centered at $(a\cos(\theta_0), a\sin(\theta_0), 0)$. in the plane $\theta = \theta_0$. Indeed,

$$\mathbf{\Phi}(\theta_0, \phi) - (a\cos(\theta_0), a\sin(\theta_0), 0)$$
$$= ((a + b\cos(\phi))\cos(\theta_0), (a + b\cos(\phi))\sin(\theta_0), b\sin(\phi)) - (a\cos(\theta_0), a\sin(\theta_0), 0)$$
$$= (b\cos(\phi))\cos(\theta_0), b\cos(\phi)\sin(\theta_0), b\sin(\phi)),$$

so that

$$||\mathbf{\Phi}(\theta_0, \phi) - (a\cos(\theta_0), a\sin(\theta_0), 0)||^2$$
$$= b^2\cos^2(\phi)\cos^2(\theta_0) + b^2\cos^2(\phi)\sin^2\theta_0) + b^2\sin^2(\phi)$$
$$= b^2\cos^2(\phi) + b^2\sin^2(\phi) = b^2.$$

$\square$

Example 14 Assume that the torus M is parametrized by

$$\mathbf{\Phi}(\theta, \phi) = ((2 + \cos(\phi))\cos(\theta), (2 + \cos(\phi))\sin(\theta), \sin(\phi)),$$

where $0 \leq \theta \leq 2\pi$ and $0 \leq \phi \leq 2\pi$.
a) Identify the coordinate curves that pass through $\mathbf{\Phi}(0, \pi/3)$.
b) Determine the tangent plane to M at $\mathbf{\Phi}(0, \pi/3)$.

Solution

a) Note that

$$\mathbf{\Phi}(0, \pi/3) = \left(\frac{5}{2}, 0, \frac{\sqrt{3}}{2}\right).$$

We have

$$\mathbf{\Phi}(\theta, \pi/3) = ((2 + \cos(\pi/3))\cos(\theta), (2 + \cos(\pi/3))\sin(\theta), \sin(\pi/3))$$
$$= \left(\frac{5}{2}\cos(\theta), \frac{5}{2}\sin(\theta), \frac{\sqrt{3}}{2}\right).$$

This is a circle of radius $5/2$ in the plane $z = \sqrt{3}/2$ that is centered at $(0, 0, \sqrt{3}/2)$.
We also have
$$\mathbf{\Phi}(0, \phi) = (2 + \cos(\phi), 0, \sin(\phi)).$$

This coordinate curve is the circle of radius 1 in the xz-plane that is centered at $(2, 0, 0)$ Indeed,

$$||\mathbf{\Phi}(0, \phi) - (2, 0, 0)||^2 = \cos^2(\phi) + \sin^2(\phi) = 1.$$

b)

$$\frac{\partial \mathbf{\Phi}}{\partial \theta}(\theta, \phi) = \frac{\partial}{\partial \theta}((2 + \cos(\phi))\cos(\theta))\mathbf{i} + \frac{\partial}{\partial \theta}((2 + \cos(\phi))\sin(\theta))\mathbf{j} + \frac{\partial}{\partial \theta}\sin(\phi)\mathbf{k}$$
$$= -(2 + \cos(\phi))\sin(\theta)\mathbf{i} + (2 + \cos(\phi))\cos(\theta)\mathbf{j},$$

$$\frac{\partial \mathbf{\Phi}}{\partial \phi}(\theta, \phi) = \frac{\partial}{\partial \phi}((2 + \cos(\phi))\cos(\theta))\mathbf{i} + \frac{\partial}{\partial \phi}((2 + \cos(\phi))\sin(\theta))\mathbf{j} + \frac{\partial}{\partial \phi}\sin(\phi)\mathbf{k}$$
$$= -\sin(\phi)\cos(\theta)\mathbf{i} - \sin(\phi)\sin(\theta)\mathbf{j} + \cos(\phi)\mathbf{k}.$$

Therefore,

$$\frac{\partial \mathbf{\Phi}}{\partial \theta}\left(0, \frac{\pi}{3}\right) = -\left(2 + \cos\left(\phi\right)\right)\sin\left(\theta\right)\mathbf{i} + \left(2 + \cos\left(\phi\right)\right)\cos\left(\theta\right)\mathbf{j}\big|_{\theta=0, \phi=\pi/3}$$

$$= -\left(2 + \frac{1}{2}\right)\sin\left(0\right)\mathbf{i} + \left(2 + \frac{1}{2}\right)\cos\left(0\right)\mathbf{j} = \frac{5}{2}\mathbf{j},$$

and

$$\frac{\partial \mathbf{\Phi}}{\partial \phi}\left(0, \frac{\pi}{3}\right) = -\sin\left(\phi\right)\cos\left(\theta\right)\mathbf{i} - \sin\left(\phi\right)\sin\left(\theta\right)\mathbf{j} + \cos\left(\phi\right)\mathbf{k}\big|_{\theta=0, \phi=\pi/3}$$

$$= -\frac{\sqrt{3}}{2}\mathbf{i} + \frac{1}{2}\mathbf{k}.$$

Thus,

$$\mathbf{N}\left(0, \frac{\pi}{3}\right) = \frac{\partial \mathbf{\Phi}}{\partial \theta}\left(0, \frac{\pi}{3}\right) \times \frac{\partial \mathbf{\Phi}}{\partial \phi}\left(0, \frac{\pi}{3}\right)$$

$$= \frac{5}{2}\mathbf{j} \times \left(-\frac{\sqrt{3}}{2}\mathbf{i} + \frac{1}{2}\mathbf{k}\right) = \frac{5\sqrt{3}}{4}\mathbf{k} + \frac{5}{4}\mathbf{i}.$$

Note that

$$\left\|\mathbf{N}\left(0, \frac{\pi}{3}\right)\right\| = \sqrt{\frac{75}{16} + \frac{25}{16}} = \frac{10}{4},$$

so that a unit vector in the direction of $\mathbf{N}\left(0, \pi/3\right)$ is

$$\mathbf{n} = \frac{1}{10}\left(5\mathbf{i} + 5\sqrt{3}\mathbf{k}\right) = \frac{1}{2}\mathbf{i} + \frac{\sqrt{3}}{2}\mathbf{k}.$$

We have

$$\mathbf{\Phi}\left(0, \frac{\pi}{3}\right) = \mathbf{\Phi}\left(0, \pi/3\right) = \left(\frac{5}{2}, 0, \frac{\sqrt{3}}{2}\right).$$

The required tangent plane consists of points $P = (x, y, z)$ such that

$$\left(x - \frac{5}{2}\right)\left(\frac{1}{2}\right) + \left(z - \frac{\sqrt{3}}{2}\right)\left(\frac{\sqrt{3}}{2}\right) = 0.$$

□

The Orientation of a Surface and an Expression for the Normal Vector (optional)

We have

$$\mathbf{N}\left(u, v\right) = \frac{\partial \mathbf{\Phi}}{\partial u}\left(u, v\right) \times \frac{\partial \mathbf{\Phi}}{\partial v}\left(u, v\right)$$

$$= \begin{vmatrix} \mathbf{i} & \mathbf{j} & \mathbf{k} \\ x_u & y_u & z_u \\ x_v & y_v & z_v \end{vmatrix}$$

$$= \begin{vmatrix} y_u & z_u \\ y_v & z_v \end{vmatrix}\mathbf{i} - \begin{vmatrix} x_u & z_u \\ x_v & z_v \end{vmatrix}\mathbf{j} + \begin{vmatrix} x_u & y_u \\ x_v & y_v \end{vmatrix}\mathbf{k}$$

$$= \left(y_u z_v - y_v z_u\right)\mathbf{i} + \left(x_v z_u - x_u z_v\right)\mathbf{j} + \left(x_u y_v - x_v y_u\right)\mathbf{k}.$$

We introduce the notation

$$\frac{\partial (y, z)}{\partial (u, v)} = \begin{vmatrix} y_u & z_u \\ y_v & z_v \end{vmatrix}, \quad \frac{\partial (z, x)}{\partial (u, v)} = \begin{vmatrix} z_u & x_u \\ z_v & x_v \end{vmatrix}, \quad \frac{\partial (x, y)}{\partial (u, v)} = \begin{vmatrix} x_u & y_u \\ x_v & y_v \end{vmatrix}$$

(read, **Jacobian of** y **and** z **with respect to** u **and** v, etc.). Thus, we can express $N(u, v)$ as

$$N(u, v) = \frac{\partial (y, z)}{\partial (u, v)} \mathbf{i} + \frac{\partial (z, x)}{\partial (u, v)} \mathbf{j} + \frac{\partial (x, y)}{\partial (u, v)} \mathbf{k}.$$

Example 15 Let ρ, ϕ and θ be spherical coordinates. For a given ϕ_0, the cone $\phi = \phi_0$ can be parametrized by the function

$$\Phi(\rho, \theta) = (\rho \sin (\phi_0) \cos (\theta), \rho \sin (\phi_0) \sin (\theta), \rho \cos (\phi_0)),$$

where $\rho \geq 0$ and $0 \leq \theta \leq 2\pi$, as in Example 11.
a) Use the above expression to determine $N(\rho, \theta)$.
b) Determine the tangent plane to the cone at $\Phi(2, \pi/2)$.

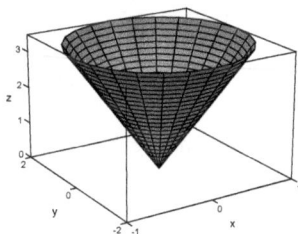

Figure 17

Solution

a) We have

$$\frac{\partial (y, z)}{\partial (\rho, \theta)} = \begin{vmatrix} \partial_\rho y (\rho, \theta) & \partial_\rho z (\rho.\theta) \\ \partial_\theta y (\rho.\theta) & \partial_\theta z (\rho, \theta) \end{vmatrix}$$

$$= \begin{vmatrix} \partial_\rho (\rho \sin (\phi_0) \sin (\theta)) & \partial_\rho (\rho \cos (\phi_0)) \\ \partial_\theta (\rho \sin (\phi_0) \sin (\theta)) & \partial_\theta (\rho \cos (\phi_0)) \end{vmatrix}$$

$$= \begin{vmatrix} \sin (\phi_0) \sin (\theta) & \cos (\phi_0) \\ \rho \sin (\phi_0) \cos (\theta) & 0 \end{vmatrix}$$

$$= -\rho \sin (\phi_0) \cos (\phi_0) \cos (\theta),$$

$$\frac{\partial (z, x)}{\partial (\rho, \theta)} = \begin{vmatrix} \partial_\rho (\rho \cos (\phi_0)) & \partial_\rho (\rho \sin (\phi_0) \cos (\theta)) \\ \partial_\theta (\rho \cos (\phi_0)) & \partial_\theta (\rho \sin (\phi_0) \cos (\theta)) \end{vmatrix}$$

$$= \begin{vmatrix} \cos (\phi_0) & \sin (\phi_0 \sin (\phi)) \\ 0 & -\rho \sin (\phi_0) \sin (\theta) \end{vmatrix}$$

$$= -\rho \sin (\phi_0) \cos (\phi_0) \sin (\theta),$$

$$\frac{\partial\left(x,y\right)}{\partial\left(\rho,\theta\right)}=\begin{vmatrix} \partial_\rho\left(\rho\sin\left(\phi_0\right)\cos\left(\theta\right)\right) & \partial_\rho\left(\rho\sin\left(\phi_0\right)\sin\left(\theta\right)\right) \\ \partial_\theta\left(\rho\sin\left(\phi_0\cos\left(\theta\right)\right)\right) & \partial_\theta\left(\rho\sin\left(\phi_0\right)\sin\left(\theta\right)\right) \end{vmatrix}$$

$$=\begin{vmatrix} \sin\left(\phi_0\right)\cos\left(\theta\right) & \sin\left(\phi_0\right)\sin\left(\theta\right) \\ -\rho\sin\left(\phi_0\right)\sin\left(\theta\right) & \rho\sin\left(\phi_0\right)\cos\left(\theta\right) \end{vmatrix}$$

$$=\rho\sin^2\left(\phi_0\right)\cos^2\left(\theta\right)+\rho\sin^2\left(\phi_0\right)\sin^2\left(\theta\right)$$

$$=\rho\sin^2\left(\phi_0\right).$$

Therefore,

$$\mathbf{N}\left(\rho,\theta\right)=\frac{\partial\left(y,z\right)}{\partial\left(\rho,\theta\right)}\mathbf{i}+\frac{\partial\left(z,x\right)}{\partial\left(\rho,\theta\right)}\mathbf{j}+\frac{\partial\left(x,y\right)}{\partial\left(\rho.\theta\right)}\mathbf{k}$$

$$=-\rho\sin\left(\phi_0\right)\cos\left(\phi_0\right)\cos\left(\theta\right)\mathbf{i}-\rho\sin(\phi_0)\cos\left(\phi_0\right)\sin\left(\theta\right)\mathbf{j}+\rho\sin^2\left(\phi_0\right)\mathbf{k}.$$

□

Let $\mathbf{T}:D^*\subset\mathbb{R}^2\to D\subset\mathbb{R}^2$ be the transformation such that

$$\mathbf{T}\left(u^*,v^*\right)=\left(u\left(u^*,v^*\right),v\left(u^*,v^*\right)\right),$$

Recall that **the Jacobian of T** is

$$\frac{\partial\left(u,v\right)}{\partial\left(u^*,v^*\right)}=\begin{vmatrix} \dfrac{\partial u}{\partial u^*} & \dfrac{\partial u}{\partial v^*} \\ \dfrac{\partial v}{\partial u^*} & \dfrac{\partial v}{\partial v^*} \end{vmatrix}.$$

Definition 2 Assume that the orientable surface S is parametrized by $\mathbf{\Phi}:D\subset\mathbb{R}^2\to\mathbb{R}^3$. We say that the function $\mathbf{\Phi}^*:D^*\subset\mathbb{R}^2\to\mathbb{R}^3$ is an **orientation-preserving parametrization** of S if there is a smooth (i.e., with continuous partial derivatives) one-one transformation $\mathbf{T}:D^*\to D$ such that $\mathbf{\Phi}^*=\mathbf{\Phi}\circ\mathbf{T}$ and the Jacobian of $\mathbf{T}$ is positive on D^*. The parametrization $\mathbf{\Phi}^*$ is said to be an **orientation-reversing parametrization** of S if the Jacobian of $\mathbf{T}$ is negative on D^*.

Problems

In the following problems, let S be the surface that is parametrized by the given function.
a) Determine the Normal to S at the given point P_0.
b) Determine the plane that is tangent to S at P_0.

1.

$$\Phi\left(u,v\right)=\left(v,3\cos\left(u\right),3\sin\left(u\right)\right),\ -4\leq v\leq 4,\ 0\leq u\leq 2\pi,$$

$$P_0=\Phi\left(\pi/6,1\right)$$

Note that S is part of the **cylinder** $y^2+z^2=9$ that has as its axis the x-axis.

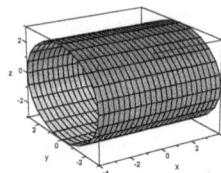

2.

$$\Phi(u, v) = \left(u^2, u\cos(v), u\sin(v)\right), \ u \geq 0, \ 0 \leq v \leq 2\pi,$$

$$P_0 = \Phi(3, \pi/3)$$

Note that S is part of the paraboloid: $x = y^2 + z^2$.

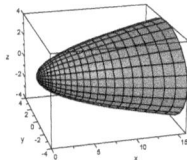

3.

$$\Phi(r, \theta) = \left(3r\cos(\theta), r^2, 2r\sin(\theta)\right), \ r \geq 0, \ 0 \leq \theta \leq \pi,$$

$$P_0 = \Phi(2, \pi/4)$$

Note that S is part of the elliptic paraboloid

$$\frac{x^2}{9} + \frac{z^2}{4} = y$$

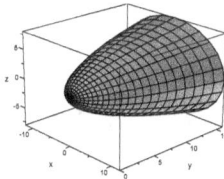

4.

$$\Phi(u, v) = (\cosh(u)\cos(v), \sinh(u), \cosh(u)\sin(v)), \ 0 \leq v \leq 2\pi, \ -2 \leq u \leq 2,$$

$$P_0 = \Phi(1, \pi/2)$$

Note that S is part of the hyperboloid

$$x^2 + z^2 - y^2 = 1$$

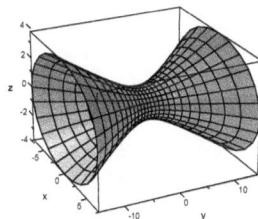

5.

$$\Phi\left(\rho,\theta\right) = \left(\rho\sin\left(\frac{\pi}{6}\right)\cos\left(\theta\right),\rho\sin\left(\frac{\pi}{6}\right)\sin\left(\theta\right),\rho\cos\left(\frac{\pi}{6}\right)\right)$$

$$= \left(\frac{1}{2}\rho\cos\left(\theta\right),\frac{1}{2}\rho\sin\left(\theta\right),\frac{\sqrt{3}}{2}\rho\right)$$

where $0 \le \rho \le 4$ and $0 \le \theta \le 2\pi$,

$$P_0 = \Phi\left(2,\pi/3\right)$$

Note that S is part of the cone $\phi = \pi/6$ (ρ, ϕ and θ are spherical coordinates).

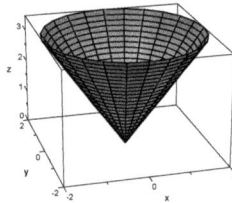

6.

$$\Phi\left(\phi,\theta\right) = \left(2\sin\left(\phi\right)\cos\left(\theta\right),\sin\left(\phi\right)\sin\left(\theta\right),\frac{1}{3}\cos\left(\phi\right)\right),$$

where $0 \le \phi \le \pi$ and $0 \le \theta \le 2\pi$,

$$P = \Phi\left(\pi/4,\pi/2\right)$$

Note that S is the ellipsoid

$$\frac{x^2}{4} + y^2 + 9z^2 = 1.$$

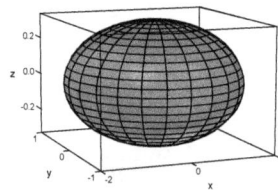

7.

$$\Omega\left(x,\theta\right) = \left(x, e^x\sin\left(\theta\right), e^x\cos\left(\theta\right)\right),$$

where $0 \le x \le 2$, $0 \le \theta \le 2\pi$,

$$P = \Omega\left(1,\pi/6\right)$$

Note that S is part of the surface of revolution that is obtained by revolving the graph of $z = e^x$ about the x-axis.

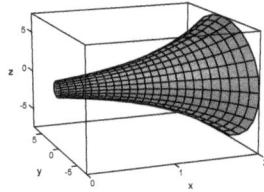

8.

$$\Omega\left(x,\theta\right)=\left(x\cos\left(\theta\right),x\sin\left(\theta\right),e^{x}\right),$$

where $0 \le x \le 2$ and $0 \le \theta \le 2\pi$,

$$P = \Omega\left(1,\pi/2\right)$$

Note that S is part of the surface of revolution that is obtained by revolving the graph of $z = e^{x}$ about the z-axis.

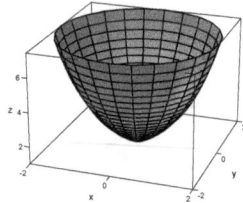

14.5 Surface Integrals

In this section we will compute the area of a parametrized surface and define the integrals of scalar functions and vector fields on such a surface. We will consider only orientable surfaces, as we discussed in Section 14.4.

Surface Area

We will motivate the definition of the area of a surface. Assume that the surface M is parametrized by the function $\mathbf{\Phi} : D \subset \mathbb{R}^{2} \to \mathbb{R}^{3}$, such that

$$\mathbf{\Phi}(u,v)=\left(x\left(u,v\right),y\left(u,v\right),z\left(u,v\right)\right),\ \left(u,v\right)\in D.$$

We assume that $\mathbf{\Phi}$ is smooth, i.e., the tangent vectors

$$\frac{\partial\mathbf{\Phi}}{\partial u}\left(u,v\right)=\frac{\partial x}{\partial u}\left(u,v\right)\mathbf{i}+\frac{\partial y}{\partial u}\left(u,v\right)\mathbf{j}+\frac{\partial z}{\partial u}\left(u,v\right)\mathbf{k}$$

and

$$\frac{\partial\mathbf{\Phi}}{\partial v}\left(u,v\right)=\frac{\partial x}{\partial v}\left(u,v\right)\mathbf{i}+\frac{\partial y}{\partial v}\left(u,v\right)\mathbf{j}+\frac{\partial z}{\partial v}\left(u,v\right)\mathbf{k}$$

are continuous on D, and the normal vector

$$\mathbf{N}\left(u,v\right)=\frac{\partial\mathbf{\Phi}}{\partial u}\left(u,v\right)\times\frac{\partial\mathbf{\Phi}}{\partial v}\left(u,v\right)\neq\mathbf{0}$$

for each $(u, v) \in D$. Let (u, v) be a point in the interior of D and assume that Δu and Δv are positive numbers that are small enough so that the rectangle $D(u, v, \Delta u, \Delta v)$ determined by the points (u, v) and $(u + \Delta u, v + \Delta v)$ is contained in the interior of D. The vectors

$$\Delta u \frac{\partial \mathbf{\Phi}}{\partial u}(u, v) = \Delta u \frac{\partial x}{\partial u}(u, v)\, \mathbf{i} + \Delta u \frac{\partial y}{\partial u}(u, v)\, \mathbf{j} + \Delta u \frac{\partial z}{\partial u}(u, v)\, \mathbf{k},$$

and

$$\Delta v \frac{\partial \mathbf{\Phi}}{\partial v}(u, v) = \Delta v \frac{\partial x}{\partial v}(u, v)\, \mathbf{i} + \Delta v \frac{\partial y}{\partial v}(u, v)\, \mathbf{j} + \Delta v \frac{\partial z}{\partial v}(u, v)\, \mathbf{k}$$

are tangent to S at $\mathbf{\Phi}(u, v)$. Since Δu and Δv are small,

$$\mathbf{\Phi}(u + \Delta u, v) - \mathbf{\Phi}(u, v) \cong \Delta u \frac{\partial \mathbf{\Phi}}{\partial u}(u, v)$$

and

$$\mathbf{\Phi}(u, v + \Delta v) - \mathbf{\Phi}(u, v) \cong \Delta v \frac{\partial \mathbf{\Phi}}{\partial v}(u, v).$$

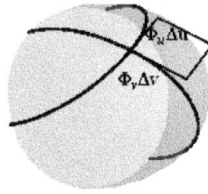

Figure 1

It is reasonable to approximate the area of the part of M corresponding to the rectangle $D(u, v, \Delta u, \Delta v)$ by the area of the part of the tangent plane at $\mathbf{\Phi}(u, v)$ that is spanned by

$$\Delta u \frac{\partial \mathbf{\Phi}}{\partial u}(u, v) \text{ and } \Delta v \frac{\partial \mathbf{\Phi}}{\partial v}(u, v).$$

Recall that this area can be expressed as

$$\left\| \Delta u \frac{\partial \mathbf{\Phi}}{\partial u}(u, v) \times \Delta v \frac{\partial \mathbf{\Phi}}{\partial v}(u, v) \right\| = \left\| \frac{\partial \mathbf{\Phi}}{\partial u}(u, v) \times \frac{\partial \mathbf{\Phi}}{\partial v}(u, v) \right\| \Delta u \Delta v$$

$$= \| \mathbf{N}(u, v) \| \Delta u \Delta v.$$

Therefore, it makes sense to define the area of the surface as the integral of the magnitude of the normal vector:

Definition 1 Assume that the surface M is parametrized by the smooth function $\mathbf{\Phi} : D \subset \mathbf{R}^2 \to \mathbf{R}^3$. The area of M is

$$\int \int_D \| \mathbf{N}(u, v) \|\, du dv = \int \int_D \left\| \frac{\partial \mathbf{\Phi}}{\partial u}(u, v) \times \frac{\partial \mathbf{\Phi}}{\partial v}(u, v) \right\| du dv.$$

Example 1 Confirm that the area of a sphere of radius ρ_0 is $4\pi\rho_0^2$.

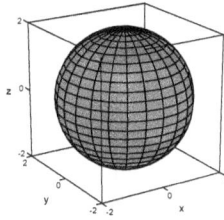

Figure 2

Solution

We can center the sphere M at the origin and parametrize it by the function

$$\boldsymbol{\Phi}\left(\phi,\theta\right) = \left(\rho_0 \sin\left(\phi\right)\cos\left(\theta\right), \rho_0 \sin\left(\phi\right)\sin\left(\theta\right), \rho_0 \cos\left(\phi\right)\right),$$

where $0 \leq \phi \leq \pi$ and $0 \leq \theta \leq 2$. As in Example 9 of Section 14.4,

$$\mathbf{N}\left(\phi,\theta\right) = \rho_0^2 \sin\left(\phi\right)\left(\sin\left(\phi\right)\cos\left(\theta\right)\mathbf{i} + \sin\left(\phi\right)\sin\left(\theta\right)\mathbf{j} + \cos\left(\phi\right)\mathbf{k}\right)$$

Therefore,

$$
\begin{aligned}
||\mathbf{N}\left(\phi,\theta\right)|| &= \rho_0^2 \sin\left(\phi\right)\sqrt{\sin^2\left(\phi\right)\cos^2\left(\theta\right) + \sin^2\left(\phi\right)\sin^2\left(\theta\right) + \cos^2\left(\phi\right)} \\
&= \rho_0^2 \sin\left(\phi\right)\sqrt{\sin^2\left(\phi\right)\left(\cos^2\left(\theta\right) + \sin^2\left(\theta\right)\right) + \cos^2\left(\phi\right)} \\
&= \rho_0^2 \sin\left(\phi\right)\sqrt{\sin^2\left(\phi\right) + \cos^2\left(\phi\right)} \\
&= \rho_0^2 \sin\left(\phi\right).
\end{aligned}
$$

Let $D = [0,\pi]\times[0,2\pi]$ denote the domain of $\boldsymbol{\Phi}$ in the $\phi\theta$-plane. The area of the sphere is

$$
\begin{aligned}
\int\int_D ||N\left(\phi,\theta\right)||\, d\phi d\theta = \int\int_D \rho_0^2 \sin\left(\phi\right)d\phi d\theta &= \rho_0^2 \int_{\phi=0}^{\phi=\pi}\int_{\theta=0}^{\theta=2\pi}\sin\left(\phi\right)d\theta d\phi \\
&= \rho_0^2 \int_{\phi=0}^{\phi=\pi} 2\pi \sin\left(\phi\right)d\phi \\
&= 2\pi\rho_0^2\left(-\cos\left(\phi\right)\big|_0^\pi\right) \\
&= 2\pi\rho_0^2\left(2\right) = 4\pi\rho_0^2,
\end{aligned}
$$

as claimed. $\square$

Remark 1 Assume that M is the graph of $f : D \to \mathbb{R}$, where D is a region in the plane. M can be parametrized by $\boldsymbol{\Phi} : D \to \mathbb{R}^3$, where

$$\boldsymbol{\Phi}\left(x,y\right) = \left(x,y,f\left(x,y\right)\right)$$

for each $(x,y) \in D$. In Section 14.4 we noted that

$$\mathbf{N}\left(x,y\right) = -\frac{\partial f}{\partial x}\left(x,y\right)\mathbf{i} - \frac{\partial f}{\partial y}\left(x,y\right)\mathbf{j} + \mathbf{k}.$$

Therefore,

$$\|\mathbf{N}(x,y)\| = \sqrt{\left(\frac{\partial f}{\partial x}\right)^2 + \left(\frac{\partial f}{\partial y}\right)^2 + 1}.$$

Therefore, the area of S can be expressed as

$$\int\int_D \sqrt{\left(\frac{\partial f}{\partial x}\right)^2 + \left(\frac{\partial f}{\partial y}\right)^2 + 1}\, dxdy.$$

$\Diamond$

Example 2 Let $f(x,y) = x^2 - y^2$, and let M be the part of the graph of f corresponding to the unit disk in the xy-plane (i.e., the disk of radius 1 centered at the origin). Determine the area of M.

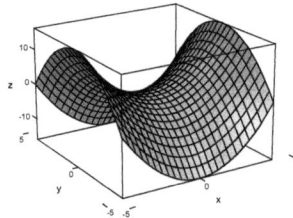

Figure 3

Solution

We have

$$\frac{\partial f}{\partial x}(x,y) = 2x \text{ and } \frac{\partial f}{\partial y}(x,y) = -2y.$$

Therefore,

$$\sqrt{\left(\frac{\partial f}{\partial x}\right)^2(x,y) + \left(\frac{\partial f}{\partial y}(x,y)\right)^2(x,y) + 1} = \sqrt{(2x)^2 + (-2y)^2 + 1} = \sqrt{4x^2 + 4y^2 + 1}$$

Thus, the area of S is

$$\int\int_D \sqrt{4x^2 + 4y^2 + 1}\, dxdy,$$

where D is the unit disk. In polar coordinates r and θ,

$$\int\int_D \sqrt{4x^2 + 4y^2 + 1}\, dxdy = \int_{\theta=0}^{2\pi}\int_{r=0}^{1} \sqrt{4r^2 + 1}\, rdrd\theta$$

$$= 2\pi \int_{r=0}^{1} \sqrt{4r^2 + 1}\, rdr$$

$$= 2\pi \left(\frac{1}{12}\left(4r^2 + 1\right)^{3/2}\Big|_0^1\right) = 2\pi\left(\frac{5}{12}\sqrt{5} - \frac{1}{12}\right)$$

Example 3 Let M be the part of the cylinder that is parametrized by

$$\boldsymbol{\Phi}(\theta, z) = (a\cos(\theta), a\sin(\theta), z),$$

where $0 \le \theta \le 2\pi$ and $0 \le z \le h$. Determine the area of S.

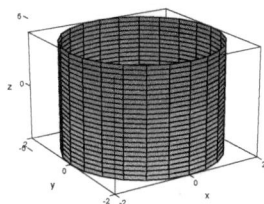

Figure 4

Solution

As in Example 11 of Section 14.4,

$$\mathbf{N}(\theta, z) = a\cos(\theta)\mathbf{i} + a\sin(\theta)\mathbf{j}.$$

Therefore,

$$||N(\theta, z)|| = \sqrt{a^2\cos^2(\theta) + a^2\sin^2(\theta)} = a.$$

Thus, the area of S is

$$\int_{\theta=0}^{2\pi}\int_{z=0}^{h} a\,dz\,d\theta = 2\pi ah,$$

as it should be (the perimeter of the base times height). $\square$

Example 4 Assume that the torus M is parametrized by

$$\boldsymbol{\Phi}(\theta, \phi) = ((a + b\cos(\phi))\cos(\theta), (a + b\cos(\phi))\sin(\theta), b\sin(\phi)),$$

where $0 \le \theta \le 2\pi$ and $0 \le \phi \le 2\pi$. Determine the area of S.

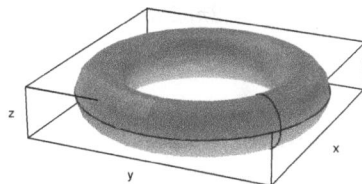

Figure 5

Solution

We have

$$\frac{\partial\boldsymbol{\Phi}}{\partial\theta}(\theta, \phi) = \frac{\partial}{\partial\theta}\left((a + b\cos(\phi))\cos(\theta)\right)\mathbf{i} + \frac{\partial}{\partial\theta}\left((a + b\cos(\phi))\sin(\theta)\right)\mathbf{j} + b\frac{\partial}{\partial\theta}\sin(\phi)\mathbf{k}$$
$$= -(a + b\cos(\phi))\sin(\theta)\mathbf{i} + (a + b\cos(\phi))\cos(\theta)\mathbf{j},$$

and

$$\frac{\partial\boldsymbol{\Phi}}{\partial\phi}(\theta, \phi) = \frac{\partial}{\partial\phi}\left((a + b\cos(\phi))\cos(\theta)\right)\mathbf{i} + \frac{\partial}{\partial\phi}\left((a + b\cos(\phi))\sin(\theta)\right)\mathbf{j} + b\frac{\partial}{\partial\phi}\sin(\phi)\mathbf{k}$$
$$= -b\sin(\phi)\cos(\theta)\mathbf{i} - b\sin(\phi)\sin(\theta)\mathbf{j} + b\cos(\phi)\mathbf{k}.$$

Therefore,

$$\mathbf{N}(\theta, \phi) = \frac{\partial \mathbf{\Phi}}{\partial \theta}(\theta, \phi) \times \frac{\partial \mathbf{\Phi}}{\partial \phi}(\theta, \phi)$$

$$= \begin{vmatrix} \mathbf{i} & \mathbf{j} & \mathbf{k} \\ -(a + b\cos(\phi))\sin(\theta) & (a + b\cos(\phi))\cos(\theta) & 0 \\ -b\sin(\phi)\cos(\theta) & -b\sin(\phi)\sin(\theta) & b\cos(\phi) \end{vmatrix}$$

$$= b(a + b\cos(\phi))(\cos(\theta)\cos(\phi)\mathbf{i} + \sin(\theta)\cos(\phi)\mathbf{j} + \sin(\phi)\mathbf{k})$$

Thus,

$$\|\mathbf{N}(\theta, \phi)\|^2 = b^2(a + b\cos(\phi))^2\left(\cos^2(\theta)\cos^2(\phi) + \sin^2(\theta)\cos^2(\phi) + \sin^2(\phi)\right)$$

$$= b^2(a + b\cos(\phi))^2\left(\cos^2(\phi) + \sin^2(\phi)\right)$$

$$= b^2(a + b\cos(\phi))^2.$$

Therefore,

$$\|\mathbf{N}(\theta, \phi)\| = b(a + b\cos(\phi))$$

Thus, the area of the torus is

$$\int\int_D \|\mathbf{N}(\theta, \phi)\| \, d\theta d\phi = \int\int_D b(a + b\cos(\phi)) \, d\theta d\phi,$$

where D is the square $[0, 2\pi] \times [0, 2\pi]$ in the $\phi\theta$-plane. Therefore,

$$\int\int_D b(a + b\cos(\phi)) \, d\theta d\phi = \int_{\phi=0}^{\phi=2\pi} \int_{\theta=0}^{\theta=2\pi} b(a + b\cos(\phi)) \, d\theta d\phi$$

$$= 2\pi b \int_{\phi=0}^{\phi=2\pi} (a + b\cos(\phi)) \, d\phi$$

$$= 2\pi b \left(a\phi + b\sin(\phi)\big|_0^{2\pi}\right)$$

$$= 2\pi b 2\pi a = 4\pi^2 ab.$$

□

Surface Integrals of Scalar Functions

Definition 2 Assume that the (orientable) surface M is parametrized by the function $\mathbf{\Phi} : D \subset \mathbb{R}^2 \to \mathbb{R}^3$, and that the scalar function f is continuous on M. We define the integral of f on M as

$$\int\int_D f(\mathbf{\Phi}(u, v)) \|\mathbf{N}(u, v)\| \, du dv.$$

We will denote the integral of f on M as

$$\int\int_M f dS.$$

Symbolically, dS is "the element of area" $\|\mathbf{N}(u, v)\| \, du dv$. You can imagine that dS represents the area of an "infinitesimal" portion of the surface M. If the surface M is a thin shell, and f is its mass density per unit area, the integral

$$\int\int_M f dS$$

yields the total mass of the shell. Soon we will discuss other contexts for the appearance of surface integrals.

Example 5 Let M be the hemisphere sphere of radius 2 that is parametrized by

$$\mathbf{\Phi}\left(\phi,\theta\right)=\left(2\sin\left(\phi\right)\cos\left(\theta\right),2\sin\left(\phi\right)\sin\left(\theta\right),2\cos\left(\phi\right)\right),$$

where $0\leq\phi\leq\pi/2$ and $0\leq\theta\leq 2\pi$ Assume that M is a thin shell with mass density

$$f\left(x,y,z\right)=4-z.$$

Determine the total mass of M.

Solution

Let $D=[0,\pi]\times[0,2\pi]$ denote the domain of $\mathbf{\Phi}$ in the $\phi\theta$-plane. As in Example 1,

$$\left\|\mathbf{N}\left(\phi,\theta\right)\right\|=2^2\sin\left(\phi\right)=4\sin\left(\phi\right).$$

Therefore, the mass of M is

$$\int\int_M f\,dS=\int\left(4-z\right)ds=\int\left(4-2\cos\left(\phi\right)\right)\left\|\mathbf{N}\left(\phi,\theta\right)\right\|d\phi d\theta$$

$$=\int\int_D\left(4-2\cos\left(\phi\right)\right)\left(4\sin\left(\phi\right)\right)d\phi d\theta$$

$$=8\int_{\phi=0}^{\phi=\pi/2}\int_{\theta=0}^{\theta=2\pi}\left(2\sin\left(\phi\right)-\cos\left(\phi\right)\sin\left(\phi\right)\right)d\phi d\theta$$

$$=16\pi\int_{\phi=0}^{\phi=\pi/2}\left(2\sin\left(\phi\right)-\cos\left(\phi\right)\sin\left(\phi\right)\right)d\phi$$

$$=16\pi\left(-2\cos\left(\phi\right)+\frac{1}{2}\cos^2\left(\phi\right)\Big|_0^{\pi/2}\right)$$

$$=16\pi\left(\frac{3}{2}\right)=24\pi.$$

(we made use of the substitution $u=\cos\left(\phi\right)$). $\square$

The integral of a scalar function on a surface M is independent of the parametrization of M. In particular, the calculation of the area of a surface does not change under an orientation preserving or reversing parametrization of M. You can find the proof at the end of this section.

Remark 2 As we noted before, if M is the graph of the function $g:D\subset\mathbb{R}^2\to\mathbb{R}$, then S can be parametrized by

$$\mathbf{\Phi}\left(x,y\right)=\left(x,y,g\left(x,y\right)\right),\ \ \left(x,y\right)\in D,$$

$$\mathbf{N}\left(x,y\right)=-\frac{\partial g}{\partial x}\left(x,y\right)\mathbf{i}-\frac{\partial g}{\partial y}\left(x,y\right)\mathbf{j}+\mathbf{k},$$

and

$$\left\|\mathbf{N}\left(x,y\right)\right\|=\sqrt{\left(\frac{\partial g}{\partial x}\right)^2+\left(\frac{\partial g}{\partial y}\right)^2+1}.$$

Therefore, the integral $\int\int_M f\,ds$ takes the form

$$\int\int_D f\left(x,y,g\left(x,y\right)\right)\left\|\mathbf{N}\left(x,y\right)\right\|dxdy=\int\int_D f\left(x,y,g\left(x,y\right)\right)\sqrt{\left(\frac{\partial g}{\partial x}\right)^2+\left(\frac{\partial g}{\partial y}\right)^2+1}dxdy.$$

The roles of the coordinates can be interchanged, of course. $\Diamond$

Example Let M be the part of the graph of $z = g(x,y) = 4 - x^2 - y^2$ over the xy-plane. Evaluate

$$\int\int_M \frac{1}{1 + 4(x^2 + y^2)}dS.$$

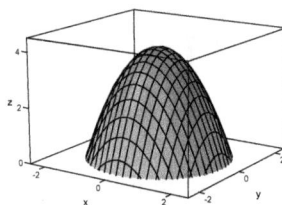

Figure 7

Solution

We have

$$4 - x^2 - y = 0 \Leftrightarrow x^2 + y^2 = 4.$$

Thus, the intersection of M with the xy-plane is the circle $x^2 + y^2 = 4$, and M is the graph of g over the disk D of radius 2 centered at the origin.
We have

$$\frac{\partial g}{\partial x}(x,y) = -2x \text{ and } \frac{\partial g}{\partial y}(x,y) = -2y.$$

Therefore,

$$\int\int_M \frac{1}{1 + 4(x^2 + y^2)}dS = \int\int_D \frac{1}{1 + 4(x^2 + y^2)}\sqrt{(-2x)^2 + (-2y^2) + 1}dxdy$$

$$= \int\int_D \frac{1}{1 + 4(x^2 + y^2)}\sqrt{4x^2 + 4y^2 + 1}dxdy$$

$$= \int\int_D \frac{1}{\sqrt{1 + 4(x^2 + y^2)}}dxdy.$$

It is convenient to transform the integral to polar coordinates:

$$\int\int_D \frac{1}{\sqrt{1 + 4(x^2 + y^2)}}dxdy = \int_{\theta=0}^{\theta=2\pi}\int_{r=0}^{r=2} \frac{1}{\sqrt{1 + 4r^2}}rdrd\theta$$

$$= 2\pi\int_{r=0}^{r=2} \frac{1}{\sqrt{1 + 4r^2}}rdr.$$

If we set $u = 1 + 4r^2$ then $du = 8rdr$, so that

$$\int_{r=0}^{r=2} \frac{1}{\sqrt{1 + 4r^2}}rdr = \frac{1}{8}\int_{u=1}^{u=17} \frac{1}{\sqrt{u}}du = \frac{1}{8}\int_{u=1}^{u=17} u^{-1/2}du$$

$$= \frac{1}{8}\left(2u^{1/2}\Big|_1^{17}\right)$$

$$= \frac{1}{4}\left(\sqrt{17} - 1\right).$$

Therefore,

$$\int\int_S \frac{1}{1+4\left(x^2+y^2\right)}dS = 2\pi \int_{r=0}^{r=2} \frac{1}{\sqrt{1+4r^2}}r\,dr$$

$$= 2\pi \left(\frac{1}{4}\left(\sqrt{17}-1\right)\right) = \frac{\pi}{2}\left(\sqrt{17}-1\right).$$

□

Flux Integrals

In order to motivate the definition of a flux integral, let us imagine that $\mathbf{v}\left(x,y,z\right)$ is the velocity of a fluid particle at $\left(x,y,z\right) \in \mathbb{R}^3$, and that M is a surface in $\mathbb{R}^3$. We would like to calculate the flow across S per unit time. Assume that a small patch of the surface is almost planar, has area dS, and $\mathbf{n}$ is the unit normal to that piece which points in the direction of the flow. The component of the velocity vector in the direction of $\mathbf{n}$ is $\mathbf{v}\cdot\mathbf{n}$. If the mass density of the fluid has the constant value δ, the mass of fluid that flows across that small patch is approximately $\delta\mathbf{v}\cdot\mathbf{n}dS$. Therefore it is reasonable to calculate the total flow across S per unit time as

$$\int\int_M \delta\mathbf{v}\cdot\mathbf{n}dS.$$

This leads to the definition of the flux integral of an arbitrary vector field over a surface:

Definition 3 Assume that $\mathbf{F}$ is a continuous vector field in $\mathbb{R}^3$ and M is a smooth surface that is parametrized by the function $\mathbf{\Phi}:D\subset\mathbb{R}^2 \to \mathbb{R}^3$. **The flux integral of F over M** is the integral of the normal component of $\mathbf{F}$ on MS, i.e.

$$\int\int_M \mathbf{F}\cdot\mathbf{n}dS$$

where $\mathbf{n}$ is the unit normal to M.

The "**vectorial element of area**" is the symbol

$$d\mathbf{S} = \mathbf{n}dS,$$

so that the flux integral of F over M can be denoted as.

$$\int\int_M \mathbf{F}\cdot d\mathbf{S}.$$

If

$$\mathbf{N}\left(u,v\right) = \frac{\partial\mathbf{\Phi}}{\partial u}\left(u,v\right) \times \frac{\partial\mathbf{\Phi}}{\partial v}\left(u,v\right),$$

then

$$\mathbf{n} = \frac{\mathbf{N}}{\|\mathbf{N}\|} \text{ and } dS = \|\mathbf{N}\|\,du\,dv.$$

Thus,

$$\int\int_M \mathbf{F}\cdot\mathbf{n}dS = \int\int_D \mathbf{F}\left(\mathbf{\Phi}\left(u,v\right)\right)\cdot\frac{\mathbf{N}\left(u,v\right)}{\|\mathbf{N}\left(u,v\right)\|}\|\mathbf{N}\left(u,v\right)\|\,du\,dv$$

$$= \int\int_D \mathbf{F}\left(\mathbf{\Phi}\left(u,v\right)\right)\cdot\mathbf{N}\left(u,v\right)\,du\,dv$$

$$= \int\int_D \mathbf{F}\left(\mathbf{\Phi}\left(u,v\right)\right)\cdot\left(\frac{\partial\mathbf{\Phi}}{\partial u}\left(u,v\right) \times \frac{\partial\mathbf{\Phi}}{\partial v}\left(u,v\right)\right)\,du\,dv.$$

Example 6 Consider the cylinder of radius 2 whose axis is along the z-axis. Let M be the portion of the cylinder between $z = 0$ and $z = 4$. As in Example 3, M can be parametrized by $\mathbf{\Phi}(\theta, z) = (2\cos(\theta), 2\sin(\theta), z)$, where $0 \leq \theta \leq 2\pi$ and $0 < z < 4$. Let $\mathbf{F}(x, y, z) = x\mathbf{i}$. Calculate the flux of $\mathbf{F}$ over M.

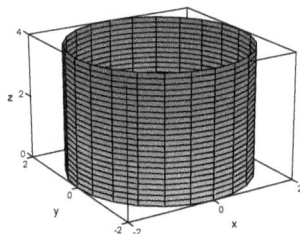

Figure 8

Solution

We have

$$\mathbf{F}(\mathbf{\Phi}(\theta, z)) = \mathbf{F}(2\cos(\theta), 2\sin(\theta), z) = 2\cos(\theta)\mathbf{i}.$$

As in Example 3,

$$\mathbf{N}(\theta, z) = 2\cos(\theta)\mathbf{i} + 2\sin(\theta)\mathbf{j}.$$

Therefore,

$$\mathbf{F}(\mathbf{\Phi}(\theta, z)) \cdot \mathbf{N}(\theta, z) = (2\cos(\theta)\mathbf{i}) \cdot (2\cos(\theta)\mathbf{i} + 2\sin(\theta)\mathbf{j}) = 4\cos^2(\theta).$$

Thus,

$$\int_M \mathbf{F} \cdot \mathbf{n}\, dS = \int \int_D \mathbf{F}(\mathbf{\Phi}(\theta, z)) \cdot \mathbf{N}(\theta, z)\, d\theta dz = \int_{z=0}^{4} \int_{\theta=0}^{2\pi} 4\cos^2(\theta)\, .d\theta dz$$

$$= 4 \int_{z=0}^{4} \left(\int_{\theta=0}^{\theta=2\pi} \cos^2(\theta)\, d\theta \right) dz$$

$$= 16 \int_{\theta=0}^{\theta=2\pi} \cos^2(\theta)\, d\theta$$

$$= 16 \left(\frac{1}{2}\cos(\theta)\sin(\theta) + \frac{1}{2}\theta \Big|_0^{2\pi} \right)$$

$$= 16\pi.$$

☐

Example 7 Let M be the sphere of radius ρ_0 that is centered at the origin. M can be parametrized by

$$\mathbf{\Phi}(\phi, \theta) = (\rho_0 \sin(\phi)\cos(\theta), \rho_0 \sin(\phi)\sin(\theta), \rho_0 \cos(\phi)),$$

where $0 \leq \phi \leq \pi$ and $0 \leq \theta \leq 2\pi$. Calculate

$$\int \int_M (x\mathbf{i} + y\mathbf{j} + z\mathbf{k}) \cdot d\mathbf{S}$$

Solution

$$\int\int_M (x\mathbf{i} + y\mathbf{j} + z\mathbf{k}) \cdot d\mathbf{S} = \int_{\theta=0}^{2\pi} \int_{\phi=0}^{\pi} \boldsymbol{\Phi}(\phi, \theta) \cdot \mathbf{N}(\phi, \theta)\, d\phi d\theta,$$

where

$$\mathbf{N}(\phi, \theta) = \rho_0^2 \sin(\phi)(\sin(\phi)\cos(\theta)\mathbf{i} + \sin(\phi)\sin(\theta)\mathbf{j} + \cos(\phi)\mathbf{k})$$
$$= \rho_0 \sin(\phi)\boldsymbol{\Phi}(\phi, \theta),$$

as in Example 1. Therefore,

$$\int_{\theta=0}^{2\pi} \int_{\phi=0}^{\pi} \boldsymbol{\Phi}(\phi, \theta) \cdot \mathbf{N}(\phi, \theta)\, d\phi d\theta\theta = \int_{\theta=0}^{2\pi} \int_{\phi=0}^{\pi} \boldsymbol{\Phi}(\phi, \theta) \cdot \rho_0 \sin(\phi)\boldsymbol{\Phi}(\phi, \theta)\, d\phi d\theta$$

$$= \rho_0 \int_{\theta=0}^{2\pi} \int_{\phi=0}^{\pi} \sin(\phi)\boldsymbol{\Phi}(\phi, \theta) \cdot \boldsymbol{\Phi}(\phi, \theta)\, d\phi d\theta$$

$$= \rho_0 \int_{\theta=0}^{2\pi} \int_{\phi=0}^{\pi} \sin(\phi) \|\boldsymbol{\Phi}(\phi, \theta)\|^2\, d\phi d\theta$$

$$= \rho_0 \int_{\theta=0}^{2\pi} \int_{\phi=0}^{\pi} \sin(\phi) \rho_0^2 d\phi d\theta$$

$$= \rho_0 (4\pi\rho_0^2) = 4\pi\rho_0^3.$$

□

Proposition 1 Assume that M is an orientable surface. The flux integral

$$\int_M \mathbf{F} \cdot \mathbf{n} dS$$

is independent of an orientation-preserving parametrization of M. The flux integral is multiplied by -1 under an orientation-reversing parametrization of M.

The proof of Proposition 2can be found at the end of this section.

Remark 3 If M is the graph of the function $g(x, y)$ over the domain D in the xy-plane, we can parametrize M by

$$\boldsymbol{\Phi}(x, y) = (x, y, g(x, y)),$$

where $(x, y) \in D$. As we noted above,

$$\mathbf{N}(x, y) = -g_x(x, y)\mathbf{i} - g_y(x, y)\mathbf{j} + \mathbf{k},$$

and

$$\|\mathbf{N}(x, y)\| = \sqrt{\left(\frac{\partial g}{\partial x}\right)^2 + \left(\frac{\partial g}{\partial y}\right)^2 + 1}.$$

Therefore,

$$
\iint_M \mathbf{F} \cdot d\mathbf{S} = \iint_M \mathbf{F} \cdot \mathbf{n} dS
$$

$$
= \iint_M \mathbf{F} \cdot \frac{\mathbf{N}}{\|\mathbf{N}\|} dS
$$

$$
= \iint_D \mathbf{F}(x, y, g(x, y)) \cdot \frac{\mathbf{N}(x, y)}{\|\mathbf{N}(x, y)\|} \|\mathbf{N}(x, y)\| \, dxdy
$$

$$
= \iint_D \mathbf{F}(x, y, g(x, y)) \cdot \mathbf{N}(x, y) \, dxdy
$$

$$
= \iint_D \mathbf{F}(x, y, g(x.y)) \cdot \left(-\frac{\partial g}{\partial x}(x, y) \mathbf{i} - \frac{\partial g}{\partial y}(x, y) \mathbf{j} + \mathbf{k} \right) dxdy..
$$

$\Diamond$

Example 8 Let $\boldsymbol{F}(x, y, z) = z\mathbf{k}$ and let M be the hemisphere of radius a centered at the origin. Calculate the flux integral of $\boldsymbol{F}$ across M in the direction of the outward normal.

Solution

The surface M is the graph of the function

$$
g(x, y) = \sqrt{a^2 - x^2 - y^2},
$$

where (x, y) is in the disk D_a of radius a centered at the origin. A normal that points outward is

$$
-g_x(x, y) \mathbf{i} - g_y(x, y) \mathbf{j} + \mathbf{k}. = \frac{x}{\sqrt{a^2 - x^2 - y^2}} \mathbf{i} + \frac{y}{\sqrt{a^2 - x^2 - y^2}} \mathbf{j} + \mathbf{k}
$$

On the hemisphere M,

$$
F(x, y, z) = F(x, y, g(x, y)) = g(x, y) \mathbf{k} = \sqrt{a^2 - x^2 - y^2} \mathbf{k}.
$$

Therefore,

$$
\iint_M \mathbf{F} \cdot d\mathbf{S} = \iint_{D_a} g(x, y) \mathbf{k} \cdot (-g_x(x, y) \mathbf{i} - g_y(x, y) \mathbf{j} + \mathbf{k}) \, dxdy
$$

$$
\iint_{D_a} \sqrt{a^2 - x^2 - y^2} \mathbf{k} \cdot \left(\frac{x}{\sqrt{a^2 - x^2 - y^2}} \mathbf{i} + \frac{y}{\sqrt{a^2 - x^2 - y^2}} \mathbf{j} + \mathbf{k} \right) dxdy
$$

$$
= \iint_{D_a} \sqrt{a^2 - x^2 - y^2} dxdy.
$$

In polar coordinates,

$$
\iint_{D_a} \sqrt{a^2 - x^2 - y^2} dxdy = \int_{\theta=0}^{2\pi} \int_{r=0}^{a} \sqrt{a^2 - r^2} r dr d\theta
$$

$$
= 2\pi \left(\int_{r=0}^{a} \sqrt{a^2 - r^2} r dr \right)
$$

$$
= 2\pi \left(\frac{a^3}{3} \right) = \frac{2\pi a^3}{3}.
$$

$\square$

Example 9 Assume that the electrostatic field due to the charge q at the origin is

$$\mathbf{F}(x, y, z) = q \frac{x\mathbf{i} + y\mathbf{j} + z\mathbf{k}}{(x^2 + y^2 + z^2)^{3/2}} = q \frac{\mathbf{r}}{||\mathbf{r}||^3}$$

Calculate the flux of $\boldsymbol{F}$ out of the sphere M of radius ρ_0 centered at the origin.

Solution

$$\int\int_M \mathbf{F} \cdot \mathbf{n} dS = \int\int_M q \frac{\mathbf{r}}{||\mathbf{r}||^3} \cdot \frac{\mathbf{r}}{||\mathbf{r}||} dS$$

$$= q \int\int_M \frac{||\mathbf{r}||^2}{||\mathbf{r}||^4} dS = q \int\int_M \frac{1}{||\mathbf{r}||^2} dS = \frac{q}{\rho_0^2} \int\int_M dS = \frac{q}{\rho_0^2} \left(4\pi\rho_0^2\right) = 4\pi q$$

□

Example 10 Let

$$\mathbf{F}(x, y, z) = \frac{-y\mathbf{i} + x\mathbf{j}}{x^2 + y^2}.$$

Determine the flux of $\mathbf{F}$ in the direction of $\mathbf{j}$ across the rectangle M in the xz-plane that has the vertices $(1, 0, 0)$ and $(3, 0, 2)$

Solution

The unit normal to M is $\mathbf{j}$. We have

$$\int\int_M \mathbf{F} \cdot \mathbf{n} dS = \int_{z=0}^2 \int_{x=1}^3 \frac{1}{x} dx dz = 2 \ln(3)$$

□

Alternative Notations and some Proofs (Optional)

At the end of Section 14.4. we saw that

$$\mathbf{N}(u, v) = \frac{\partial(y, z)}{\partial(u, v)} \mathbf{i} + \frac{\partial(z, x)}{\partial(u, v)} \mathbf{j} + \frac{\partial(x, y)}{\partial(u, v)} \mathbf{k},$$

where

$$\frac{\partial(y, z)}{\partial(u, v)} = \begin{vmatrix} y_u & z_u \\ y_v & z_v \end{vmatrix}, \quad \frac{\partial(z, x)}{\partial(u, v)} = \begin{vmatrix} z_u & x_u \\ z_v & x_v \end{vmatrix}, \quad \frac{\partial(x, y)}{\partial(u, v)} = \begin{vmatrix} x_u & y_u \\ x_v & y_v \end{vmatrix}$$

Therefore, we can express **the area of the surface** that is parametrized by $(x(u, v), y(u, v), z(u, v))$, where $(u, v) \in D$, as

$$\int\int_D ||\mathbf{N}(u, v)|| \, dudv = \int\int_D \sqrt{\left(\frac{\partial(y, z)}{\partial(u, v)}\right)^2 + \left(\frac{\partial(z, x)}{\partial(u, v)}\right)^2 + \left(\frac{\partial(x, y)}{\partial(u, v)}\right)^2} \, dudv.$$

Example 11 Consider the cone that is parametrized by

$$\mathbf{\Phi}(\rho, \theta) = (\rho \sin(\phi_0) \cos(\theta), \rho \sin(\phi_0) \sin(\theta)), \rho \cos(\phi_0).$$

Here ϕ_0 is a given angle, $\rho \geq 0$ and $0 \leq \theta \leq 2\pi$. Let M be the part of the cone of slant height l, i.e., $0 \leq \rho \leq l$. Determine the area of M.

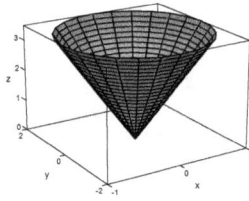

Figure 9

Solution

As in Example 14 of Section 14.4,

$$\mathbf{N}\left(\rho,\theta\right) = \frac{\partial\left(y,z\right)}{\partial\left(\rho,\theta\right)}\mathbf{i} + \frac{\partial\left(z,x\right)}{\partial\left(\rho,\theta\right)}\mathbf{j} + \frac{\partial\left(x,y\right)}{\partial\left(\rho.\theta\right)}\mathbf{k}$$
$$= -\rho\sin\left(\phi_0\right)\cos\left(\phi_0\right)\cos\left(\theta\right)\mathbf{i} - \rho\sin(\phi_0)\cos\left(\phi_0\right)\sin\left(\theta\right)\mathbf{j} + \rho\sin^2\left(\phi_0\right)\mathbf{k}.$$

Therefore,

$$\left\|\mathbf{N}\left(\rho,\theta\right)\right\| = \sqrt{\rho^2\sin^2\left(\phi_0\right)\cos^2\left(\phi_0\right)\cos^2\left(\theta\right) + \rho^2\sin^2\left(\phi_0^2\right)\cos^2\left(\phi_0\right)\sin^2\left(\theta\right) + \rho^2\sin^4\left(\phi_0\right)}$$
$$= \sqrt{\rho^2\sin^2\left(\phi_0\right)\cos^2\left(\phi_0\right) + \rho^2\sin^4\left(\phi_0\right)}$$
$$= \rho\sin\left(\phi_0\right)\sqrt{\cos^2\left(\phi_0\right) + \sin^2\left(\phi\right)} = \rho\sin\left(\phi_0\right)$$

Thus, the area of S is

$$\int_{\theta=0}^{2\pi}\int_{\rho=0}^{l}\rho\sin\left(\phi_0\right)d\rho d\theta = 2\pi\sin\left(\phi_0\right)\int_{\rho=0}^{l}\rho d\rho = 2\pi\sin\left(\phi_0\right)\left(\frac{l^2}{2}\right)$$
$$= \pi l^2\sin\left(\phi_0\right).$$

Note that the radius of the base is $l\sin\left(\phi_0\right)$, so that the above expression is $\pi\times$ slant height $\times$ radius of the base. $\square$

Proposition 2 The surface integral

$$\int\int_M f dS$$

is independent of an orientation-preserving or orientation reversing parametrization of the surface M.

Proof

Assume that M is parametrized by $\mathbf{\Phi}$,

$$\mathbf{\Phi}\left(u,v\right) = \left(x\left(u,v\right),y\left(u,v\right).z\left(u,v\right)\right), \text{ for each } \left(u,v\right) \in D,$$

and

$$u = u\left(u^*,v^*\right),\ v = v\left(u^*,v^*\right)$$

for each $\left(u^*,v^*\right) \in D^*$. Set

$$x\left(u^*,v^*\right) = x\left(u\left(u^*,v^*\right),v\left(u^*,v^*\right)\right),\ y\left(u^*,v^*\right) = y\left(u\left(u^*,v^*\right),v\left(u^*,v^*\right)\right)$$
$$z\left(u^*,v^*\right) = z\left(u\left(u^*,v^*\right),v\left(u^*,v^*\right)\right),$$

and
$$\mathbf{\Phi}^*\left(u^*,v^*\right)=\left(x\left(u^*,v^*\right),y\left(u^*,v^*\right).z\left(u^*,v^*\right)\right),\text{ where }\left(u^*,v^*\right)\in D^*.$$

If we make use of the parametrization $\mathbf{\Phi}^*$, the surface integral can be expressed as

$$\iint_{D^*} f\left(\mathbf{\Phi}^*\left(u^*,v^*\right)\right)\left\|\mathbf{N}\left(u^*,v^*\right)\right\|du^*dv^*$$
$$=\iint_{D^*} f\left(\mathbf{\Phi}^*\left(u^*,v^*\right)\right)\sqrt{\left(\frac{\partial\left(y,z\right)}{\partial\left(u^*,v^*\right)}\right)^2+\left(\frac{\partial\left(z,x\right)}{\partial\left(u^*,v^*\right)}\right)^2+\left(\frac{\partial\left(x,y\right)}{\partial\left(u^*,v^*\right)}\right)^2}\,du^*dv^*$$

As a consequence of the chain rule,

$$\frac{\partial\left(y,z\right)}{\partial\left(u^*,v^*\right)}=\frac{\partial\left(y,z\right)}{\partial\left(u,v\right)}\frac{\partial\left(u,v\right)}{\partial\left(u^*,v^*\right)},$$
$$\frac{\partial\left(z,x\right)}{\partial\left(u^*,v^*\right)}=\frac{\partial\left(z,x\right)}{\partial\left(u,v\right)}\frac{\partial\left(u,v\right)}{\partial\left(u^*,v^*\right)},$$
$$\frac{\partial\left(x,y\right)}{\partial\left(u^*,v^*\right)}=\frac{\partial\left(x,y\right)}{\partial\left(u,v\right)}\frac{\partial\left(u,v\right)}{\partial\left(u^*,v^*\right)}.$$

Therefore,

$$\iint_{D^*} f\left(\mathbf{\Phi}^*\left(u^*,v^*\right)\right)\left\|\mathbf{N}\left(u^*,v^*\right)\right\|du^*dv^*$$
$$=\iint_{D^*}\int_{D^*} f\left(\mathbf{\Phi}^*\left(u^*,v^*\right)\right)\sqrt{\left(\frac{\partial\left(y,z\right)}{\partial\left(u^*,v^*\right)}\right)^2+\left(\frac{\partial\left(z,x\right)}{\partial\left(u^*,v^*\right)}\right)^2+\left(\frac{\partial\left(x,y\right)}{\partial\left(u^*,v^*\right)}\right)^2}\,du^*dv^*$$
$$=\iint_{D^*}\int_{D^*} f\left(\mathbf{\Phi}^*\left(u^*,v^*\right)\right)\sqrt{\left(\frac{\partial\left(y,z\right)}{\partial\left(u,v\right)}\right)^2+\left(\frac{\partial\left(z,x\right)}{\partial\left(u,v\right)}\right)^2+\left(\frac{\partial\left(x,y\right)}{\partial\left(u,v\right)}\right)^2}\left|\frac{\partial\left(u,v\right)}{\partial\left(u^*,v^*\right)}\right|du^*dv^*$$
$$=\iint_{D} f\left(\mathbf{\Phi}\left(u,v\right)\right)\sqrt{\left(\frac{\partial\left(y,z\right)}{\partial\left(u,v\right)}\right)^2+\left(\frac{\partial\left(z,x\right)}{\partial\left(u,v\right)}\right)^2+\left(\frac{\partial\left(x,y\right)}{\partial\left(u,v\right)}\right)^2}\,dudv,$$

by the rule for the change of variables in double integrals. Note that

$$\sqrt{\left(\frac{\partial\left(y,z\right)}{\partial\left(u,v\right)}\right)^2+\left(\frac{\partial\left(z,x\right)}{\partial\left(u,v\right)}\right)^2+\left(\frac{\partial\left(x,y\right)}{\partial\left(u,v\right)}\right)^2}=\left\|\mathbf{N}\left(u,v\right)\right\|.$$

Therefore,

$$\iint_{D^*}\left\|\mathbf{N}\left(u^*,v^*\right)\right\|du^*dv^*=\iint_{D}\left\|\mathbf{N}\left(u,v\right)\right\|dudv.$$

Thus, the surface integral

$$\iint_{M} fdS$$

can be calculated by making use of either parametrization of S, as claimed .∎

The Differential Form Notation for a Flux Integral

As we saw before, if $\mathbf{\Phi}\left(u,v\right)=\left(x\left(u,v\right),y\left(u,v\right),z\left(u,v\right)\right)$ then

$$\mathbf{N}\left(u,v\right)=\frac{\partial\left(y,z\right)}{\partial\left(u,v\right)}\mathbf{i}+\frac{\partial\left(z,x\right)}{\partial\left(u,v\right)}\mathbf{j}+\frac{\partial\left(x,y\right)}{\partial\left(u,v\right)}\mathbf{k}.$$

Therefore, if $\mathbf{F}\left(x,y,z\right)=M\left(x,y,z\right)\mathbf{i}+N\left(x,y,z\right)\mathbf{j}+P\left(x,y,z\right)\mathbf{k}$ then

$$\mathbf{F}\left(\mathbf{\Phi}\left(u,v\right)\right)\cdot\mathbf{N}\left(u,v\right)=M\left(\mathbf{\Phi}\left(u,v\right)\right)\frac{\partial\left(y,z\right)}{\partial\left(u,v\right)}+N\left(\mathbf{\Phi}\left(u,v\right)\right)\frac{\partial\left(z,x\right)}{\partial\left(u,v\right)}+P\left(\mathbf{\Phi}\left(u,v\right)\right)\frac{\partial\left(x,y\right)}{\partial\left(u,v\right)}$$

Therefore,

$$\int\int_{M}\mathbf{F}\cdot\mathbf{n}dS=\int\int_{D}\mathbf{F}\left(\mathbf{\Phi}\left(u,v\right)\right)\cdot\mathbf{N}\left(u,v\right)dudv$$

$$=\int\int_{D}\left(M\left(\mathbf{\Phi}\left(u,v\right)\right)\frac{\partial\left(y,z\right)}{\partial\left(u,v\right)}+N\left(\mathbf{\Phi}\left(u,v\right)\right)\frac{\partial\left(z,x\right)}{\partial\left(u,v\right)}+P\left(\mathbf{\Phi}\left(u,v\right)\right)\frac{\partial\left(x,y\right)}{\partial\left(u,v\right)}\right)dudv.$$

Symbolically, we can set

$$dydz=\frac{\partial\left(y,z\right)}{\partial\left(u,v\right)}dudv,\ dzdx=\frac{\partial\left(z,x\right)}{\partial\left(u,v\right)}dudv,\ dxdy=\frac{\partial\left(x,y\right)}{\partial\left(u,v\right)}dudv,$$

and write

$$\int\int_{D}\left(M\left(\mathbf{\Phi}\left(u,v\right)\right)\frac{\partial\left(y,z\right)}{\partial\left(u,v\right)}+N\left(\mathbf{\Phi}\left(u,v\right)\right)\frac{\partial\left(z,x\right)}{\partial\left(u,v\right)}+P\left(\mathbf{\Phi}\left(u,v\right)\right)\frac{\partial\left(x,y\right)}{\partial\left(u,v\right)}\right)dudv$$

$$=\int\int_{M}Mdydz+Ndzdx+Pdxdy.$$

This is the differential form notation for the flux integral of

$$\mathbf{F}\left(x,y,z\right)=M\left(x,y,z\right)\mathbf{i}+N\left(x,y,z\right)\mathbf{j}+P\left(x,y,z\right)\mathbf{k}$$

over the surface M.

Example 12 Let M be the torus that is parametrized by

$$\mathbf{\Phi}\left(\theta,\phi\right)=\left(\left(a+b\cos\left(\phi\right)\right)\cos\left(\theta\right),\left(a+b\cos\left(\phi\right)\right)\sin\left(\theta\right),b\sin\left(\phi\right)\right),$$

where $a>b>0$, $0\leq\theta\leq2\pi$ and $0\leq\phi\leq2\pi$. Determine the flux integral

$$\int\int_{M}xdydz+ydzdx+zdxdy$$

Solution

We have

$$xdydz=\left(a+b\cos\left(\phi\right)\right)\cos\left(\theta\right)\frac{\partial\left(y,z\right)}{\partial\left(\theta,\phi\right)}d\theta d\phi$$

$$=\left(a+b\cos\left(\phi\right)\right)\cos\left(\theta\right)b\left(a+b\cos\left(\phi\right)\right)\cos\left(\theta\right)\cos\left(\phi\right)d\theta d\phi$$

$$=b\left(a+b\cos\left(\phi\right)\right)^{2}\cos^{2}\left(\theta\right)\cos\left(\phi\right)d\theta d\phi,$$

$$ydzdx=\left(a+b\cos\left(\phi\right)\right)\sin\left(\theta\right)\frac{\partial\left(z,x\right)}{\partial\left(\theta,\phi\right)}d\theta d\phi$$

$$=b\left(a+b\cos\left(\phi\right)\right)\sin\left(\theta\right)\left(a+b\cos\left(\phi\right)\right)\sin\left(\theta\right)\cos\left(\phi\right)d\theta d\phi$$

$$=b\left(a+b\cos\left(\phi\right)\right)^{2}\sin^{2}\left(\theta\right)\cos\left(\phi\right)d\theta d\phi,$$

and

$$zdxdy = z\frac{\partial(x,y)}{\partial(u,v)}d\theta d\phi$$
$$= b\sin(\phi)(a + b\cos(\phi))\sin(\phi)\,d\theta d\phi$$
$$= b(a + b\cos(\phi))\sin^2(\phi)\,d\theta d\phi.$$

Therefore,

$xdydz + ydzdx + zdxdy$
$$= b(a + b\cos(\phi))\left((a + b\cos(\phi))\cos^2(\theta)\cos(\phi) + (a + b\cos(\phi))\sin^2(\theta)\cos(\phi) + \sin^2(\phi)\right)d\theta d\phi$$
$$= b(a + b\cos(\phi))\left((a + b\cos(\phi))\cos(\phi) + \sin^2(\phi)\right)d\theta d\phi$$

Thus

$$\int\int_S xdydz + ydzdx + zdxdy$$
$$= \int_{\theta=0}^{\theta=2\pi}\int_{\phi=0}^{\phi=\pi} b(a + b\cos(\phi))\left((a + b\cos(\phi))\cos(\phi) + \sin^2(\phi)\right)d\theta d\phi$$
$$= 2\pi\left(b^2 a\pi + \frac{1}{2}ba\pi\right) = 2ab^2\pi^2 + \pi^2 ab.$$

$\square$

The Proof of Proposition 2:

Assume that M is an orientable surface. The flux integral

$$\int_M \mathbf{F}\cdot\mathbf{n}dS$$

is independent of an orientation-preserving parametrization of M. The flux integral is multiplied by -1 under an orientation-reversing parametrization of M.

As in the proof of Proposition 1, assume that M is parametrized by $\mathbf{\Phi}$,

$$\mathbf{\Phi}(u,v) = (x(u,v), y(u,v).z(u,v)), \text{ for each } (u,v) \in D,$$

and

$$u = u(u^*,v^*), \ v = v(u^*,v^*)$$

for each $(u^*, v^*) \in D^*$. Set

$$x(u^*,v^*) = x(u(u^*,v^*), v(u^*,v^*)), \ y(u^*,v^*) = y(u(u^*,v^*), v(u^*,v^*))$$
$$z(u^*,v^*) = z(u(u^*,v^*), v(u^*,v^*)),$$

and

$$\mathbf{\Phi}^*(u^*,v^*) = (x(u^*,v^*), y(u^*,v^*).z(u^*,v^*)), \text{ where } (u^*,v^*) \in D^*.$$

Let's assume that the transformation $(u^*, v^*) \to (u, v)$ is orientation preserving so that

$$\frac{\partial(u,v)}{\partial(u^*,v^*)} > 0$$

on D^*. We have

$$
\iint_{D^*} (M\left(\boldsymbol{\Phi}^*\left(u^*,v^*\right)\right) \frac{\partial\left(y,z\right)}{\partial\left(u^*,v^*\right)} + N\left(\boldsymbol{\Phi}^*\left(u^*,v^*\right)\right) \frac{\partial\left(z,x\right)}{\partial\left(u^*,v^*\right)}
$$
$$
+ P\left(\boldsymbol{\Phi}^*\left(u^*,v^*\right)\right) \frac{\partial\left(x,y\right)}{\partial\left(u^*,v^*\right)})du^*dv^*
$$
$$
= \iint_{D^*} (M\left(\boldsymbol{\Phi}^*\left(u^*,v^*\right)\right) \frac{\partial\left(y,z\right)}{\partial\left(u,v\right)} \frac{\partial\left(u,v\right)}{\partial\left(u^*,v^*\right)} + N\left(\boldsymbol{\Phi}^*\left(u^*,v^*\right)\right) \frac{\partial\left(z,x\right)}{\partial\left(u,v\right)} \frac{\partial\left(u,v\right)}{\partial\left(u^*,v^*\right)}
$$
$$
+ P\left(\boldsymbol{\Phi}^*\left(u^*,v^*\right)\right) \frac{\partial\left(x,y\right)}{\partial\left(u,v\right)} \frac{\partial\left(u,v\right)}{\partial\left(u^*,v^*\right)})du^*dv^*
$$
$$
= \iint_{D^*} \left(M\left(\boldsymbol{\Phi}^*\left(u^*,v^*\right)\right) \frac{\partial\left(y,z\right)}{\partial\left(u,v\right)} + N\left(\boldsymbol{\Phi}^*\left(u^*,v^*\right)\right) \frac{\partial\left(z,x\right)}{\partial\left(u,v\right)} + \left(\boldsymbol{\Phi}^*\left(u^*,v^*\right)\right) \frac{\partial\left(x,y\right)}{\partial\left(u,v\right)} \right) \frac{\partial\left(u,v\right)}{\partial\left(u^*,v^*\right)}du^*dv^*
$$
$$
= \iint_D \left(M\left(\boldsymbol{\Phi}\left(u,v\right)\right) \frac{\partial\left(y,z\right)}{\partial\left(u,v\right)} + N\left(\boldsymbol{\Phi}\left(u,v\right)\right) \frac{\partial\left(z,x\right)}{\partial\left(u,v\right)} + P\left(\boldsymbol{\Phi}\left(u,v\right)\right) \frac{\partial\left(x,y\right)}{\partial\left(u,v\right)} \right) dudv
$$
$$
= \iint_M \mathbf{F} \cdot \mathbf{n}dS
$$

by the rule for the change of variables in double integrals.
If the transformation is orientation reversing so that

$$
\frac{\partial\left(u,v\right)}{\partial\left(u^*,v^*\right)} < 0
$$

on D^*, we have

$$
\iint_{D^*} \left(M\left(\boldsymbol{\Phi}^*\left(u^*,v^*\right)\right) \frac{\partial\left(y,z\right)}{\partial\left(u,v\right)} + N\left(\boldsymbol{\Phi}^*\left(u^*,v^*\right)\right) \frac{\partial\left(z,x\right)}{\partial\left(u,v\right)} + \left(\boldsymbol{\Phi}^*\left(u^*,v^*\right)\right) \frac{\partial\left(x,y\right)}{\partial\left(u,v\right)} \right) \frac{\partial\left(u,v\right)}{\partial\left(u^*,v^*\right)}du^*dv^*
$$
$$
= -\iint_{D^*} \left(M\left(\boldsymbol{\Phi}^*\left(u^*,v^*\right)\right) \frac{\partial\left(y,z\right)}{\partial\left(u,v\right)} + N\left(\boldsymbol{\Phi}^*\left(u^*,v^*\right)\right) \frac{\partial\left(z,x\right)}{\partial\left(u,v\right)} + \left(\boldsymbol{\Phi}^*\left(u^*,v^*\right)\right) \frac{\partial\left(x,y\right)}{\partial\left(u,v\right)} \right) \left| \frac{\partial\left(u,v\right)}{\partial\left(u^*,v^*\right)} \right| du^*d
$$
$$
= -\iint_D \left(M\left(\boldsymbol{\Phi}\left(u,v\right)\right) \frac{\partial\left(y,z\right)}{\partial\left(u,v\right)} + N\left(\boldsymbol{\Phi}\left(u,v\right)\right) \frac{\partial\left(z,x\right)}{\partial\left(u,v\right)} + P\left(\boldsymbol{\Phi}\left(u,v\right)\right) \frac{\partial\left(x,y\right)}{\partial\left(u,v\right)} \right) dudv
$$
$$
= -\int_M \mathbf{F} \cdot \mathbf{n}dS.
$$

■

Problems

In problems 1-5, determine the area of the surface that is parametrized by the given function (leave the relevant integral in a simplified form, if that is indicated)

1 (a cylinder)

$$
\Phi\left(u,v\right) = \left(v, 3\cos\left(u\right), 3\sin\left(u\right)\right), \ -4 \le v \le 4, \ 0 \le u \le 2\pi.
$$

2 (a paraboloid)

$$
\Phi\left(u,v\right) = \left(u^2, u\cos\left(v\right), u\sin\left(v\right)\right), \ 0 \le u \le 3, \ 0 \le v \le 2\pi.
$$

3 (a cone)

$$\Phi\left(\rho,\theta\right) = \left(\frac{1}{2}\rho\cos\left(\theta\right),\frac{1}{2}\rho\sin\left(\theta\right),\frac{\sqrt{3}}{2}\rho\right)$$

where $0 \le \rho \le 4$ and $0 \le \theta \le 2\pi$.

4 (an ellipsoid) Let

$$\Phi\left(\phi,\theta\right) = \left(2\sin\left(\phi\right)\cos\left(\theta\right),\sin\left(\phi\right)\sin\left(\theta\right),\frac{1}{3}\cos\left(\phi\right)\right),$$

where $0 \le \phi \le \pi$ and $0 \le \theta \le 2\pi$. Express the area of the surface as an integral. Simplify the expression as much as possible. Do not attempt to evaluate the integral.

5 (a surface of revolution)

$$\Omega\left(x,\theta\right) = \left(x,e^x\sin\left(\theta\right),e^x\cos\left(\theta\right)\right),$$

where $0 \le x \le 2$, $0 \le \theta \le 2\pi$.

6. Let S be the hemisphere that is parametrized by

$$\Phi\left(\phi,\theta\right) = \left(3\sin\left(\phi\right)\cos\left(\theta\right),3\sin\left(\phi\right)\sin\left(\theta\right),3\cos\left(\phi\right)\right),$$

where $0 \le \phi \le \pi$ and $0 \le \theta \le \pi$ Evaluate

$$\int\int_S \left(x^2+y^2\right)dS$$

7. Let S be the surface that is parametrized by

$$\Phi\left(u,v\right) = \left(v,2\cos\left(u\right),2\sin\left(u\right)\right),\quad -4 \le v \le 4,\ 0 \le u \le \pi.$$

Evaluate

$$\int\int_S y^2 dS$$

(this is part of the cylinder $y^2+z^2=1$).

8. Let S be the helicoid that is parametrized by

$$\Phi\left(r,\theta\right) = \left(r\cos\left(\theta\right),r\sin\left(\theta\right),\theta\right),\ 0 \le r \le 4,\ 0 \le \theta \le 2\pi.$$

Evaluate

$$\int\int_S \sqrt{x^2+y^2}dS$$

9. Let S be the part of the plane

$$x+y+z=1$$

in the first octant (i.e., $x \ge 0$, $y \ge 0$ and $z \ge 0$). Evaluate

$$\int\int_S x\,dS.$$

10. Let S be the part of the sphere

$$x^2 + y^2 + z^2 = 1$$

that is in the first octant (i.e., $x \geq 0$, $y \geq 0$ and $z \geq 0$). Evaluate

$$\int\int_S z^2 dS.$$

11. Let S be the surface that is parametrized by

$$\Phi(u, v) = (\cos(u), v, \sin(u)), \ -2 \leq v \leq 2, \ 0 \leq u \leq \pi$$

(this is part of the cylinder $x^2 + z^2 = 1$). Evaluate

$$\int\int_S (x\mathbf{i} + z\mathbf{k}) \cdot d\mathbf{S}.$$

12. Let S be the planar surface that is parametrized by

$$\Phi(y, z) = (2, y, z), \ 0 \leq y \leq 3, \ 0 \leq z \leq 4.$$

Evaluate

$$\int\int_S (y\mathbf{i} + x\mathbf{j} + z\mathbf{k}) \cdot d\mathbf{S}.$$

13. Let S be the hemisphere of radius 3 that is parametrized by

$$\Phi(\phi, \theta) = (3\sin(\phi)\cos(\theta), 3\sin(\phi)\sin(\theta), 3\cos(\phi)),$$

where $0 \leq \phi \leq \pi$ and $0 \leq \theta \leq \pi$. Evaluate

$$\int\int_S \frac{x\mathbf{i} + y\mathbf{j} + z\mathbf{k}}{(x^2 + y^2 + z^2)} \cdot d\mathbf{S}.$$

14. Let S be the portion of the plane

$$z = 1 - \frac{1}{2}x - \frac{1}{2}y$$

in the first octant. Evaluate

$$\int\int_S (x\mathbf{i} - z\mathbf{k}) \cdot d\mathbf{S},$$

if S is oriented so that the normal points upward.

15. Let S be the hemisphere

$$z = \sqrt{4 - x^2 - y^2}.$$

Evaluate

$$\int\int_S (x\mathbf{i} + y\mathbf{j} + z\mathbf{k}) \cdot d\mathbf{S},$$

if S is oriented so that the normal points upward.

16. Let S be the portion of the paraboloid

$$z = 1 - x^2 - y^2$$

above the xy-plane. Evaluate

$$\int\int_S (y\mathbf{j} + \mathbf{k}) \cdot d\mathbf{S}$$

if S is oriented so that the normal points upward.

17. Assume that S consists of the hemisphere S_1

$$z = \sqrt{4 - x^2 - y^2}$$

and the disk

$$S_2 = \{(x, y, 0) : x^2 + y^2 \le 4\}$$

Evaluate

$$\int\int_S (yz\mathbf{i} + xz\mathbf{j} + xy\mathbf{k}) \cdot d\mathbf{S},$$

if S is oriented so that the normal points towards the exterior of the region enclosed by S.

14.6   Green's Theorem

In this section we will discuss Green's Theorem that establishes a link between a line integral on a closed curve and a double integral and some of the applications of the theorem. For example, Green's theorem will be used to show that the necessary condition for the existence of a potential for a vector field is also sufficient if the vector field is smooth in a domain without holes.

Green's Theorem in Simply Connected Regions

Assume that the boundary of a region D in the plane is **a simple closed curve** C, i.e., a closed curve without self intersections, and that C is oriented so that D is to the left as C is traversed. In this case we will say that C is **the positively oriented boundary of** D and refer to D as **the interior of** C. The region D is said to be **simply connected. There are no holes in** D. You can imagine that any simple closed curve that lies in the interior of D can be shrunk to a point in D without leaving D.

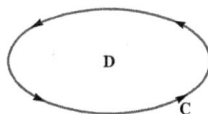

Figure 1: A simply connected region D and its positively oriented boundary C

Green's Theorem relates a line integral on C to a double integral on D:

Theorem 1 (Green's Theorem) Let the simple closed curve C be the positively oriented boundary of the region D. We have

$$\int_C M\,dx + N\,dy = \int\int_D \left(\frac{\partial N}{\partial x} - \frac{\partial M}{\partial y}\right) dx\,dy,$$

provided that the partial derivatives of M and N are continuous in $D \cup C$.

We will not give the proof of Green's Theorem for an arbitrary region, but we will confirm the statement of the theorem in the case of a region that is both of type I and type II. We will refer to such a region as a **simple region**. The theorem will be confirmed by showing that

$$\int_C M dx = - \int\int_D \frac{\partial M}{\partial y} dx dy,$$

and

$$\int_C N dy = \int\int_D \frac{\partial N}{\partial x} dx dy.$$

Let's establish the first equality. Since D is of type I, D can be expressed as the set of point (x, y) such that $a \le x \le b$ and $f(x) \le y \le g(x)$, as illustrated in Figure 2.

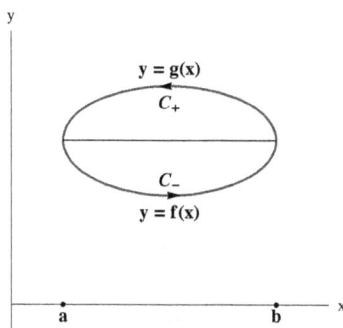

Figure 2

By Fubini's Theorem,

$$\int\int_D \frac{\partial M}{\partial y} dx dy = \int_a^b \left(\int_{y=f(x)}^{y=g(x)} \frac{\partial M}{\partial y} (x, y) \, dy \right) dx.$$

By the Fundamental Theorem of Calculus,

$$\int_{y=f(x)}^{y=g(x)} \frac{\partial M}{\partial y} (x, y) \, dy = M(x, g(x)) - M(x, f(x)).$$

Therefore,

$$\int\int_D \frac{\partial M}{\partial y} dx dy = \int_a^b (M(x, g(x)) - M(x, f(x))) \, dx$$

$$= \int_a^b M(x, g(x)) \, dx - \int_a^b M(x, f(x)) \, dx.$$

Note that

$$\int_a^b M(x, g(x)) \, dx = - \int_{C_+} M dx,$$

and

$$\int_a^b M(x, f(x)) \, dx = \int_{C_-} M dx.$$

Therefore,

$$\int_C M\,dx = \int_{C_+} M\,dx + \int_{C_-} M\,dx$$

$$= -\int_a^b M\left(x, g\left(x\right)\right) dx + \int_a^b M\left(x, f\left(x\right)\right) dx$$

$$= -\left(\int_a^b M\left(x, g\left(x\right)\right) dx - \int_a^b M\left(x, f\left(x\right)\right) dx\right)$$

$$= -\int\int_D \frac{\partial M}{\partial y}\,dx\,dy,$$

as claimed.

Similarly, since D is also of type II, we can express D as the set of points (x, y) such that $c \le y < d$ and $F\left(y\right) \le x < G\left(y\right)$, as illustrated in Figure 3.

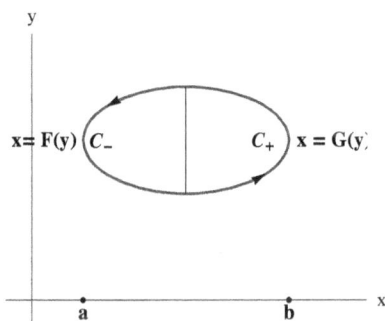

Figure 3

By Fubini's Theorem,

$$\int\int_D \frac{\partial N}{\partial x}\,dx\,dy = \int_c^d \left(\int_{x=F(x)}^{x=G(y)} \frac{\partial N}{\partial x}\left(x, y\right) dx\right) dy.$$

By the Fundamental Theorem of Calculus,

$$\int_{x=F(x)}^{x=G(y)} \frac{\partial N}{\partial x}\left(x, y\right) dx = N\left(G(y), y\right) - N\left(F(y), y\right).$$

Therefore,

$$\int\int_D \frac{\partial N}{\partial x}\,dx\,dy = \int_c^d \left(N\left(G(y), y\right) - N\left(F(y), y\right)\right) dy$$

$$= \int_c^d N\left(G(y), y\right) dy - \int_c^d N\left(F(y), y\right) dy$$

Note that

$$\int_c^d N\left(G(y), y\right) dy = \int_{C_+} N\,dy,$$

and

$$-\int_c^d N\left(F(y), y\right) dy = \int_{C_-} N\,dy.$$

Therefore,

$$
\int_C Ndy \;=\; \int_{C_+} Ndy + \int_{C_-} Ndy
$$

$$
=\; \int_c^d N\left(G(y),y\right) dy - \int_c^d N\left(F(y),y\right) dy = \int\int_D \frac{\partial N}{\partial x}\,dxdy,
$$

as claimed. ∎

Remark   Note that

$$
\int_C Mdx + Ndy = \int_C (M\mathbf{i} + N\mathbf{j}) \cdot d\boldsymbol{\sigma}.
$$

Thus, Green's Theorem can be stated as

$$
\int_C \mathbf{F}\cdot d\boldsymbol{\sigma} = \int\int_D \left(\frac{\partial N}{\partial x} - \frac{\partial M}{\partial y}\right) dxdy,
$$

where $\mathbf{F} = M\mathbf{i} + N\mathbf{j}$.

Example 1 Let D be the unit disk $\{(x,y) : x^2 + y^2 \le 1\}$ and let C be its positively oriented boundary. Confirm Green's Theorem for the line integral

$$
\int_C ydx + \ln\left(x^2 + y^2 + 1\right) dy.
$$

Solution

Let's evaluate the line integral directly. We can parametrize the unit circle C by $\boldsymbol{\sigma}(t) = (\cos(t), \sin(t))$, $t \in [0, 2\pi]$. Thus,

$$
\int_C ydx + \ln\left(x^2 + y^2 + 1\right) dy = \int_0^{2\pi} \left(\sin(t)\left(-\sin t\right) + \ln(2)\cos(t)\right) dt
$$

$$
= \int_0^{2\pi} \left(-\sin^2(t) + \ln(2)\cos(t)\right) dt = -\pi.
$$

By Green's Theorem,

$$
\int_C ydx + \ln\left(x^2 + y^2 + 1\right) dy = \int\int_D \left(\frac{\partial}{\partial x}\ln\left(x^2 + y^2 + 1\right) - \frac{\partial}{\partial y}(y)\right) dxdy
$$

$$
= \int\int_D \left(\frac{1}{x^2 + y^2 + 1}(2x) - 1\right) dxdy.
$$

In polar coordinates,

$$
\int\int_D \left(\frac{1}{x^2 + y^2 + 1}(2x) - 1\right) dxdy = \int_{\theta=0}^{2\pi}\int_{r=0}^1 \left(\frac{2r\cos(\theta)}{r^2 + 1} - 1\right) rdrd\theta
$$

$$
= \int_{r=0}^1 \frac{2r^2}{r^2 + 1}\int_{\theta=0}^{2\pi}\cos(\theta)\,d\theta dr - \int_{\theta=0}^{\theta=2\pi}\int_{r=0}^1 rdrd\theta
$$

$$
= 0 - 2\pi\left(\frac{1}{2}\right) = -\pi.
$$

Thus, we have confirmed that

$$
\int_C ydx + \ln\left(x^2 + y^2 + 1\right) dy = \int\int_D \left(\frac{1}{x^2 + y^2 + 1}(2x) - 1\right) dxdy.
$$

□

Green's Theorem in Multiply Connected Regions

Assume that D is a region whose boundary consists of the closed curves C_1 and C_2, as in Figure 4, with the orientations of the curves as indicated (if you imagine that you walk along the boundary you should see the region D to your left). Such a region is **doubly connected**: You can consider that there is a hole in the interior of C_1 that is bounded by the inner curve C_2.

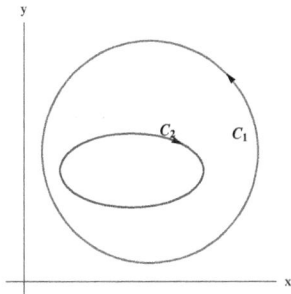

Figure 4

In this case Green's Theorem takes the following form:

$$\int_{C_1} M\,dx + N\,dy + \int_{C_2} M\,dx + N\,dy = \int\int_D \left(\frac{\partial N}{\partial x} - \frac{\partial M}{\partial y}\right) dx\,dy.$$

We can provide a plausibility argument as follows: Imagine that a cut is made in D so that the resulting region D' is bounded by the simple closed curve $C_1 + C_3 + C_2 + C_4$, as in Figure 5 (you can imagine that the cut has width that can be made arbitrarily small, so that C_3 and C_4 correspond to the same curve traversed in opposite directions).

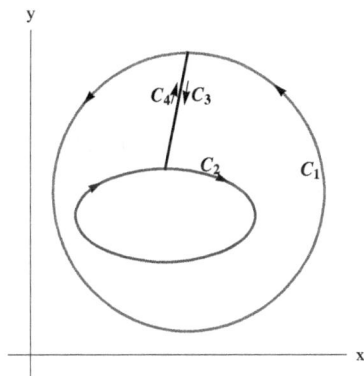

Figure 5

Since Green's Theorem is valid in the simply connected domain D', we have

$$\int_{C_1} M\,dx + N\,dy + \int_{C_3} M\,dx + N\,dy + \int_{C_2} M\,dx + N\,dy + \int_{C_4} M\,dx + N\,dy = \int\int_{D'} \left(\frac{\partial N}{\partial x} - \frac{\partial M}{\partial y}\right) dx\,dy$$

Since

$$\int_{C_3} M\,dx + N\,dy + \int_{C_4} M\,dx + N\,dy = 0$$

due to the opposite orientations of C_3 and C_4.

$$\int_{C_1} M\,dx + N\,dy + \int_{C_2} M\,dx + N\,dy = \int\int_{D'} \left(\frac{\partial N}{\partial x} - \frac{\partial M}{\partial y}\right) dx\,dy$$

Since we can imagine that the width of the cut tends to 0, we have can replace the double integral on D' by the double integral on D. Thus

$$\int_{C_1} M\,dx + N\,dy + \int_{C_2} M\,dx + N\,dy = \int\int_D \left(\frac{\partial N}{\partial x} - \frac{\partial M}{\partial y} \right) dx\,dy$$

as claimed. ∎

The above version of Green's theorem can be generalized to **"multiply connected regions"** with more than one hole by making cuts that enable us to use Green's Theorem for a simply connected region. For example, assume that the region D is bounded by the curves C_1, C_2 and C_3, with the orientations that are indicated the picture:

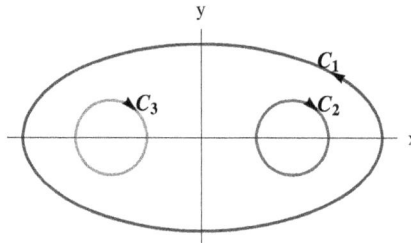

Figure 6

Then

$$\int_{C_1} M\,dx + N\,dy + \int_{C_2} M\,dx + N\,dy + \int_{C_3} M\,dx + N\,dy = \int\int_D \left(\frac{\partial N}{\partial x} - \frac{\partial M}{\partial y} \right) dx\,dy.$$

Example 2 Let

$$\mathbf{F}(x,y) = -\frac{y}{x^2 + y^2}\mathbf{i} + \frac{x}{x^2 + y^2}\mathbf{j}$$

Assume that C_1 is a simple closed curve that is the positively oriented boundary of the region D and that the positively oriented (counterclockwise) circle C_2 of radius a is in D.

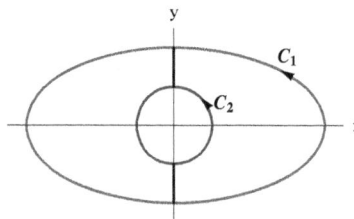

Figure 7

a) Use Green's Theorem to show that

$$\int_{C_1} \mathbf{F} \cdot d\boldsymbol{\sigma} = \int_{C_2} \mathbf{F} \cdot d\boldsymbol{\sigma}$$

b) Use the result of part a) to evaluate $\int_{C_1} \mathbf{F} \cdot d\boldsymbol{\sigma}$.

Solution

a) We need to consider the negative orientation of the inner boundary C_2 in order to apply Green's Theorem. Thus

$$\int_{C_1} \mathbf{F} \cdot d\boldsymbol{\sigma} - \int_{C_2} \mathbf{F} \cdot d\boldsymbol{\sigma} = \int \int_D \left(\frac{\partial}{\partial x} \left(\frac{x}{x^2 + y^2} \right) - \frac{\partial}{\partial y} \left(-\frac{y}{x^2 + y^2} \right) \right) dx dy$$

$$= \int \int_D \left(-\frac{x^2 - y^2}{(x^2 + y^2)^2} + \frac{x^2 - y^2}{(x^2 + y^2)^2} \right) dx dy = 0.$$

Therefore

$$\int_{C_1} \mathbf{F} \cdot d\boldsymbol{\sigma} = \int_{C_2} \mathbf{F} \cdot d\boldsymbol{\sigma},$$

as claimed.

b) We can parametrize C_2 by $\sigma(t) = (a \cos(t), a \sin(t))$, $t \in [0, 2\pi]$. By part a)

$$\int_{C_1} \mathbf{F} \cdot d\boldsymbol{\sigma} = \int_{C_2} \mathbf{F} \cdot d\boldsymbol{\sigma}$$

$$= \int_{C_2} -\frac{y}{x^2 + y^2} dx + \frac{x}{x^2 + y^2} dy$$

$$= \int_0^{2\pi} \left(-\frac{a \sin(t)}{a^2} (-a \sin(t)) + \frac{a \cos(t)}{a^2} (a \cos(t)) \right) dt$$

$$= \int_0^{2\pi} \left(\sin^2(t) + \cos^2(t) \right) dt$$

$$= \int_0^{2\pi} 1 dt = 2\pi.$$

$\square$

Green's Theorem can be used to express the area of a region in terms of a line integral on its boundary:

Proposition 1 Assume that D is the interior of C and that C is positively oriented. Then

$$\text{Area of } D = \frac{1}{2} \int_C x dy - y dx.$$

Proof

$$\frac{1}{2} \int_C x dy - y dx = \frac{1}{2} \int \int_D \left(\frac{\partial}{\partial x} (x) + \frac{\partial}{\partial y} (y) \right) dx dy$$

$$= \frac{1}{2} \int \int_D (2) dx dy = \text{ Area of } D.$$

$\blacksquare$

Interpretations of Green's Theorem

In Section 14.1 we defined the **curl** of a two-dimensional vector field $\mathbf{F}(x, y) = M(x, y) \mathbf{i} + N(x, y) \mathbf{j}$ as

$$\boldsymbol{\nabla} \times \mathbf{F}(x, y) = \boldsymbol{\nabla} \times (M(x, y) \mathbf{i} + N(x, y) \mathbf{j} + 0\mathbf{k}) = \left(\frac{\partial N}{\partial x} - \frac{\partial M}{\partial y} \right) \mathbf{k}.$$

Therefore, we can express Green's Theorem,

$$\int_C M\,dx + N\,dy = \int\int_D \left(\frac{\partial N}{\partial x} - \frac{\partial M}{\partial y}\right) dx\,dy,$$

as

$$\int_C \mathbf{F} \cdot d\boldsymbol{\sigma} = \int\int_D (\boldsymbol{\nabla} \times \mathbf{F}) \cdot \mathbf{k}\, dx\,dy.$$

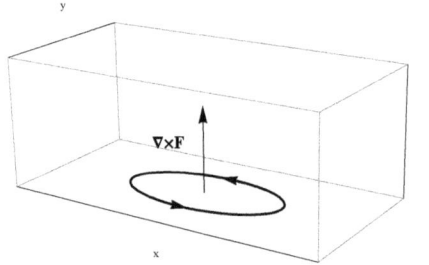

Figure 10

If $\mathbf{F}(x, y)$ is the velocity of a fluid particle at (x, y), the line integral is **the circulation around** C. Therefore, **the circulation of the velocity field around C is equal to the integral of the normal component of its curl over its interior D.** We can think of **the normal component of curl of F at (x_0, y_0) as the average circulation of F at (x_0, y_0):**

Proposition 2 Assume that C_r is a circle of radius r centered at (x_0, y_0) that is oriented counterclockwise and D_r is the disk bounded by C_r. We have

$$\lim_{r \to 0} \frac{1}{\pi r^2} \int_{C_r} \mathbf{F} \cdot d\boldsymbol{\sigma} = (\boldsymbol{\nabla} \times \mathbf{F})\,(x_0, y_0) \cdot \mathbf{k}$$

(assuming that $\boldsymbol{\nabla} \times \mathbf{F}$ is continuous).

A plausibility argument for Proposition 2:

By Green's Theorem,

$$\int_{C_r} \mathbf{F} \cdot d\boldsymbol{\sigma} = \int\int_{D_r} \boldsymbol{\nabla} \times \mathbf{F} \cdot \mathbf{k}\, dx\,dy.$$

By the continuity of $\boldsymbol{\nabla} \times \mathbf{F}$, if r is small,

$$\int\int_{D_r} \boldsymbol{\nabla} \times \mathbf{F} \cdot \mathbf{k}\, dx\,dy \cong (\boldsymbol{\nabla} \times \mathbf{F}\,(x_0, y_0) \cdot \mathbf{k})\,(\text{area of } D_r) = (\boldsymbol{\nabla} \times \mathbf{F}\,(x_0, y_0) \cdot \mathbf{k})\,(\pi r^2).$$

Thus,

$$\frac{1}{\pi r^2} \int_{\partial D_r} \mathbf{F} \cdot d\boldsymbol{\sigma} = \frac{1}{\pi r^2} \int\int_{D_r} \boldsymbol{\nabla} \times \mathbf{F} \cdot \mathbf{k}\, dS \cong \frac{1}{\pi r^2} (\boldsymbol{\nabla} \times \mathbf{F}\,(x_0, y_0) \cdot \mathbf{k})\,(\pi r^2) = \boldsymbol{\nabla} \times \mathbf{F}\,(x_0, y_0) \cdot \mathbf{k}.$$

This lends support to the assertion that

$$\lim_{r \to 0} \frac{1}{\pi r^2} \int_{\partial D_r} \mathbf{F} \cdot d\boldsymbol{\sigma} = \boldsymbol{\nabla} \times \mathbf{F}\,(x_0, y_0) \cdot \mathbf{k}.$$

■

Example 3 Let
$$\mathbf{F}(x, y) = -\omega y \mathbf{i} + \omega x \mathbf{j},$$
so that the vector field represents circulation around the origin.

In Section 14.1 we saw that

$$\nabla \times \mathbf{F}(x, y) = 2\omega \mathbf{k} \Rightarrow \nabla \times \mathbf{F}(x, y) \cdot \mathbf{k} = 2\omega.$$

Let C_r be the circle of radius r that is centered at the origin. We have

$$\frac{1}{\pi r^2} \int_{C_r} \mathbf{F} \cdot d\boldsymbol{\sigma} = \frac{\omega}{\pi r^2} \int_0^{2\pi} r^2 \left(-\sin(t)\, i + \cos(t)\, j\right) \cdot \left(-\sin(t) + \cos(t)\right) dt$$

$$= \frac{\omega}{\pi} \int_0^{2\pi} \left(\sin^2(t) + \cos^2(t)\right) dt$$

$$= \frac{\omega}{\pi} (2\pi) = 2\omega,$$

as well. $\square$

We can also interpret Green's Theorem in terms of the flux of a vector field across a simple closed curve C. If $\mathbf{F}(x, y) = M(x, y)\mathbf{i} + N(x, y)\mathbf{j}$, the divergence of $\mathbf{F}$ at (x, y) is

$$\nabla \cdot \mathbf{F}(x, y) = \frac{\partial M}{\partial x}(x, y) + \frac{\partial N}{\partial y}(x, y)$$

By Green's Theorem,

$$\int_C -N dx + M dy = \int\int_D \left(\frac{\partial N}{\partial y} + \frac{\partial M}{\partial x}\right) dx dy = \int\int_D \nabla \cdot \mathbf{F}(x, y)\, dA$$

Let us reexamine the expression on the left hand side. Assume that C is parametrized by $\sigma(t) = (x(t), y(t))$, where $t \in [a, b]$:

$$\int_C -N dx + M dy = \int_a^b \left(-N(x(t), y(t))\right) \frac{dx}{dt} + M(x(t), y(t)) \frac{dy}{dt}\, dt$$

$$= \int_a^b (M(x, y)\mathbf{i} + N(x, y)\mathbf{j}) \cdot \left(\frac{dy}{dt}\mathbf{i} - \frac{dx}{dt}\mathbf{j}\right) dt$$

$$= \int_a^b \mathbf{F}(x(t), y(t)) \cdot \left(\frac{dy}{dt}\mathbf{i} - \frac{dx}{dt}\mathbf{j}\right) dt$$

$$= \int_a^b \mathbf{F}(x(t), y(t)) \cdot \frac{\left(\frac{dy}{dt}\mathbf{i} - \frac{dx}{dt}\mathbf{j}\right)}{\sqrt{\left(\frac{dx}{dt}\right)^2 + \left(\frac{dy}{dt}\right)^2}} \sqrt{\left(\frac{dx}{dt}\right)^2 + \left(\frac{dy}{dt}\right)^2}\, dt.$$

Note that

$$\mathbf{n}(t) = \frac{\left(\frac{dy}{dt}\mathbf{i} - \frac{dx}{dt}\mathbf{j}\right)}{\sqrt{\left(\frac{dx}{dt}\right)^2 + \left(\frac{dy}{dt}\right)^2}}$$

is orthogonal to the tangent

$$\frac{d\boldsymbol{\sigma}}{dt} = \frac{dx}{dt}\mathbf{i} + \frac{dy}{dt}\mathbf{j},$$

$\|\mathbf{n}(t)\| = 1$, and $\boldsymbol{n}(t)$ points towards the exterior of C. We will refer to $\mathbf{n}(t)$ as **the unit normal** at $(x(t), y(t))$ that points towards the exterior of C. Thus,

$$\int_C -N\,dx + M\,dy = \int_a^b \mathbf{F}(x(t), y(t)) \cdot \frac{\left(\dfrac{dy}{dt}\mathbf{i} - \dfrac{dx}{dt}\mathbf{j}\right)}{\sqrt{\left(\dfrac{dx}{dt}\right)^2 + \left(\dfrac{dy}{dt}\right)^2}} \sqrt{\left(\frac{dx}{dt}\right)^2 + \left(\frac{dy}{dt}\right)^2}\, dt$$

$$= \int_a^b \mathbf{F}(x(t), y(t)) \cdot \mathbf{n}(t)\, ds$$

$$= \int_C \mathbf{F} \cdot \mathbf{n}\, ds$$

The quantity

$$\int_C \mathbf{F} \cdot \mathbf{n}\, ds.$$

is **the outward flux of** $\mathbf{F}$ across C. By Green's Theorem,

$$\int_C \mathbf{F} \cdot \mathbf{n}\, ds = \int\int_D \boldsymbol{\nabla} \cdot \mathbf{F}(x, y)\, dx\, dy$$

Thus, .the outward flux of $\mathbf{F}$ across C is equal to the integral of the divergence of F over the interior of C.

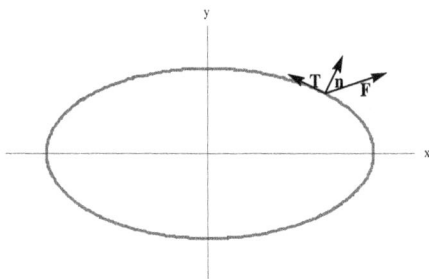

Figure 11

We can interpret the divergence of F at (x_0, y_0) as the flux per unit area of F at (x_0, y_0).

Proposition 3 Assume that C_r is a circle of radius r centered at (x_0, y_0) that is oriented counterclockwise and D_r is the disk bounded by C_r. We have

$$\lim_{r \to 0} \frac{1}{\pi r^2} \int_{C_r} \mathbf{F} \cdot \mathbf{n}\, ds = \boldsymbol{\nabla} \cdot \mathbf{F}(x_0, y_0)$$

(assuming that $\boldsymbol{\nabla} \cdot \mathbf{F}$ is continuous).

A plausibility argument for Proposition2:

We have

$$\int_{C_r} \mathbf{F} \cdot \mathbf{n}\, ds = \int\int_{D_r} \boldsymbol{\nabla} \cdot \mathbf{F}(x, y)\, dx\, dy$$

By continuity, if r is small

$$\int\int_{D_r} \boldsymbol{\nabla} \cdot \mathbf{F}(x, y)\, dx\, dy = \boldsymbol{\nabla} \cdot \mathbf{F}(x_0, y_0) \times (\text{area of } D_r) = \boldsymbol{\nabla} \cdot \mathbf{F}(x_0, y_0) \times \left(\pi r^2\right).$$

Therefore

$$\frac{1}{\pi r^2} \int_{C_r} \mathbf{F} \cdot \mathbf{n} ds \cong \frac{1}{\pi r^2} \boldsymbol{\nabla} \cdot \mathbf{F}(x_0, y_0) \times (\pi r^2) = \boldsymbol{\nabla} \cdot \mathbf{F}(x_0, y_0)$$

if r is small. This supports the claim that

$$\lim_{r \to 0} \frac{1}{\pi r^2} \int_{C_r} \mathbf{F} \cdot \mathbf{n} ds = \boldsymbol{\nabla} \cdot \mathbf{F}(x_0, y_0).$$

∎

Irrotational and Incompressible Flow

If $\mathbf{F}$ is the velocity field of a fluid in motion, we say that the flow is **incompressible** If $\boldsymbol{\nabla} \cdot \mathbf{F}(x, y) = 0$ for each (x, y) in a region D that is relevant to the flow. The flow is **irrotational** if we have $\boldsymbol{\nabla} \times \mathbf{F}(x, y) = 0$ for each $(x, y) \in D$. Assume that C is simply connected. If C is a closed curve that lies in D and $\mathbf{F} = M\mathbf{i} + N\mathbf{j}$, then

$$\int_C -N dx + M dy = \int_C \mathbf{F} \cdot \mathbf{n} ds = \int\int_D \boldsymbol{\nabla} \cdot \mathbf{F}(x, y) \, dA = 0.$$

Therefore, the vector field $-N\mathbf{i} + M\mathbf{j}$ is the gradient of a scalar function, say $g(x, y)$. Thus,

$$-N(x, y) = \frac{\partial g}{\partial x}(x, y) \text{ and } M(x, y) = \frac{\partial g}{\partial y}(x, y).$$

Since we also have $\boldsymbol{\nabla} \times \mathbf{F}(x, y) = 0$ for each $(x, y) \in D$, F is the gradient of scalar function f:

$$M(x, y) = \frac{\partial f}{\partial x}(x, y) \text{ and } N(x, y) = \frac{\partial f}{\partial y}(x, y).$$

Therefore,

$$\frac{\partial f}{\partial x}(x, y) = \frac{\partial g}{\partial y}(x, y) \text{ and } \frac{\partial f}{\partial y}(x, y) = -\frac{\partial g}{\partial x}(x, y)$$

Also note that

$$\frac{\partial^2 f}{\partial x^2} + \frac{\partial^2 f}{\partial y^2} = \frac{\partial M}{\partial x} + \frac{\partial N}{\partial y} = \frac{\partial^2 g}{\partial x \partial y} + \frac{\partial}{\partial y}\left(-\frac{\partial g}{\partial x}\right) = 0,$$

$$\frac{\partial^2 g}{\partial x^2} + \frac{\partial^2 g}{\partial y^2} = \frac{\partial}{\partial x}(-N) + \frac{\partial M}{\partial y} = -\frac{\partial^2 f}{\partial x \partial y} + \frac{\partial}{\partial y}\left(\frac{\partial f}{\partial x}\right) = 0.$$

Thus, if a vector field $\mathbf{F}$ is irrotational and conservative, a potential function f satisfies Laplace's equation. Such functions are said to be **harmonic**. The stream function, i.e., a potential g for $-N\mathbf{i} + M\mathbf{j}$ also satisfies Laplace's equation.

Sufficient Conditions for the Existence of a Potential Function

In Section 14.3 we saw that the integral of a gradient field is independent of path, and that a necessary condition for $\mathbf{F}(x, y) = M(x, y)\mathbf{i} + N(x, y)\mathbf{j}$ to be a gradient field in a region D is that

$$\frac{\partial M}{\partial y}(x, y) = \frac{\partial N}{\partial x}(x, y)$$

for each $(x, y) \in D$. Green's Theorem enables us to discuss the sufficiency of this condition for a vector field to be conservative:

Theorem 2 Assume that D is a simply connected region, $\mathbf{F}(x,y) = M(x,y)\mathbf{i} + N(x,y)\mathbf{j}$ and

$$\frac{\partial M}{\partial y}(x,y) = \frac{\partial N}{\partial x}(x,y)$$

for each $(x,y) \in D$, and that these partial derivatives are continuous on D. Then the vector field $\mathbf{F}$ is conservative.

Proof

Step 1. To begin with, let us note that **the line integral of a vector field is independent of path if its integral around any simple closed curve is 0**. We will not give a proof of this statement under general conditions. For example, if C_1 and C_2 join P_1 to P_2 as in Figure 8, then $C_1 - C_2$ is a simple closed curve.

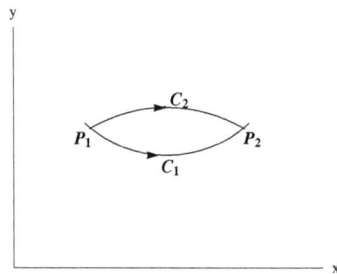

Figure 8

We have

$$\int_{C_1 - C_2} \mathbf{F} \cdot d\boldsymbol{\sigma} = 0,$$

so that

$$\int_{C_1} \mathbf{F} \cdot d\boldsymbol{\sigma} - \int_{C_2} \mathbf{F} \cdot d\boldsymbol{\sigma} = 0 \Rightarrow \int_{C_1} \mathbf{F} \cdot d\boldsymbol{\sigma} = \int_{C_2} \mathbf{F} \cdot d\boldsymbol{\sigma}.$$

Step 2. If $\boldsymbol{F}(x,y) = M(x,y)\mathbf{i} + N(x,y)\mathbf{j}$ and

$$\frac{\partial M}{\partial y}(x,y) = \frac{\partial N}{\partial x}(x,y)$$

for each $(x,y) \in D$, **the line integrals of $\boldsymbol{F}$ are independent of path**. By **Step 1**, it is sufficient to show that

$$\int_C \mathbf{F} \cdot d\boldsymbol{\sigma} = 0$$

if C is a simple closed curve in D,. The interior of C is entirely in D since D is assumed to be simply connected. By Green's Theorem,

$$\int_C \mathbf{F} \cdot d\boldsymbol{\sigma} = \int_C M\,dx + N\,dy = \int\int_{Int(C)} \left(\frac{\partial N}{\partial x} - \frac{\partial M}{\partial y}\right) dx\,dy = 0.$$

Step 3. Let's fix a point $P_0 = (x_0, y_0)$ in D and define $f(P) = f(x,y)$ as

$$\int_{C(P_0, P_1)} \mathbf{F} \cdot d\boldsymbol{\sigma},$$

where $C(P_0, P_1)$ is an arbitrary curve in D that joins P_0 to P. The definition of $f(x,y)$ makes sense since the line integrals of $\boldsymbol{F}$ are independent of path, by **Step 2**.

Claim:

$$\boldsymbol{F}(x,y) = \boldsymbol{\nabla} f(x,y)$$

for each $(x,y) \in D$.

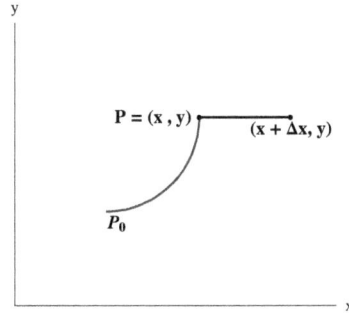

Figure 9

Let $|\Delta x|$ be small enough so that any point of the form $(x + \Delta x, y) \in D$. By independence of path, we can set

$$f(x+\Delta x, y) - f(x,y) = \int_{C(P_0,P)+C_{\Delta x}} \mathbf{F} \cdot d\boldsymbol{\sigma} - \int_{C(P_0,P)} \mathbf{F} \cdot d\boldsymbol{\sigma} = \int_{C_{\Delta x}} \mathbf{F} \cdot \boldsymbol{\sigma}$$

where $C_{\Delta x}$ is the line segment from $P = (x,y)$ to $(x + \Delta x, y)$. We can parametrize $C_{\Delta x}$ by $\sigma(t) = (x + t\Delta x, y)$, where $0 \le t \le 1$. Then,

$$\int_{C_{\Delta x}} \mathbf{F} \cdot d\boldsymbol{\sigma} = \int_{C_{\Delta x}} Mdx + Ndy = \int_0^1 \left(M(x+t\Delta x, y)\frac{dx}{dt} + N(x+t\Delta x)\frac{dy}{dt} \right) dt$$

$$= \int_0^1 M(x+t\Delta x, y)\, \Delta x\, dt$$

$$= \Delta x \int_0^1 M(x+t\Delta x, y)\, dt$$

Therefore,

$$\lim_{\Delta x \to 0} \frac{f(x+\Delta x, y) - f(x,y)}{\Delta x} = \lim_{\Delta x \to 0} \int_0^1 M(x+t\Delta x, y)\, dt$$

$$= \lim_{\Delta x \to 0} \left(\int_0^1 M(x+t\Delta x, y) - M(x,y) \right) + M(x,y)$$

$$= \lim_{\Delta x \to 0} \left(\int_0^1 M(x+t\Delta x, y) - \int_0^1 M(x,y)\, dt \right) + M(x,y)$$

$$= \lim_{\Delta x \to 0} \int_0^1 \left(M(x+t\Delta x)\ \ M(x,y) \right) dt + M(x,y)$$

$$= M(x,y),$$

since

$$\lim_{\Delta x \to 0} \int_0^1 \left(M(x+t\Delta x) - M(x,y) \right) dt = \int_0^1 \lim_{\Delta x \to 0} \left(M(x+t\Delta x) - M(x,y) \right) dt = 0,$$

by the continuity of M at (x,y).

Thus, we have shown that

$$\frac{\partial f}{\partial x}(x,y) = \lim_{\Delta x \to 0} \frac{f(x+\Delta x, y) - f(x,y)}{\Delta x} = M(x,y).$$

Similarly, by considering the vertical line that connects (x,y) to $(x, y+\Delta y)$, we can show that

$$\frac{\partial f}{\partial y}(x,y) = N(x,y).$$

Therefore,,

$$\mathbf{F}(x,y) = M(x,y)\mathbf{i} + N(x,y)\mathbf{j} = \frac{\partial f}{\partial x}(x,y)\mathbf{i} + \frac{\partial f}{\partial y}(x,y)\mathbf{j} = \boldsymbol{\nabla} f(x,y),$$

as claimed. ∎

Remark 1 The assumption that D is simply connected is essential. If

$$\mathbf{F}(x,y) = -\frac{y}{x^2+y^2}\mathbf{i} + \frac{x}{x^2+y^2}\mathbf{j},$$

as in Example 2, we have

$$\frac{\partial}{\partial y}\left(-\frac{y}{x^2+y^2}\right) = \frac{\partial}{\partial x}\left(\frac{x}{x^2+y^2}\right)$$

for each $(x,y) \neq (0,0)$. But $\mathbf{F}$ is *not* conservative in $D = \{(x,y) \in \mathbb{R}^2 : (x,y) \neq (0,0)\}$, since we showed that

$$\int_C \mathbf{F} \cdot d\boldsymbol{\sigma} = 2\pi$$

if C is any simple closed curve that contains the origin in its interior (if $\mathbf{F}$ were conservative in D such a line integral would have been 0). There is no contradiction here, since F does not satisfy the condition

$$\frac{\partial}{\partial y}\left(-\frac{y}{x^2+y^2}\right) = \frac{\partial}{\partial x}\left(\frac{x}{x^2+y^2}\right)$$

at the origin. D has a hole at the origin, so that it is not simply connected. ◊

Problems

In problems 1-4
a) Use Green's Theorem to express the given line integral as a double integral,
b) Evaluate the double integral.

1.

$$\int_C y^3 dx - x^3 dy,$$

where C is the positively oriented boundary of the disk bounded by the circle $x^2 + y^2 = 4$.

2.

$$\int_C xy^2 dx + 2x^2 y dy,$$

where C is the positively oriented boundary of the triangular region with vertices $(0,0)$, $(2,0)$ and $(2,4)$.

3.

$$\int_C \cos(y)\, dx + x^2 \sin(y)\, dy,$$

where C is the positively oriented boundary of the rectangular region with vertices $(0,0)$, $(5,0)$, $(5,\pi)$ and $(0,\pi)$.

4.

$$\int_C xe^{-2x}\,dx + \left(x^4 + 2x^2y^2\right) dy,$$

where C is the positively oriented boundary of the annular region bounded by the circles $x^2+y^2 = 1$ an $x^2 + y^2 = 4$.

In problems 5 and 6,
a) Use Green's Theorem to express

$$\int_C \mathbf{F} \cdot d\boldsymbol{\sigma}$$

as a double integral,
b) Evaluate the double integral.

5.

$$\mathbf{F}(x,y) = y^2 \cos(x)\,\mathbf{i} + \left(x^2 + 2y\sin(x)\right)\mathbf{j}$$

and $C = C_1 + C_2 + C_3$, where C_1 is the line segment from $(0,0)$ to $(2,6)$, C_2 is the line segment from $(2,6)$ to $(2,0)$, and C_3 is the line segment from $(2,0)$ to $(0,0)$ (pay attention to orientation).

6.

$$\mathbf{F}(x,y) = \left(e^x + x^2y\right)\mathbf{i} + \left(e^y - xy^2\right)\mathbf{j}$$

and C is the positively oriented boundary of the disk bounded by the circle $x^2 + y^2 = 25$.

7. Let

$$\mathbf{F}(x,y) = \frac{x}{x^2 + y^2}\mathbf{i} + \frac{y}{x^2 + y^2}\mathbf{j}$$

Assume that C_1 is a positively oriented simple closed curve such that the circle of radius 1 centered at the origin is contained in its interior. Calculate

$$\int_{C_1} \mathbf{F} \cdot d\boldsymbol{\sigma}$$

Caution: Green's Theorem is not applicable in the interior of C_1 since $\mathbf{F}$ is not defined at the origin.

In problems 8 and 9, make use of the appropriate form of Green's Theorem to calculate

$$\int_C \mathbf{F} \cdot \mathbf{n}\,ds,$$

where $\mathbf{n}$ is the unit normal that points towards the exterior of C:

8.

$$\mathbf{F}(x,y) = x^2y^3\mathbf{i} - xy\mathbf{j}$$

and C is the positively oriented boundary of the triangular region with vertices $(0,0)$, $(1,0)$ and $(1,2)$.

9.

$$\mathbf{F}(x,y) = x^3\mathbf{i} + y^3\mathbf{j}$$

and C is the positively oriented boundary of the disk bounded by the circle $x^2 + y^2 = 4$.

14.7   Stokes' Theorem

In this section we will discuss Stokes' Theorem which is a generalization of Green's Theorem to $\mathbb{R}^3$.

The Meaning and Plausibility of Stokes' Theorem

Assume that M is a smooth orientable surface that is parametrized by the function $\mathbf{\Phi} : D \subset \mathbb{R}^2 \to \mathbb{R}^3$. Thus, $\mathbf{N} = \partial_u\mathbf{\Phi} \times \partial_v\mathbf{\Phi}$ is continuous on D, and let's assume that $\mathbf{N}$ specifies the positive orientation of M. Also assume that the boundary curve ∂M is parametrized by a function $\phi : J \subset \mathbb{R} \to \mathbb{R}^3$. Informally, the boundary of M is said to be positively oriented if a person who walks along the boundary in the direction of the tangent determined by ϕ and whose head is in the direction of the normal N to the surface sees M to his/her left.

Recall that we can express Green's Theorem as follows:

$$\int_C \mathbf{F} \cdot d\boldsymbol{\sigma} = \int_C \mathbf{F} \cdot \mathbf{T} ds = \int\int_D (\boldsymbol{\nabla} \times \mathbf{F}) \cdot \mathbf{k} dS$$

if C is the positively oriented boundary of the simply connected planar region D. Stokes' Theorem generalizes Green's Theorem to surfaces than span a curve in space:

Theorem 1 Assume that M is a smooth orientable surface, ∂M is its positively oriented boundary and F is a smooth vector field (i.e., the components of F has continuous partial derivatives). Then,

$$\int_{\partial M} \mathbf{F} \cdot d\boldsymbol{\sigma} = \int_{\partial M} \mathbf{F} \cdot \mathbf{T} ds = \int\int_M (\boldsymbol{\nabla} \times \mathbf{F}) \cdot \mathbf{n} dS$$

You can find a discussion of the plausibility of Stokes' Theorem at the end of this section.

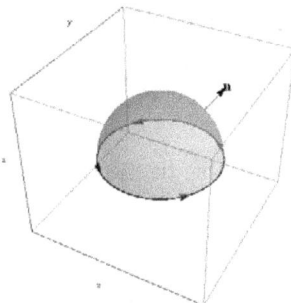

Figure 1

Example 1 Let M be the image of $z = g(x,y) = 4 - x^2 - y^2$, where $x^2 + y^2 \leq 4$, and let

$$\mathbf{F}(x,y,z) = 2z\mathbf{i} + x\mathbf{j} + y^2\mathbf{k}$$

Use Stokes' Theorem in order to evaluate

$$\int\int_M (\nabla \times \mathbf{F}) \cdot \mathbf{n} dS,$$

where $\mathbf{n}$ denotes the unit normal that points upwards.

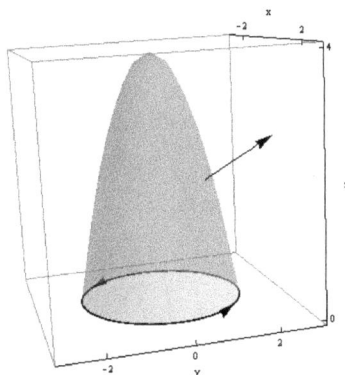

Figure 2

Solution

The circle $x^2 + y^2 = 4$ is the boundary of M and can be parametrized by $\boldsymbol{\sigma}(t) = 2\cos(t)\,\mathbf{i} + 2\sin(t)\,\mathbf{j} + 0\mathbf{k}$, $0 \le t \le 2\pi$ (note that $\boldsymbol{\sigma}$ provides the boundary ∂M with the required orientation). By Stokes' Theorem,

$$
\iint_M (\nabla \times \mathbf{F}) \cdot \mathbf{n}\, dS = \int_{\partial M} \mathbf{F} \cdot d\boldsymbol{\sigma}
$$

$$
= \int_{\partial M} \left(2z\mathbf{i} + x\mathbf{j} + y^2\mathbf{k}\right) \cdot d\mathbf{r}
$$

$$
= \int_0^{2\pi} \left((0)\,\mathbf{i} + 2\cos(t)\,\mathbf{j} + 4\sin^2(t)\,\mathbf{k}\right) \cdot \left(-2\sin(t)\,\mathbf{i} + 2\cos(t)\,\mathbf{j}\right) dt
$$

$$
= \int_0^{2\pi} 4\cos^2(t)\, dt = 4\pi
$$

$\square$

Stokes' Theorem leads to a nice interpretation of the curl of a vector field:

Proposition 1 Assume that Π is a plane that contains the point (x_0, y_0, z_0) and let $\mathbf{n}$ denote a unit vector that is orthogonal to Π. Let D_r be the disk of radius r that is in the plane Π and centered at (x_0, y_0, z_0). Let ∂D_r be the boundary of D_r that has positive orientation with respect to the orientation of the plane as determined by $\mathbf{n}$ (as in Figure 3).Then

$$
\lim_{r \to 0} \frac{1}{\pi r^2} \int_{\partial D_r} \mathbf{F} \cdot d\boldsymbol{\sigma} = \lim_{r \to 0} \frac{1}{\pi r^2} \int_{\partial D_r} \mathbf{F} \cdot \mathbf{T}\, ds = (\boldsymbol{\nabla} \times \mathbf{F})(x_0, y_0, z_0) \cdot \mathbf{n}.
$$

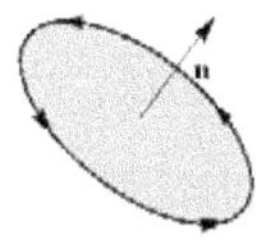

Figure 3

Note that the integral $\int_{\partial D_r} \mathbf{F} \cdot \mathbf{T} ds$ is the circulation of the vector field around ∂D_r. Therefore, Proposition 1 says that the limiting case of the circulation of $\mathbf{F}$ as the disk shrinks to the point (x_0, y_0, z_0) is the component of the curl of $\mathbf{F}$ that is normal to the plane of the disks. Thus, we can interpret $(\boldsymbol{\nabla} \times \mathbf{F})(x_0, y_0, z_0) \cdot \mathbf{n}$ as the circulation of the restriction of $\mathbf{F}$ to that plane at the point (x_0, y, z).

A plausibility argument for Proposition1:

By Stokes Theorem,

$$\int_{\partial D_r} \mathbf{F} \cdot d\boldsymbol{\sigma} = \int\int_{D_r} \boldsymbol{\nabla} \times \mathbf{F} \cdot \mathbf{n} \, dS.$$

If r is small,

$$\int\int_{D_r} \boldsymbol{\nabla} \times \mathbf{F} \cdot \mathbf{n} \, dS \cong (\boldsymbol{\nabla} \times \mathbf{F}(x_0, y_0, z_0) \cdot \mathbf{n})(\text{area of } D_r) = (\boldsymbol{\nabla} \times \mathbf{F}(x_0, y_0, z_0) \cdot \mathbf{n})(\pi r^2).$$

Thus,

$$\frac{1}{\pi r^2} \int_{\partial D_r} \mathbf{F} \cdot d\boldsymbol{\sigma} = \frac{1}{\pi r^2} \int\int_{D_r} \boldsymbol{\nabla} \times \mathbf{F} \cdot \mathbf{n} \, dS \cong \frac{1}{\pi r^2}(\boldsymbol{\nabla} \times \mathbf{F}(x_0, y_0, z_0) \cdot \mathbf{n})(\pi r^2) = \boldsymbol{\nabla} \times \mathbf{F}(x_0, y_0, z_0) \cdot \mathbf{n}.$$

This supports to the assertion that

$$\lim_{r \to 0} \frac{1}{\pi r^2} \int_{\partial D_r} \mathbf{F} \cdot d\boldsymbol{\sigma} = \boldsymbol{\nabla} \times \mathbf{F}(x_0, y_0, z_0) \cdot \mathbf{n}.$$

$\blacksquare$

Remark In Section 14.1 we noted that the condition $\boldsymbol{\nabla} \times \mathbf{F} = \mathbf{0}$ is necessary for the field to be conservative. **Stokes Theorem can be used to show that the condition is also sufficient for F to be conservative in certain types of regions.** For example, **convex regions** are those that have the property that any two points in the region can be connected to each other by a line segment that stays in the region. Interiors of spheres and ellipsoids are examples of convex regions. The condition is sufficient for $\mathbf{F}$ to be conservative in a convex region. $\Diamond$

The proof of the above assertion is left to a post-calculus course in vector analysis.

A Plausibility Argument for Stokes' Theorem (Optional)

Let's express Stokes' Theorem in terms of the components of the vector field $\mathbf{F}$ and in the differential form notations for the line and surface integrals. If

$$F(x, y, z) = M(x, y, z)\,i + N(x, y, z)\,j + P(x, y, z),$$

then,

$$\nabla \times F = \left(\frac{\partial P}{\partial y} - \frac{\partial N}{\partial z}\right)\mathbf{i} - \left(\frac{\partial P}{\partial x} - \frac{\partial M}{\partial z}\right)\mathbf{j} + \left(\frac{\partial N}{\partial x} - \frac{\partial M}{\partial y}\right)\mathbf{k},$$

$$\int_{\partial M} \mathbf{F} \cdot d\boldsymbol{\sigma} = \int_{\partial M} M\,dx + N\,dy + P\,dz,$$

and

$$\int\int_M (\boldsymbol{\nabla} \times \mathbf{F}) \cdot \mathbf{n} \, dS = \int\int_M \left(\frac{\partial P}{\partial y} - \frac{\partial N}{\partial z}\right)dydz + \left(\frac{\partial M}{\partial z} - \frac{\partial P}{\partial x}\right)dzdx + \left(\frac{\partial N}{\partial x} - \frac{\partial M}{\partial y}\right)dxdy.$$

Therefore, Stokes' Theorem takes the following form:

$$\int_{\partial M} M dx + N dy + P dz = \int\int_M \left(\frac{\partial P}{\partial y} - \frac{\partial N}{\partial z} \right) dy dz + \left(\frac{\partial M}{\partial z} - \frac{\partial P}{\partial x} \right) dz dx + \left(\frac{\partial N}{\partial x} - \frac{\partial M}{\partial y} \right) dx dy.$$

It is enough to show that

$$\int_{\partial M} M dx = \int\int_M \frac{\partial M}{\partial z} dz dx - \frac{\partial M}{\partial y} dx dy$$

(for $F = M\mathbf{i}$). The cases $F = N\mathbf{j}$ and $F = P\mathbf{k}$ are similar.

We will assume that $\Phi : D \subset \mathbb{R}^2 \to \mathbb{R}^3$ parametrizes the surface M and that the restriction of Φ to the boundary of D, to be denoted by ∂D, parametrizes ∂M with the positive orientation. Let's set

$$\Phi(u, v) = (x(u, v), y(u, v), z(u, v))$$

and

$$F(x, y, z) = M(x, y, z) i + N(x, y, z) j + P(x, y, z) k.$$

We have

$$\int_{\partial M} M dx = \int_{\partial D} M \left(\frac{\partial x}{\partial u} du + \frac{\partial x}{\partial v} dv \right),$$

and

$$\int\int_M \frac{\partial M}{\partial z} dz dx - \frac{\partial M}{\partial y} dx dy = \int\int_D \left(\frac{\partial M}{\partial z} \frac{\partial(z, x)}{\partial(u, v)} - \frac{\partial M}{\partial y} \frac{\partial(x, y)}{\partial(u, v)} \right) du dv.$$

Therefore,, we need to show that

$$\int_{\partial D} M \left(\frac{\partial x}{\partial u} du + \frac{\partial x}{\partial v} dv \right) = \int\int_D \left(\frac{\partial M}{\partial z} \frac{\partial(z, x)}{\partial(u, v)} - \frac{\partial M}{\partial y} \frac{\partial(x, y)}{\partial(u, v)} \right) du dv.$$

By Green's Theorem,

$$\int_{\partial D} M \left(\frac{\partial x}{\partial u} du + \frac{\partial x}{\partial v} dv \right)$$

$$= \int\int_D \left(\frac{\partial}{\partial u} \left(M \frac{\partial x}{\partial v} \right) - \frac{\partial}{\partial v} \left(M \frac{\partial x}{\partial u} \right) \right) du dv$$

$$= \int\int_D \left(\left(\frac{\partial M}{\partial u} \frac{\partial x}{\partial v} + M \frac{\partial^2 x}{\partial u \partial v} \right) - \left(\frac{\partial M}{\partial v} \frac{\partial x}{\partial u} + M \frac{\partial^2 x}{\partial v \partial u} \right) \right) du dv$$

$$= \int\int_D \left(\frac{\partial M}{\partial u} \frac{\partial x}{\partial v} - \frac{\partial M}{\partial v} \frac{\partial x}{\partial u} \right) du dv$$

$$= \int\int_D \left(\left(\frac{\partial M}{\partial x} \frac{\partial x}{\partial u} + \frac{\partial M}{\partial y} \frac{\partial y}{\partial u} + \frac{\partial M}{\partial z} \frac{\partial z}{\partial u} \right) \frac{\partial x}{\partial v} - \left(\frac{\partial M}{\partial x} \frac{\partial x}{\partial v} + \frac{\partial M}{\partial y} \frac{\partial y}{\partial v} + \frac{\partial M}{\partial z} \frac{\partial z}{\partial u} \right) \frac{\partial x}{\partial u} \right) du dv$$

$$= \int\int_D \left(\frac{\partial M}{\partial x} \left(\frac{\partial x}{\partial u} \frac{\partial x}{\partial v} - \frac{\partial x}{\partial v} \frac{\partial x}{\partial u} \right) + \frac{\partial M}{\partial y} \left(\frac{\partial y}{\partial u} \frac{\partial x}{\partial v} - \frac{\partial y}{\partial v} \frac{\partial x}{\partial u} \right) + \frac{\partial M}{\partial z} \left(\frac{\partial z}{\partial u} \frac{\partial x}{\partial v} - \frac{\partial z}{\partial v} \frac{\partial x}{\partial u} \right) \right) du dv$$

$$= \int\int_D \left(\frac{\partial M}{\partial z} \frac{\partial(z, x)}{\partial(u, v)} - \frac{\partial M}{\partial y} \frac{\partial(x, y)}{\partial(u, v)} \right) du dv.$$

∎

Problems

In problems 1 and 2 use Stokes' Theorem to express

$$\int\int_M (\nabla \times \mathbf{F}) \cdot \mathbf{n} dS$$

as a line integral on C. Evaluate the line integral:

1.

$$\mathbf{F}(x, y, z) = y\mathbf{i} + 2x\mathbf{j} + z\mathbf{k},$$

M is the graph of

$$z = \sqrt{16 - x^2 - y^2} \text{ where } x^2 + y^2 \le 16.$$

(M is a hemisphere of radius 4 centered at the origin). C is the circle of radius 4 in the xy-plane centered at the origin. M is oriented so that the normal points upward and C is oriented so that it is traversed counterclockwise.

2.

$$\mathbf{F}(x, y, z) = y\mathbf{i} - x\mathbf{j} + z^2\mathbf{k},$$

M is the portion of the sphere

$$x^2 + y^2 + (z - 1)^2 = 2$$

above the xy-plane. C is the circle of radius 1 in the xy-plane centered at the origin. M is oriented so that the normal points upward and C is oriented so that it is traversed counterclockwise.

In problems 3 and 4 use Stokes Theorem to express the line integral

$$\int_C \mathbf{F} \cdot d\boldsymbol{\sigma}$$

as a surface integral on M. Evaluate the surface integral:

3.

$$\mathbf{F}(x, y, z) = z\mathbf{i} + x^3\mathbf{j} + y^2\mathbf{k}$$

M is the part of the plane $z = 4$ within the cylinder $x^2 + y^2 = 9$ and the normal points upward. C is the intersection of the plane and the cylinder. The orientation of C is counterclockwise (a person whose head is in the direction of the normal to the plane and traverses C sees the plane on his/her left).

4.

$$\mathbf{F}(x, y, z) = z\mathbf{i} - x\mathbf{j} + 2y\mathbf{k}.$$

M is the part of the plane $z = y + 2$ that is inside the cylinder $x^2 + y^2 = 4$ and the normal points upward. C is the intersection of the plane and the cylinder. The orientation of C is counterclockwise (a person whose head is in the direction of the normal to the plane and traverses C sees the plane on his/her left).

14.8   Gauss' Theorem

The Meaning and Plausibility of Gauss' Theorem

Gauss' Theorem establishes an important link between the flux of a vector field across a closed surface such a sphere and the divergence of the vector field in the interior of the surface:

Theorem 1 Assume that W is a region in $\mathbb{R}^3$ that is bounded by the smooth orientable closed surface ∂W. If $\mathbf{n}$ is the unit normal that points towards the exterior of W then

$$\int\int_{\partial W} \mathbf{F} \cdot \mathbf{n} dS = \int\int\int_W \boldsymbol{\nabla} \cdot \mathbf{F} dV$$

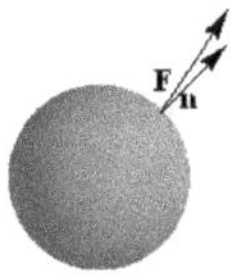

Figure 1

Gauss' Theorem is also referred to as the **Divergence Theorem** since the statement involves the divergence of the vector field that is being discussed. You can find a plausibility argument for Gauss' Theorem at the end of this section.

Example 1 Let D be the interior of the sphere $x^2 + y^2 + z^2 = 4$ and let

$$\mathbf{F}(x, y, z) = 2x^3\mathbf{i} + 2y^3\mathbf{j} + 2z^3\mathbf{k}.$$

Use the Divergence Theorem in order to determine

$$\int\int_{\partial D} \mathbf{F} \cdot \mathbf{n} dS,$$

where $\mathbf{n}$ is the unit normal to ∂D that points towards the exterior of D.

Solution

We have

$$(\boldsymbol{\nabla} \cdot \mathbf{F})(x, y, z) = \frac{\partial}{\partial x}\left(2x^3\right) + \frac{\partial}{\partial y}\left(2y^3\right) + \frac{\partial}{\partial z}\left(2z^3\right) = 6\left(x^2 + y^2 + z^2\right).$$

Therefore,

$$\begin{aligned}
\int\int_{\partial D} F \cdot \mathbf{n} dS &= \int\int\int_D \left(\boldsymbol{\nabla} \cdot \mathbf{F}\right) dx dy dz \\
&= \int\int\int_D 6\left(x^2 + y^2 + z^2\right) dx dy dz \\
&= 6\int_{\rho=0}^{\rho=2} \int_{\phi=0}^{\pi} \int_{\theta=0}^{\theta=2\pi} \rho^2 \rho^2 \sin(\phi)\, d\theta d\phi d\rho \\
&= 6\int_{\rho=0}^{\rho=2} \int_{\phi=0}^{\pi} \int_{\theta=0}^{\theta=2\pi} \rho^4 \sin(\phi)\, d\theta d\phi d\rho \\
&= \frac{768\pi}{5}
\end{aligned}$$

□

Gauss' Theorem leads to an interpretation of divergence as outward flux per unit volume at a point:

Proposition 1 Let D_r be the ball of radius r that is centered at (x_0, y_0, z_0) and let ∂D_r be the sphere that bounds D_r. If $V(D_r)$ denotes the volume of D_r then

$$\lim_{r \to 0} \frac{1}{V(D_r)} \int\int_{\partial D_r} \mathbf{F} \cdot \mathbf{n} dS = (\nabla \cdot \mathbf{F})(x_0, y_0, z_0)$$

(assuming that the divergence of F is continuous).

A plausibility argument for the Proposition:

By Gauss' Theorem

$$\int\int_{\partial D_r} \mathbf{F} \cdot \mathbf{n} dS = \int\int\int_{D_r} \nabla \cdot \mathbf{F} dV.$$

If r is small,

$$\int\int\int_{D_r} \nabla \cdot \mathbf{F} dV \cong (\nabla \cdot \mathbf{F})(x_0, y_0, z_0) V(D_r)$$

so that

$$\frac{1}{V(D_r)} \int\int_{\partial D_r} \mathbf{F} \cdot \mathbf{n} dS \cong (\nabla \cdot \mathbf{F})(x_0, y_0, z_0).$$

This makes it plausible that

$$\lim_{r \to 0} \frac{1}{V(D_r)} \int\int_{\partial D_r} \mathbf{F} \cdot \mathbf{n} dS = (\nabla \cdot \mathbf{F})(x_0, y_0, z_0)$$

■

Thus, we can interpret $(\nabla \cdot \mathbf{F})(x_0, y_0, z_0)$ as the **flux per unit volume at the point** (x_0, y_0, z_0).

Example 2 (Another law by Gauss) Let W be a region in $\mathbb{R}^3$ and assume that the origin is not on the boundary of W (∂W). If $\mathbf{r}(x, y, x) = x\mathbf{i} + y\mathbf{j} + z\mathbf{k}$ is the position vector of the point (x, y, z) then

$$\int\int_{\partial W} \frac{\mathbf{r} \cdot \mathbf{n}}{\|\mathbf{r}\|^3} dS = 4\pi$$

if the origin is in W. Otherwise the integral is 0.

Indeed, if the origin is not in W,

$$\int\int_{\partial W} \frac{\mathbf{r} \cdot \mathbf{n}}{\|\mathbf{r}\|^3} dS = \int\int\int_W \nabla \cdot \left(\frac{\mathbf{r}}{\|\mathbf{r}\|^3}\right) dV$$

by Gauss' Theorem. Now,

$$\nabla \cdot \left(\frac{\mathbf{r}}{\|\mathbf{r}\|^3}\right) = \nabla \cdot \left(\frac{1}{(x^2 + y^2 + z^2)^{3/2}}(x\mathbf{i} + y\mathbf{j} + z\mathbf{k})\right)$$

$$= \frac{\partial}{\partial x}\left(\frac{x}{(x^2 + y^2 + z^2)^{3/2}}\right) + \frac{\partial}{\partial y}\left(\frac{y}{(x^2 + y^2 + z^2)^{3/2}}\right) + \frac{\partial}{\partial z}\left(\frac{z}{(x^2 + y^2 + z^2)^{3/2}}\right).$$

We have

$$\frac{\partial}{\partial x}\left(\frac{x}{\left(x^2+y^2+z^2\right)^{3/2}}\right) = \frac{\left(x^2+y^2+z^2\right)^{3/2} - x\left(3/2\right)\left(x^2+y^2+z^2\right)^{1/2}\left(2x\right)}{\left(x^2+y^2+z^2\right)^3}$$

$$= \frac{\left(x^2+y^2+z^2\right)^{1/2}\left(\left(x^2+y^2+z^2\right)-3x^2\right)}{\left(x^2+y^2+z^2\right)^3}$$

$$= \frac{\left(-2x^2+y^2+z^2\right)}{\left(x^2+y^2+z^2\right)^{5/2}}.$$

Thus,

$$\nabla\cdot\left(\frac{\mathbf{r}}{||\mathbf{r}||^3}\right) = \frac{\left(-2x^2+y^2+z^2\right)}{\left(x^2+y^2+z^2\right)^{5/2}} + \frac{\left(-2y^2+x^2+z^2\right)}{\left(x^2+y^2+z^2\right)^{5/2}} + \frac{\left(-2z^2+x^2+y^2\right)}{\left(x^2+y^2+z^2\right)^{5/2}} = 0.$$

Therefore,

$$\iint_{\partial W}\frac{\mathbf{r}\cdot\mathbf{n}}{||\mathbf{r}||^3}dS = \iiint_W \nabla\cdot\left(\frac{\mathbf{r}}{||\mathbf{r}||^3}\right)dV = 0.$$

If $(0,0,0)$ is in W, we apply the divergence theorem to the region W' between ∂W and ∂B_a, where B is the ball of sufficiently small radius a centered at $(0,0,0)$:

$$\iint_{\partial W}\frac{\mathbf{r}\cdot\mathbf{n}}{||\mathbf{r}||^3}dS - \iint_{\partial B_a}\frac{\mathbf{r}\cdot\mathbf{n}}{||\mathbf{r}||^3}dS = \iiint_{W'} \nabla\cdot\left(\frac{\mathbf{r}}{||\mathbf{r}||^3}\right)dV = 0.$$

The $(-)$ sign in front of the integral on the integral on the inner boundary ∂B is due to the fact that $-\mathbf{n}$ points towards the interior of B_a which is the exterior of W'. Thus,

$$\iint_{\partial W}\frac{\mathbf{r}\cdot\mathbf{n}}{||\mathbf{r}||^3}dS = \iint_{\partial B_a}\frac{\mathbf{r}}{||\mathbf{r}||^3}\cdot\mathbf{n}dS = \iint_{\partial B_a}\frac{\mathbf{r}}{||\mathbf{r}||^3}\cdot\frac{\mathbf{r}}{||\mathbf{r}||}dS = \iint_{\partial B_a}\frac{||\mathbf{r}||^2}{||\mathbf{r}||^4}dS = \frac{1}{a^2}\iint_{\partial B_a}dS = \frac{1}{a^2}\left(4\pi a^2\right) = 4\pi.$$

$\square$

Example 3 (The Continuity Equation) Let $\rho(x,y,z,t)$ be the density of the matter (for example water) at a point (x,y,z) and time t. Let $\mathbf{v}(x,y,z,t)$ be the corresponding velocity. The amount of material in a region W at the instant t is

$$\iiint_W \rho\left(x,y,z,t\right)dV.$$

This changes at the rate

$$\frac{d}{dt}\iiint_W \rho\left(x,y,z,t\right)dV = \iiint_W \frac{\partial\rho}{\partial t}dV.$$

The rate at which the material flows through the boundary ∂W is

$$\iint_{\partial W}\rho\mathbf{v}\cdot\mathbf{n}dS$$

If material is conserved, the amount of matter in W decreases at the rate at which matter flows outward across ∂W. Thus

$$\iiint_W \frac{\partial\rho}{\partial t}dV = -\iint_{\partial W}\rho\mathbf{v}\cdot\mathbf{n}dS$$

By **Gauss' Theorem**,

$$\iint_{\partial W} \rho \mathbf{v} \cdot \mathbf{n} dS = \iiint_W \boldsymbol{\nabla} \cdot (\rho \mathbf{v})\, dV$$

Therefore,

$$\iiint_W \frac{\partial \rho}{\partial t} dV = -\iiint_W \boldsymbol{\nabla} \cdot (\rho \mathbf{v})\, dV$$

for every W. This implies that

$$\frac{\partial \rho}{\partial t} + \nabla \cdot (\rho \mathbf{v}) = 0$$

for each (x, y, z, t) (this can be justified by considering the equality of the integrals for a sequence of regions that shrink towards a given point (x, y, z) at a given time t). This is **the continuity equation.**of continuum mechanics.

For example, if the density ρ is constant we have

$$\nabla \cdot \mathbf{v} = 0.$$

Thus, **if the fluid is incompressible, the divergence of the velocity field is 0.** $\square$

Example 4 (Electrostatics) Gauss' law for electrostatics can be expressed in the following form:

$$\iint_{\partial W} \mathbf{F} \cdot \mathbf{n} dS = \frac{1}{\in_0} \iiint_W \rho dV.$$

Here $\in_0$is a universal constant and ρ denotes the charge density. Thus, **the flux across a closed surface is equal to the total charge in the interior.** By Gauss' Theorem,

$$\iint_{\partial W} \mathbf{F} \cdot \mathbf{n} dS = \iiint_W \boldsymbol{\nabla} \cdot \mathbf{F} dV.$$

Therefore,

$$\iiint_W \nabla \cdot \mathbf{F} dV = \frac{1}{\in_0} \iiint_W \rho dV$$

for every W. Thus we must have

$$\boldsymbol{\nabla} \cdot \mathbf{F}\,(x, y, z) = \frac{1}{\in_0} \rho\,(x, y, z)$$

at each (x, y, z). If $\mathbf{F}$ is conservative and $-\Phi$ is a potential so that

$$\mathbf{F} = -\boldsymbol{\nabla}\Phi,$$

then

$$\boldsymbol{\nabla} \cdot (\boldsymbol{\nabla}\Phi) = -\frac{\rho}{\in_0}$$

We have

$$\nabla \cdot (\nabla\Phi) = \Delta\boldsymbol{\Phi},$$

where

$$\Delta\boldsymbol{\Phi} = \frac{\partial^2 \Phi}{\partial x^2} + \frac{\partial^2 \Phi}{\partial y^2} + \frac{\partial^2 \Phi}{\partial z^2}.$$

Thus

$$\Delta\boldsymbol{\Phi} = -\frac{\rho}{\in_0}$$

at each point (x, y, z). This is **Poisson's Equation** for $-\rho/\in_0$. $\square$

A plausibility argument for Gauss' Theorem (Optional)

In terms of the components of $\mathbf{F} = M\mathbf{i} + N\mathbf{j} + P\mathbf{k}$, and the differential form notation,

$$\iint_{\partial W} \mathbf{F} \cdot \mathbf{n}dS = \iint_{\partial W} Mdydz + Ndzdx + Pdxdy.$$

The integral over W is

$$\iiint_{W} \boldsymbol{\nabla} \cdot \mathbf{F}dV = \iiint_{W} \left(\frac{\partial M}{\partial x} + \frac{\partial N}{\partial y} + \frac{\partial P}{\partial z} \right) dxdydz.$$

Therefore, Gauss' Theorem reads as follows in the differential form notation:

$$\iint_{\partial W} Mdydz + Ndzdx + Pdxdy = \iiint_{W} \left(\frac{\partial M}{\partial x} + \frac{\partial N}{\partial y} + \frac{\partial P}{\partial z} \right) dxdydz$$

For example, let us confirm that

$$\iint_{\partial W} Pdxdy = \iiint_{W} \frac{\partial P}{\partial z}dxdydz$$

if W is xy-simple, i.e., W can be expressed as

$$\{(x, y, z) : \varphi_1(x, y) \leq z \leq \varphi_2(x, y) \text{ ,, for each } (x, y) \in D, \}$$

where D is a region in the xy-plane. In this case,

$$\iiint_{W} \frac{\partial P}{\partial z}dxdydz = \iint_{D} \left(\int_{z=\varphi_1(x,y)}^{z=\varphi_2(x,y)} \frac{\partial P}{\partial z}(x, y, z)\, dz \right) dxdy$$

$$= \iint_{D} (P(x, y, \varphi_2(x, y)) - P(x, y, \varphi_1(x, y)))\, dxdy,$$

by the Fundamental Theorem of Calculus. Note that

$$\iint_{D} (P(x, y, \varphi_2(x, y)) - P(x, y, \varphi_1(x, y)))\, dxdy$$

$$= \iint_{D} P(x, y, \varphi_2(x, y))\, dxdy - \iint_{D} P(x, y, \varphi_1(x, y))\, dxdy,$$

The first integral is the part of the surface integral

$$\iint_{\partial W} Pdxdy$$

that corresponds to the upper part of the surface. Indeed, that part of ∂W is the graph G_2 of the function φ_2 on the region D, the outward unit normal is $\mathbf{k}$, and

$$\iint_{G_2} Pdxdy = \iint_{G_2} P\mathbf{i} \cdot \mathbf{k}dS =$$

$$\iint_{D} P\mathbf{i} \cdot \mathbf{k}dxdy = \iint_{D} P(x, y, \varphi_2(x, y))\, dxdy.$$

Similarly, the integral

$$-\iint_{D} P(x, y, \varphi_1(x, y))\, dxdy$$

is the part of the surface integral

$$\iint_{\partial W} P dx dy$$

that corresponds to the lower surface that is the graph G_1 of the function φ_1. There, the outward unit normal is $-\mathbf{k}$, and

$$\iint_{G_1} P dx dy = \iint_{G_1} P\mathbf{i} \cdot (-\mathbf{k}) \, dS = -\iint_D P\mathbf{i} \cdot \mathbf{k} dx dy = -\iint_D P(x, y, \varphi_2(x, y)) \, dx dy.$$

The part of the integral

$$\iint_{\partial W} P dx dy$$

that corresponds to the lateral boundary S of W add up to 0, since

$$\iint_S P dx dy = \iint_S P\mathbf{i} \cdot \mathbf{n} dS,$$

where the unit normal $\mathbf{n}$ is orthogonal is parallel to the xy-plane. In particular $\mathbf{i} \cdot \mathbf{n} = 0$. Thus,

$$\iint_S P\mathbf{i} \cdot \mathbf{n} dS.$$

Therefore, we confirmed the equality

$$\iiint_W \frac{\partial P}{\partial z} dx dy dz = \iint_{\partial W} P dx dy.$$

Problems

In problems 1-4 evaluate the flux

$$\iint_M \mathbf{F} \cdot \mathbf{n} dS,$$

where M is a closed surface that encloses the region W by using Gauss' Theorem ($\mathbf{n}$ points towards the exterior of W):

1.

$$\mathbf{F}(x, y, z) = x\mathbf{i} + y\mathbf{j} + z\mathbf{k},$$

M is the sphere of radius 4 centered at the origin, and W is the interior of M.

2.

$$\mathbf{F}(x, y, z) = z^3\mathbf{k}$$

M is the sphere of radius 2 centered at the origin, and W is the interior of M.

3.

$$\mathbf{F}(x, y, z) = x\mathbf{i} + y\mathbf{j} + z\mathbf{k},$$

and M is the surface of the cube

$$W = \{(x, y, z) : 0 \le x \le 4, \ 0 \le y \le 4, \ 0 \le z \le 4\}.$$

4.

$$\mathbf{F}(x, y, z) = x^2\mathbf{i}$$

and M is the surface of the cube

$$W = \{(x, y, z) : 0 \le x \le 1, \ 0 \le y \le 1, \ 0 \le z \le 1\}.$$

Appendix K

Answers to Some Problems

Answers of Some Problems of Section 11.1

1.

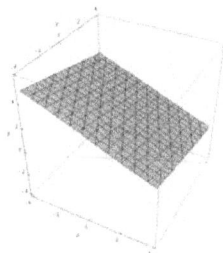

If $x = c$ then

$$2c + y + 4z = 8 \Rightarrow y + 4z = 8 - 2c.$$

These are lines.
If $y = c$ then

$$2x + c + 4z = 8 \Rightarrow 2x + 4z = 8 - c.$$

These are lines.
If $z = c$ then

$$2x + y + 4c = 8 \Rightarrow 2x + y = 8 - 4c.$$

These are lines.

3.

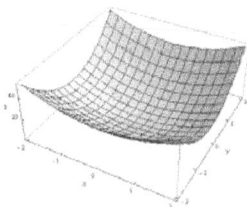

If $x = c$ then

$$z = 4c^2 + 9y^2$$

These are parabolas.
If $y = c$ then

$$z = 4x^2 + 9c^2$$

These are parabolas.
If $z = c$ then

$$c = 4x^2 + 9y^2$$

These are ellipses if $c > 0$. If $c = 0$ the curve is reduced to a single point, the origin. The surface does not have any points below the xy-plane corresponding to $c < 0$.

5.

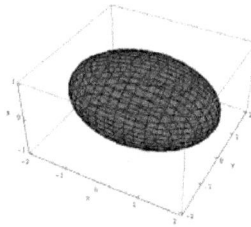

If $x = c$ then

$$c^2 + 2y^2 + 4z^2 = 4 \Rightarrow 2y^2 + 4z^2 = 4 - c^2.$$

These are ellipses if $-2 < c < 2$. The curve is reduced to a single point if $c = \pm 2$. The surface does not have any points corresponding to $x < -2$ or $x > 2$.
If $y = c$ then

$$x^2 + 2c^2 + 4z^2 = 4 \Rightarrow x^2 + 4z^2 = 4 - 2c^2$$

These are ellipses if $-\sqrt{2} < c < \sqrt{2}$. The curve is reduced to a single point if $c = \pm\sqrt{2}$. The surface does not have any points corresponding to $y > \sqrt{2}$ or $y < -\sqrt{2}$.
If $z = c$ then

$$x^2 + 2y^2 + 4c^2 = 4 \Rightarrow x^2 + 2y^2 = 4 - 4c^2$$

These are ellipses if $-1 < c < 1$. The curve is reduced to a single point if $c = \pm 1$. The surface does not have any points corresponding to $z > 1$ or $z < -1$.

7.

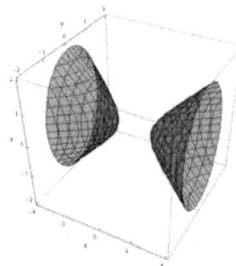

If $x = c$ then

$$c^2 - 9y^2 - 4z^2 = 1 \Rightarrow 9y^2 + 4z^2 = c^2 - 1$$

These are ellipses if $c < -1$ or $c > 1$. The curve is reduced to a single point if $c = \pm 1$. The surface does not have any points corresponding to $-1 < x < 1$.

If $y = c$ then

$$x^2 - 9c^2 - 4z^2 = 1 \Rightarrow x^2 - 4z^2 = 1 + 9c^2$$

These are hyperbolas.
If $z = c$ then

$$x^2 - 9y^2 - 4c^2 = 1 \Rightarrow x^2 - 9y^2 = 1 + 4c^2$$

These are hyperbolas.

9.

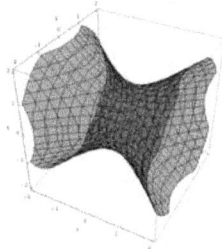

If $x = c$ then

$$y^2 - c^2 + z^2 = 1 \Rightarrow y^2 + z^2 = 1 + c^2$$

These are circles.
If $y = c$ then

$$c^2 - x^2 + z^2 = 1 \Rightarrow -x^2 + z^2 = 1 - c^2$$

These are hyperbolas.
If $z = c$ then

$$y^2 - x^2 + c^2 = 1 \Rightarrow y^2 - x^2 = 1 - c^2$$

These are hyperbolas.

Answers of Some Problems of Section 11.2

1.
a) $\mathbf{v} = (2, 3)$

b) $Q_2 = (4, 4)$

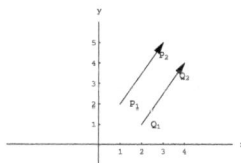

3.
a) $\mathbf{v} = (2, 5)$

b) $Q_2 = (5, 7)$

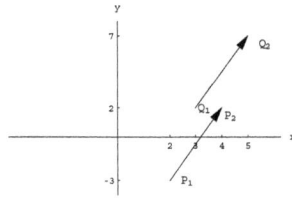

5.
a) $\mathbf{v} + \mathbf{w} = (5, 5)$
b)

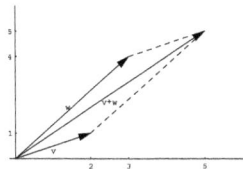

7.
a) $\mathbf{v} + \mathbf{w} = (0, 3)$
b)

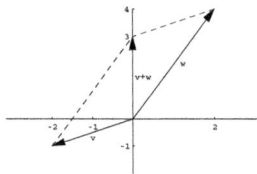

9.
a) $\mathbf{v} + \mathbf{w} = (1, 6)$
b)

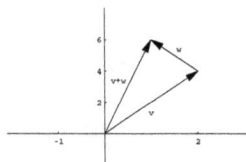

11.
a) $\mathbf{v} - \mathbf{w} = (4, 2)$

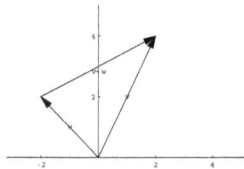

13. $2\mathbf{v} - 3\mathbf{w} = (12, -17)$

15.
a)

$$\mathbf{u} = \left(\frac{3}{5}, \frac{4}{5} \right).$$

b)

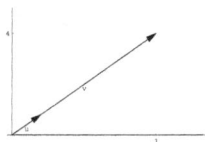

17.
a) $\mathbf{v} = 3\mathbf{i} + 2\mathbf{j}$, $\mathbf{w} = -2\mathbf{i} + 4\mathbf{j}$
b) $2\mathbf{v} - 3\mathbf{w} = 12\mathbf{i} - 8\mathbf{j}$

19.
a) $\mathbf{v} = -2\mathbf{i} + 3\mathbf{j} + 6\mathbf{k}$, $\mathbf{w} = 4\mathbf{i} - 2\mathbf{j} + \mathbf{k}$
b) $-\mathbf{v} + 4\mathbf{w} = 18\mathbf{i} - 11\mathbf{j} - 2\mathbf{k}$

Answers of Some Problems of Section 11.3

1.
a)

$$\|\mathbf{v}\| = 1, \ \|\mathbf{w}\| = \sqrt{2}, \ \mathbf{v} \cdot \mathbf{w} = 1.$$

b)

$$\theta = \frac{\pi}{4}.$$

3.
a)

$$\|\mathbf{v}\| = \sqrt{2}, \ \|\mathbf{w}\| = \sqrt{2}, \ \mathbf{v} \cdot \mathbf{w} = \sqrt{3}$$

b)

$$\theta = \frac{\pi}{6}$$

5.
a)

$$\|\mathbf{v}\| = \sqrt{2}, \ \|\mathbf{w}\| = \sqrt{2}, \ \mathbf{v} \cdot \mathbf{w} = 1$$

b)

$$\theta = \frac{\pi}{3}.$$

7.

a)

$$\|\mathbf{v}\| = \sqrt{5}, \ \|\mathbf{w}\| = \sqrt{2}, \ \mathbf{v} \cdot \mathbf{w} = -1$$

b)

$$\cos(\theta) = -\frac{1}{\sqrt{10}}.$$

c)

$$\theta \cong 1.89255$$

9.

a)

$$\|\mathbf{v}\| = \sqrt{5}, \ \|\mathbf{w}\| = \sqrt{10}, \ \mathbf{v} \cdot \mathbf{w} = -1$$

b)

$$\cos(\theta) = -\frac{1}{\sqrt{50}}.$$

c)

$$\theta \cong 1.71269$$

11.

a)

$$\mathbf{u} = \left(\frac{2}{\sqrt{13}}, \frac{3}{\sqrt{13}} \right)$$

b) The direction cosines of **v** are

$$\frac{2}{\sqrt{13}} \text{ and } \frac{3}{\sqrt{13}}$$

13.

a)

$$\mathbf{u} = \left(-\frac{3}{5}, \frac{4}{5} \right)$$

b) The direction cosines of **v** are

$$-\frac{3}{5} \text{ and } \frac{4}{5}$$

15.

a)

$$\mathbf{u} = \left(\frac{6}{\sqrt{40}}, \frac{2}{\sqrt{40}} \right)$$

b)

$$comp_{\mathbf{w}}\mathbf{v} = \frac{26}{\sqrt{40}}$$

c)

$$P_{\mathbf{w}}\mathbf{v} = \left(\frac{39}{10}, \frac{13}{10} \right)$$

d)

$$\mathbf{v_2} = \left(-\frac{9}{10}, \frac{27}{10} \right)$$

17.

a)

$$\mathbf{u} = \left(\frac{2}{\sqrt{5}}, -\frac{1}{\sqrt{5}} \right)$$

b)

$$comp_{\mathbf{w}}\mathbf{v} = \frac{4}{\sqrt{5}}$$

c)

$$P_{\mathbf{w}}\mathbf{v} = \left(\frac{8}{5}, -\frac{4}{5} \right)$$

d)

$$\mathbf{v_2} = \left(-7, -\frac{6}{5} \right)$$

Answers of Some Problems of Section 11.4

1.
$$\mathbf{v} \times \mathbf{w} = 5\mathbf{k}$$

3.
$$\mathbf{v} \times \mathbf{w} = -11\mathbf{i} - 11\mathbf{j} + 11\mathbf{k}$$

5. 12

7. 7

9.
$$3x + 2y + z = 13$$

11.
$$-x + y + 2z = -8$$

13.
$$-5 + 2y + 5z = 9$$

Answers of Some Problems of Section 12.1

1.

a)

$$\boldsymbol{\sigma}'(t) = \mathbf{i} + 8\sin(4t)\cos(4t)\mathbf{j}, \ \boldsymbol{\sigma}'\left(\frac{\pi}{3}\right) = \mathbf{i} + 2\sqrt{3}\mathbf{j}, \ \mathbf{T}\left(\frac{\pi}{3}\right) = \frac{1}{\sqrt{13}}\mathbf{i} + \frac{2\sqrt{3}}{\sqrt{13}}\mathbf{j}$$

b)

$$\mathbf{L}(u) = \left(\frac{\pi}{3} + u, \frac{3}{4} + 2\sqrt{3}u\right), \quad -\infty < u \le +\infty.$$

3.

a)

$$\boldsymbol{\sigma}'(t) = \left(\frac{-3t^2 + 3}{(t^2 + 1)^2}, \frac{6t}{(t^2 + 1)^2}\right), \quad \boldsymbol{\sigma}'(1) = \left(0, \frac{3}{2}\right), \quad \mathbf{T}(1) = \mathbf{j}$$

b)

$$\mathbf{L}(u) = \left(\frac{3}{2}, \frac{3}{2} + \frac{3}{2}u\right), \quad u \in \mathbb{R}.$$

5.

$$\mathbf{v}(t) = \boldsymbol{\sigma}'(t) = e^{-t}\left(\cos(t) - \sin(t)\right)\mathbf{i} - e^{-t}\left(\cos(t) + \sin(t)\right)\mathbf{j}$$

$$\mathbf{v}(\pi) = \left(-e^{-\pi}, e^{-\pi}\right) = -e^{-\pi}\mathbf{i} + e^{-\pi}\mathbf{j}$$

Speed at $t = \pi$ is

$$\|\mathbf{v}(\pi)\| = \sqrt{2}e^{-\pi}.$$

Answers of Some Problems of Section 12.2

1.

$$\mathbf{a}(t) = -48\cos(4t)\mathbf{i} - 48\sin(4t)\mathbf{j}$$

$$\mathbf{a}(\pi/12) = -24\mathbf{i} - 24\sqrt{3}\mathbf{j}$$

3.

$$\mathbf{a}(t) = -\frac{2t}{(t^2 + 1)^2}\mathbf{j}.$$

$$\mathbf{a}(1) = -\frac{2}{4}\mathbf{j} = -\frac{1}{2}\mathbf{j}.$$

5.

$$\mathbf{T}(\pi/6) = \frac{1}{2}\left(-1, \sqrt{3}\right).$$

$$\mathbf{N}(\pi/6) = -\frac{1}{2}\left(\sqrt{3}, 1\right).$$

7.

a)

$$s(t) = \int_0^t \sqrt{4\sin^2(\tau) + 9\cos^2(\tau)}d\tau.$$

b)

$$s(\pi) \cong 7.93272$$

9.

$$\kappa(\pi/2) = \frac{3}{4}.$$

11. The tangential component of the acceleration is always 0. The normal component of the acceleration is 8.

Answers of Some Problems of Section 12.3

In the plots of the level curves for problems 1-5, the smaller values of f are indicated by the darker color.

1.
a)

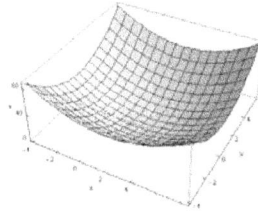

b)

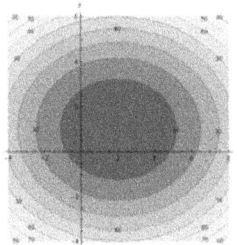

3.
a)

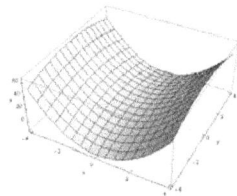

b)

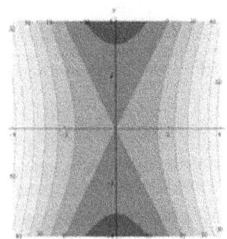

5.

a)

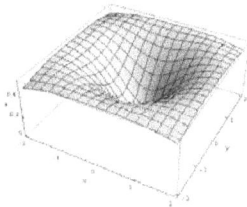

b)

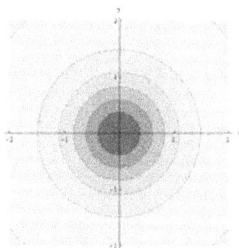

7.

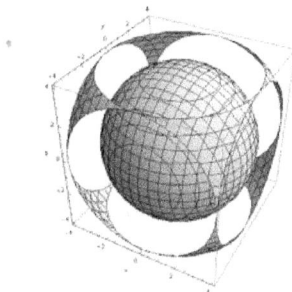

The level surfaces are spheres. In the pictures, the outer sphere is shown partially in order to show the inner sphere.

9. The pictures show two level surfaces of f.

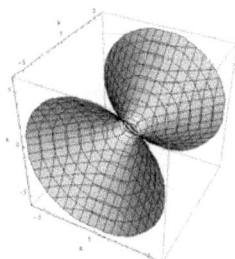

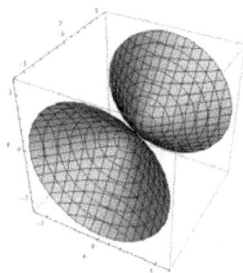

Answers of Some Problems of Section 12.4

1.

$$\frac{\partial}{\partial x}\left(4x^2 + 9y^2\right) = 8x, \ \frac{\partial}{\partial y}\left(4x^2 + 9y^2\right) = 18y$$

3.

$$\frac{\partial}{\partial x}\sqrt{2x^2 + y^2} = \frac{2x}{\sqrt{2x^2 + y^2}}, \ \frac{\partial}{\partial y}\sqrt{2x^2 + y^2} = \frac{y}{\sqrt{2x^2 + y^2}}$$

5.

$$\frac{\partial}{\partial x}e^{-x^2-y^2} = -2xe^{-x^2-y^2}, \ \frac{\partial}{\partial y}e^{-x^2-y^2} = -2ye^{-x^2-y^2}$$

7.

$$\frac{\partial}{\partial x}\ln\left(x^2 + y^2\right) = \frac{2x}{x^2 + y^2}, \ \frac{\partial}{\partial y}\ln\left(x^2 + y^2\right) = \frac{2y}{x^2 + y^2}$$

9.

$$\frac{\partial}{\partial x}\arctan\left(\frac{y}{\sqrt{x^2 + y^2}}\right) = -\frac{xy}{\left(x^2 + 2y^2\right)\sqrt{x^2 + y^2}},$$

$$\frac{\partial}{\partial y}\arctan\left(\frac{y}{\sqrt{x^2 + y^2}}\right) = \frac{x^2}{\left(x^2 + 2y^2\right)\sqrt{x^2 + y^2}}$$

11.

$$\frac{\partial}{\partial x}\left(\frac{1}{\sqrt{x^2 + y^2 + z^2}}\right) = -\frac{x}{\left(x^2 + y^2 + z^2\right)^{3/2}}$$

Similarly,

$$\frac{\partial}{\partial y}\left(\frac{1}{\sqrt{x^2 + y^2 + z^2}}\right) = -\frac{z}{\left(x^2 + y^2 + z^2\right)^{3/2}}$$

13.

$$\frac{\partial}{\partial x}\arcsin\left(\frac{1}{\sqrt{x^2 + y^2 + z^2}}\right) = -\frac{x}{\sqrt{x^2 + y^2 + z^2 - 1}\left(x^2 + y^2 + z^2\right)}$$

15.
a)

$$\mathbf{i} + \frac{3}{\sqrt{10}}\mathbf{k} \text{ and } \mathbf{j} + \frac{1}{\sqrt{10}}\mathbf{k}$$

b)

$$\mathbf{L}_1\left(u\right) = \left(3 + u, 1, \sqrt{10} + \frac{3}{\sqrt{10}}u\right),$$

and

$$\mathbf{L}_2\left(u\right) = \left(3, 1+u, \sqrt{10} + \frac{1}{\sqrt{10}}u\right), \ u \in \mathbb{R}.$$

17.

a)

$$\mathbf{i} + 2e\mathbf{k} \text{ and } \mathbf{j}$$

b)

$$\mathbf{L}_1\left(u\right) = \left(1+u, 0, e+2eu\right),$$

and

$$\mathbf{L}_2\left(u\right) = \left(1, u, e\right), \ u \in \mathbb{R}.$$

19.

$$\frac{\partial f}{\partial x}\left(x, y\right) = 8x, \ \frac{\partial f}{\partial y}\left(x, y\right) = 18y, \ \frac{\partial^2 f}{\partial x^2}\left(x, y\right) = 0$$

21.

$$\frac{\partial f}{\partial x}\left(x, y\right) = -2xe^{-x^2+y^2}, \ \frac{\partial^2 f}{\partial x^2}\left(x, y\right) = -2e^{-x^2+y^2} + 4x^2 e^{-x^2+y^2}, \ \frac{\partial^2 f}{\partial y \partial x}\left(x, y\right) = -4xye^{-x^2+y^2}$$

23.

$$\frac{\partial f}{\partial z}\left(x, y, z\right) = xye^{x-y+2z} + 2xyze^{x-y+2z},$$

$$\frac{\partial^2 f}{\partial x \partial z}\left(x, y, z\right) = e^{x-y+2z}\left(y + xy + 2yz + 2xyz\right)$$

$$\frac{\partial^2 f}{\partial y \partial z}\left(x, y, z\right) = e^{x-y+2z}\left(x - xy + 2xz - 2xyz\right)$$

25.

$$f_{xy}\left(x, y\right) = -\frac{16xy}{\left(x^2 + 4y^2\right)^2} = f_{yx}\left(x, y\right)$$

Answers of Some Problems of Section 12.5

1.

a)

$$L\left(x, y\right) = 125 + 45\left(x - 3\right) + 60(y - 4).$$

b)

$$f\left(3.1, 3.9\right) \cong L\left(3.1, 3.9\right) = 123.5$$

c) According to a calculator, $f\left(3.1, 3.9\right) \cong 123.652$. The absolute error is

$$\left|f\left(3.1, 3.9\right) - 123.5\right| \cong 0.152.$$

3.

a)

$$L\left(x, y\right) = \frac{\pi}{6} - \frac{1}{4}\left(x - \sqrt{3}\right) + \frac{\sqrt{3}}{4}\left(y - 1\right).$$

b)
$$f(1.8, 0.8) \cong L(1.8, 0.8) \cong 0.420\,009$$

c) According to a calculator $f(1.8, 0.8) \cong 0.418\,224$. The absolute error is
$$|f(1.8, 0.8) - L(1.8, 0.8)| \cong 1.8 \times 10^{-3}$$

5.

a)
$$L(x, y, z) = \sqrt{14} + \frac{1}{\sqrt{14}}(x - 1) + \frac{2}{\sqrt{14}}(y - 2) + \frac{3}{\sqrt{14}}(z - 3).$$

b)
$$f(0.9, 2.2, 2.9) \cong L(0.9, 2.2, 2.9) \cong 3.741\,66$$

c). According to a calculator $f(0.9, 2.2, 2.9) \cong f(0.9, 2.2, 2.9)$. The absolute error is
$$|f(0.9, 2.2, 2.9) - L(0.9, 2.2, 2.9)| \cong 8 \times 10^{-3}$$

7.

a)
$$df = \frac{yz}{2\sqrt{xyz}}dx + \frac{xz}{2\sqrt{xyz}}dy + \frac{xy}{2\sqrt{xyz}}dz.$$

b)
$$f(0.9, 2.9, 3.1) - f(1, 3, 3) \cong df(1, 3, 3, -0.1, -0.1, 0.1) = -0.15$$

Therefore,
$$f(0.9, 2.9, 3.1) \cong f(1, 3, 3) - 0.15 = 3 - 0.15 = 2.85$$

c) According to a calculator $f(0.9, 2.9, 3.1) \cong 2.844\,47$. The absolute error is
$$|f(0.9, 2.9, 3.1) - 2.85| \cong 5.5 \times 10^{-3}$$

Answers of Some Problems of Section 12.6

1.

a)
$$\frac{d}{dt}f(x(t), y(t)) = \frac{dz}{dt} = \frac{\partial z}{\partial x}\frac{dx}{dt} + \frac{\partial z}{\partial y}\frac{dy}{dt}$$

$$= \left(\frac{x}{\sqrt{x^2 + y^2}}\right)\cos(t) + \left(\frac{y}{\sqrt{x^2 + y^2}}\right)(-2\sin(t))$$

$$= -\frac{3\sin(t)\cos(t)}{\sqrt{\sin^2(t) + 4\cos^2(t)}}$$

b) We have
$$f(x(t), y(t)) = f(\sin(t), 2\cos(t)) = \sqrt{\sin^2(t) + 4\cos^2(t)}.$$

Therefore,
$$\frac{d}{dt}f(x(t), y(t)) = \frac{d}{dt}\sqrt{\sin^2(t) + 4\cos^2(t)} = -\frac{3\sin(t)\cos(t)}{\sqrt{\sin^2(t) + 4\cos^2(t)}}$$

3.

a)

$$\frac{\partial}{\partial u} f\left(x\left(u,v\right)\right) = \frac{\partial z}{\partial u} = \frac{dz}{dx}\frac{\partial x}{\partial u} = \left(\frac{1}{x}\right)\left(-\frac{u}{\left(u^2+v^2\right)^{3/2}}\right) = -\frac{u}{u^2+v^2}$$

and

$$\frac{\partial}{\partial v} f\left(x\left(u,v\right)\right) = \frac{\partial z}{\partial v} = \frac{dz}{dx}\frac{\partial x}{\partial v} = \left(\frac{1}{x}\right)\left(-\frac{v}{\left(u^2+v^2\right)^{3/2}}\right) = -\frac{v}{u^2+v^2}.$$

b) We have

$$f\left(x\left(u,v\right)\right) = f\left(\frac{1}{\sqrt{u^2+v^2}}\right) = \ln\left(\sqrt{u^2+v^2}\right).$$

Therefore.

$$\frac{\partial}{\partial u} f\left(x\left(u,v\right)\right) = \frac{\partial}{\partial u} \ln\left(\frac{1}{\sqrt{u^2+v^2}}\right) = -\frac{u}{u^2+v^2},$$

and

$$\frac{\partial}{\partial v} f\left(x\left(u,v\right)\right) = \frac{\partial}{\partial v} \ln\left(\frac{1}{\sqrt{u^2+v^2}}\right) = -\frac{v}{u^2+v^2}.$$

5.

a)

$$\frac{\partial}{\partial u} f\left(x\left(u,v\right),y\left(u,v\right)\right) = \frac{\partial z}{\partial u} = \frac{dz}{dx}\frac{\partial x}{\partial u} + \frac{dz}{dy}\frac{\partial y}{\partial u}$$

$$= \left(-\frac{y}{x^2+y^2}\right)\cos\left(v\right) + \left(\frac{x}{x^2+y^2}\right)\sin\left(v\right) = 0,$$

$$\frac{\partial}{\partial v} f\left(x\left(u,v\right),y\left(u,v\right)\right) = \frac{\partial z}{\partial v} = \frac{dz}{dx}\frac{\partial x}{\partial v} + \frac{dz}{dy}\frac{\partial y}{\partial v}$$

$$= \left(-\frac{y}{x^2+y^2}\right)\left(-u\sin\left(v\right)\right) + \left(\frac{x}{x^2+y^2}\right)\left(u\cos\left(v\right)\right) = 1.$$

b) We have

$$f\left(x\left(u,v\right),y\left(u,v\right)\right) = f\left(u\cos\left(v\right),u\sin\left(v\right)\right) = \arcsin\left(\sin\left(v\right)\right) = v.$$

Therefore,

$$\frac{\partial}{\partial u} f\left(x\left(u,v\right),y\left(u,v\right)\right) = 0 \text{ and } \frac{\partial}{\partial v} f\left(x\left(u,v\right),y\left(u,v\right)\right) = 1.$$

9.

b)

$$u\left(x,0\right) = \sin\left(x\right),\ \ u\left(x,2\right) = \cos\left(x\right),\ \ u\left(x,4\right) = -\cos\left(x\right)$$

c)

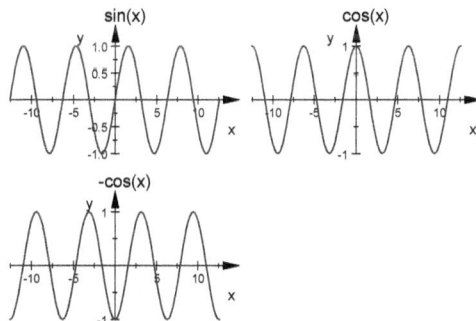

11.
a)
$$\frac{\partial z}{\partial x} = -\frac{x}{z}, \ \frac{\partial z}{\partial y} = \frac{y}{z}$$

b)
$$\frac{\partial z}{\partial x}\Big|_{x=3,y=6,z=6} = -\frac{3}{6} = -\frac{1}{2}, \ \frac{\partial z}{\partial y}\Big|_{x=3,y=3,z=6} = \frac{6}{6} = 1.$$

Therefore, the tangent plane is the graph of the equation

$$z = 6 - \frac{1}{2}(x-3) + (y-6).$$

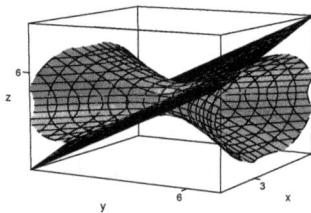

Answers of Some Problems of Section 12.7

1.
a)
$$\nabla f(x,y) = 8x\mathbf{i} + 18y\mathbf{j}$$

b)
$$D_{\mathbf{u}}f(3,4) = \frac{24\sqrt{5}}{5}$$

3.
a)
$$\nabla f(x,y) = 2xe^{x^2-y^2}\mathbf{i} - 2ye^{x^2-y^2}\mathbf{j}$$

b)
$$D_{\mathbf{u}}f(2,1) = -\sqrt{10}e^3$$

5.
a)
$$\nabla f(x,y,z) = 2x\mathbf{i} - 2y\mathbf{j} + 4z\mathbf{k}.$$

b)
$$D_{\mathbf{u}}f(1,-1,2) = \frac{8}{\sqrt{3}}$$

7.
a)
$$\mathbf{v} = \nabla f(2,3) = -\frac{2}{13^{3/2}}\mathbf{i} - \frac{3}{13^{3/2}}\mathbf{j}$$

The corresponding rate of increase of f is

$$\|\mathbf{v}\| = \frac{1}{13}$$

b)

$$\mathbf{w} = -\nabla f(2,3) = \frac{2}{13^{3/2}}\mathbf{i} + \frac{3}{13^{3/2}}\mathbf{j}$$

The corresponding rate of decrease of f is $1/13$.

9.

a)

$$\frac{d}{dt}f(\boldsymbol{\sigma}(t))\bigg|_{t=\pi/6} = \nabla f\left(\sigma\left(\frac{\pi}{6}\right)\right) \cdot \frac{d\sigma}{dt}\left(\frac{\pi}{6}\right) = \left(2\sqrt{3}e^2\mathbf{i} - 2e^2\mathbf{j}\right) \cdot \left(-\mathbf{i} + \sqrt{3}\mathbf{j}\right) = -4\sqrt{3}e.$$

b)

$$D_{d\sigma/dt}f\left(\sigma\left(\pi/6\right)\right) = -2\sqrt{3}e^2$$

11.

a) Let $f(x,y) = 2x^2 + 3y^2$.

$$\nabla f(x,y) = 4x\mathbf{i} + 6y\mathbf{j}.$$
$$\nabla f(2,3) = 8\mathbf{i} + 18\mathbf{j}$$

is orthogonal to the curve $f(x,y) = 35$ at $(2,3)$.

b) The tangent line is the graph of the equation

$$8(x-2) + 18(y-3) = 0$$

13.

a) Let

$$f(x,y) = e^{25-x^2-y^2}.$$
$$\nabla f(3,4) = -6\mathbf{i} - 8\mathbf{j}$$

is orthogonal to the curve $f(x,y) = 1$ at $(3,4)$.

b) The tangent line is the graph of the equation

$$-6(x-3) - 8(y-4) = 0.$$

15.

a) Let $f(x,y,z) = x^2 - y^2 + z^2$.

$$\nabla f(2,2,1) = 4\mathbf{i} - 4\mathbf{j} + 2\mathbf{k}$$

is orthogonal to the surface at $(2,2,1)$.

b) The plane that is tangent to the surface at $(2,2,1)$ is the graph of the equation

$$4(x-2) - 4(y-2) + 2(z-1) = 0.$$

17.

Let $f(x,y,z) = x - \sin(y)\cos(z)$.

$$\nabla f(1, \pi/2, 0) = \mathbf{i}$$

is orthogonal to the surface at $(1, \pi/2, 0)$.

b) The plane that is tangent to the surface at $(1, \pi/2, 0)$ is the graph of the equation

$$x - 1 = 0 \Leftrightarrow x = 1.$$

Answers of Some Problems of Section 12.8

1. $(-1/5, -3/5)$ is the only critical point. The function has a saddle point at $(-1/5, -3/5)$.

3. Any point on the line $y = -x$ is a critical point. The function attains its absolute maximum or minimum on the line $y = -x$.

5. $(-18/5, -11/15)$.is the only critical point. The function has a local (and absolute) minimum at $(-18/5, -11/15)$.

7. The critical points are $(0,0)$, $(1, -1)$ and $(-1, 1)$. The function has a saddle pioint at $(0, 0)$, local maxima at $(1, -1)$ and $(-1, 1)$.

Answers of Some Problems of Section 12.9

1. The maximum value of f on the circle $x^2 + y^2 = 4$ is $f\left(\sqrt{2}, \sqrt{2}\right) = 2\sqrt{2}$ and the minimum value is $f\left(-\sqrt{2}, -\sqrt{2}\right) = -2\sqrt{2}$.

3. The minimum value of $f(x, y)$ subject to $4x^2 + y^2 = 8$ is $--2$, the maximum value is 2.

5. The minimum value of f in D is 0, and its maximum value in D is 8.

Answers of Some Problems of Section 13.1

1.
$$\frac{81}{2}$$

3.
$$\frac{1}{3}e^{12} - e^4 - \frac{1}{3}e^6 + e^2$$

5.
$$\frac{8}{3}\ln(2)$$

7.
$$\frac{\pi}{12}$$

Answers of Some Problems of Section 13.2

1. The integral is 32.

3.

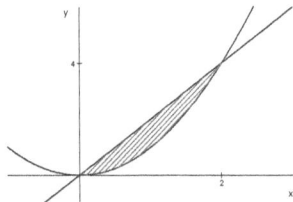

The integral is 64/3.

5.

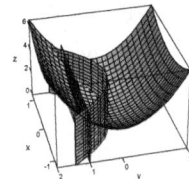

The integral is
$$\frac{1}{2}e - 1$$

7.

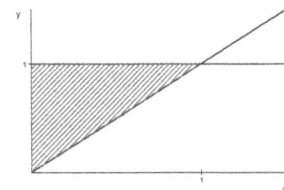

9.

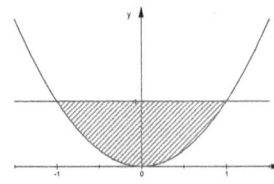

The integral is
$$\frac{76}{35}$$

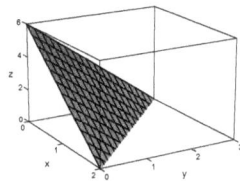

The integral is 6.

11.

a)

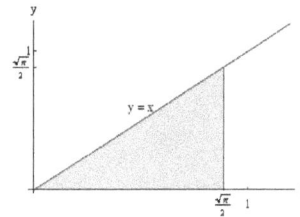

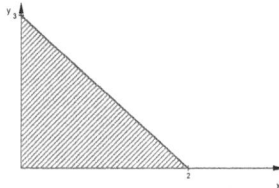

b) The integral is

$$\frac{\sqrt{2}}{4}$$

Answers of Some Problems of Section 13.3

1.

a)

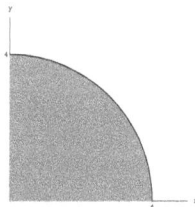

b) The integral is 32.

3.

a)

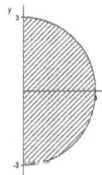

b) The integral is 9π.

5.

a)

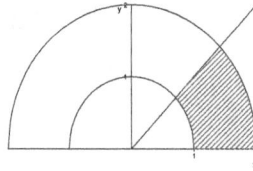

b) The integral is

$$\frac{3\pi^2}{64}$$

7.

a)

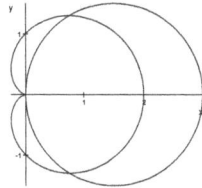

b) The area of D is $\pi/4$.

9. The volume of D is

$$\frac{4\pi}{3}12^{3/2}$$

11.

a)

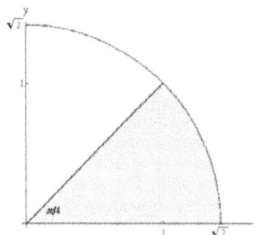

b) The integral is 2/3.

Answers of Some Problems of Section 13.5

1. The volume of the region is 81π.

3.

$$\iiint_D x^2 dx dy dz = \frac{1}{3}$$

5.

$$\iiint_D ze^{x+y} dx dy dz = 2\left(e-1\right)^2.$$

7.

$$\iiint_D z dx dy dz = \frac{1}{2}\iint_R dx dy = \frac{\pi}{8}.$$

9.

$$\iiint_D xy\,dx\,dx\,ydz = \frac{3}{28}$$

Answers of Some Problems of Section 13.6

1.
$$\left(\sqrt{2}, \sqrt{2}, 1\right)$$

3.
$$\left(-\frac{3}{2}, \frac{3\sqrt{3}}{2}, 4\right)$$

5.
$$r = \sqrt{2}, \ \theta = \frac{3\pi}{4}, \ z = 4$$

7.
$$r = \sqrt{2}, \ \theta = -\frac{\pi}{4}, \ z = 2$$

9.
$$\iiint_D \sqrt{x^2 + y^2}\,dx\,dy\,dz = 384\pi$$

11.
$$\iiint_D e^z\,dx\,dy\,dz = \pi\left(e^6 - 5 - e\right)$$

13. The volume of the region is
$$\frac{4\pi}{3}\left(8 - 3^{3/2}\right)$$

15.
$$\left(\frac{\sqrt{2}}{2}, \frac{\sqrt{6}}{2}, \sqrt{2}\right)$$

17.
$$\left(-\sqrt{6}, \sqrt{6}, 2\right)$$

19.
$$\rho = \sqrt{2}, \ \phi = \frac{3\pi}{4}, \ \theta = -\frac{\pi}{2}$$

21.
$$\rho = 2\sqrt{2}, \ \phi = \frac{\pi}{6}, \ \theta = \frac{3\pi}{4}$$

23.
$$\iiint_D z\,dx\,dy\,dz = \frac{15}{16}\pi$$

25.
$$\iiint_D x^2\,dx\,dy\,dz = \frac{1562}{15}\pi$$

27.
$$\iiint_D dx\,dy\,dz = 9\left(\sqrt{3} - 1\right)\pi$$

29. The volume of the region D is
$$\frac{8\sqrt{2}}{3}\pi$$

Answers of Some Problems of Section 14.1

1.

a)

$$\nabla \cdot \mathbf{F}(x, y, z) = yz$$

b)

$$\nabla \times \mathbf{F}(x, y, z) = -x^2 \mathbf{i} + 3xy \mathbf{i} - xz \mathbf{k}$$

3.

a)

$$\nabla \cdot \mathbf{F}(x, y, z) = 0.$$

b)

$$\nabla \times \mathbf{F}(x, y, z) = (-x \cos(xy) + x \sin(xz)) \mathbf{i} + y \cos(xy) \mathbf{j} - z \sin(xz) \mathbf{k}$$

5.

a)

$$\nabla \cdot \mathbf{F}(x, y, z) = 2y + 2z$$

b)

$$\nabla \times \mathbf{F}(x, y, z) = \mathbf{0}$$

Answers of Some Problems of Section 14.2

1.

$$\int_C y^3 ds = \frac{1}{54} \left(145^{3/2} - 1\right)$$

3.

$$\int_C xe^y ds = \frac{e^5}{2} - \frac{e}{2}$$

5.

$$\int_C (x - 2) e^{(y-3)(z-4)} ds = \frac{\sqrt{14}}{12} \left(e^6 - 1\right)$$

7.

$$\int_C \left(x^2 \mathbf{i} + y^2 \mathbf{j}\right) \cdot d\boldsymbol{\sigma} = -\frac{\pi}{2}$$

9.

$$\int_C (\cos(y) \mathbf{i} + \sin(z) \mathbf{j} + x \mathbf{k}) \cdot d\boldsymbol{\sigma} = \frac{3}{4}$$

11.

$$\int_C (-2xy \mathbf{i} + (y + 1) \mathbf{j}) \cdot \mathbf{T} ds = \frac{28}{3}$$

13.

a)

$$\int_C \mathbf{F} \cdot d\boldsymbol{\sigma} = \int_C -xy dx + \frac{1}{x^2 + 1} dy.$$

b)

$$\int_C -xy dx + \frac{1}{x^2 + 1} dy = \ln\left(\frac{2}{17}\right) + \frac{255}{4}$$

15.

$$\int_C 3x^2 dx - 2y^3 dy = -\frac{3}{2}$$

17.

$$\int_C -\sin(x) dx + \cos(x) dy = -6.$$

Answers of Some Problems of Section 14.3

1.
b)
$$f(x, y) = x^2 - 3yx + g(y) = x^2 - 3yx + 2y^2 - 8y + K$$
is a potential for **F** (K is an arbitrary constant).

3.
b)
$$f(x, y) = e^x \sin(y) + K$$
is a is a potential for **F** (K is an arbitrary constant).

5.
a)
$$f(x, y) = \frac{1}{2}x^2 y^2$$
is a potential for **F**.
b)
$$\int_C \mathbf{F} \cdot d\boldsymbol{\sigma} = 2.$$

7.
a)
$$f(x, y, z) = xyz + z^2$$
is a potential for **F**.
b)
$$\int_C \mathbf{F} \cdot d\boldsymbol{\sigma} = 77$$

9.
$$\int_C \frac{y^2}{1 + x^2} dx + 2y \arctan(x) \, dy = \int_C df,$$
where
$$f(x, y) = y^2 \arctan(x).$$
$$\int_C \frac{y^2}{1 + x^2} dx + 2y \arctan(x) \, dy = \pi.$$

Answers of Some Problems of Section 14.4

1.
a)
$$\mathbf{N}\left(\frac{\pi}{6}, 1\right) = \frac{3\sqrt{3}}{2}\mathbf{j} + \frac{3}{2}\mathbf{k}$$
b)
$$\frac{3\sqrt{3}}{2}\left(y - \frac{3\sqrt{3}}{2}\right) + \frac{3}{2}\left(z - \frac{3}{2}\right) = 0$$

3.
a)
$$\mathbf{N}\left(2, \frac{\pi}{4}\right) = 8\sqrt{2}\mathbf{i} - \left(3\sqrt{2} + 6\right)\mathbf{j} + 12\mathbf{k}$$
b)
$$8\sqrt{2}\left(x - 3\sqrt{2}\right) - \left(3\sqrt{2} + 6\right)(y - 4) + 12\left(z - 2\sqrt{2}\right) = 0.$$

5.

a)

$$N\left(2, \frac{\pi}{3}\right) = -\frac{\sqrt{3}}{4}i - \frac{3}{4}j + \frac{1}{2}k$$

b)

$$-\frac{\sqrt{3}}{4}\left(x - \frac{1}{2}\right) - \frac{3}{4}\left(y - \frac{\sqrt{3}}{2}\right) + \frac{1}{2}\left(z - \sqrt{3}\right) = 0.$$

7.

a)

$$N\left(1, \frac{\pi}{6}\right) = -e^2i + \frac{1}{2}ej + \frac{\sqrt{3}}{2}k$$

b)

$$-e^2(x - 1) + \frac{1}{2}e\left(y - \frac{1}{2}e\right) + \frac{\sqrt{3}}{2}\left(z - \frac{\sqrt{3}}{2}e\right) = 0.$$

Answers of Some Problems of Section 14.5

1 The area of the surface is 48π.

3. The area of the surface is 8π.

5 The area of the surface is

$$2\pi\left(\frac{1}{2}e^2\sqrt{(e^4 + 1)} + \frac{1}{2}\operatorname{arcsinh}\left(e^2\right) - \frac{1}{2}\sqrt{2} - \frac{1}{2}\operatorname{arcsinh}(1)\right)$$

7.

$$\int\int_S y^2 dS = 8\pi$$

9.

$$\int\int_S x dS = -4\pi$$

13.

$$\int_S \frac{xi + yj + zk}{(x^2 + y^2 + z^2)} \cdot dS = 12\pi$$

15.

$$\int\int_S (xi + yj + zk) \cdot dS = 16\pi$$

Answers of Some Problems of Section 14.6

1.

a)

$$\int_C y^3 dx - x^3 dy = -3\int\int_D \left(x^2 + y^2\right) dx dy$$

b)

$$-3\int\int_D \left(x^2 + y^2\right) dx dy = -24\pi$$

3.

a) Let D be the rectangular region with vertices $(0,0)$, $(5,0)$, $(5,\pi)$ and $(0,\pi)$. Then

$$\int_C \cos(y)\,dx + x^2 \sin(y)\,dy = \int\int_D (2x+1)\sin(y)\,dxdy$$

b)
$$\int\int_D (2x+1)\sin(y)\,dxdy = 60$$

5.

a) Let D be the triangular region with vertices $(0,0), (2,6)$ and $(2,0)$.

$$\int_C \mathbf{F}\cdot d\boldsymbol{\sigma} = -\int\int_D 2x\,dxdy$$

b)
$$-\int\int_D 2x\,dxdy = -16$$

7.
$$\int_{C_1} \mathbf{F}\cdot d\boldsymbol{\sigma} = 0.$$

9.
$$\int_C \mathbf{F}\cdot \mathbf{n}ds = 24\pi$$

Answers of Some Problems of Section 14.7

1.
$$16\pi$$

3.
$$\frac{243}{4}\pi$$

Answers of Some Problems of Section 14.8

1. 3

3. 192

Appendix L

Basic Differentiation and Integration formulas

Basic Differentiation Formulas

1. $\dfrac{d}{dx} x^r = r x^{r-1}$

2. $\dfrac{d}{dx} \sin(x) = \cos(x)$

3. $\dfrac{d}{dx} \cos(x) = -\sin(x)$

4. $\dfrac{d}{dx} \sinh(x) = \cosh(x)$

5. $\dfrac{d}{dx} \cosh(x) = \sinh(x)$

6. $\dfrac{d}{dx} \tan(x) = \dfrac{1}{\cos^2(x)}$

7. $\dfrac{d}{dx} a^x = \ln(a) a^x$

8. $\dfrac{d}{dx} \log_a(x) = \dfrac{1}{x \ln(a)}$

9. $\dfrac{d}{dx} \arcsin(x) = \dfrac{1}{\sqrt{1-x^2}}$

10. $\dfrac{d}{dx} \arccos(x) = -\dfrac{1}{\sqrt{1-x^2}}$

11. $\dfrac{d}{dx} \arctan(x) = \dfrac{1}{1+x^2}$

Basic Antidifferentiation Formulas

C denotes an arbitrary constant.

1. $\displaystyle\int x^r dx = \dfrac{1}{r+1} x^{r+1} + C \ (r \neq -1)$

2. $\displaystyle\int \dfrac{1}{x} dx = \ln(|x|) + C$

3. $\displaystyle\int \sin(x)\, dx = -\cos(x) + C$

4. $\displaystyle\int \cos(x)\, dx = \sin(x) + C$

5. $\int \sinh(x) dx = \cosh(x) + C$

6. $\int \cosh(x) dx = \sinh(x) + C$

7. $\displaystyle\int e^x dx = e^x + C$

8. $\displaystyle\int a^x dx = \dfrac{1}{\ln(a)} a^x + C \ (a > 0)$

9. $\displaystyle\int \dfrac{1}{1+x^2} dx = \arctan(x) + C$

10. $\int \dfrac{1}{\sqrt{1-x^2}} dx = \arcsin(x) + C$

Index

Absolute extrema, 122
Acceleration, 46
Arc length, 52

Binormal, 49

Cartesian Coordinates, 1
Chain rule, 85
Circulation of a vector field, 268
Conservative vector fields, 215
Continuity, 66
Continuity equation, 283
Curvature, 53
    radius of curvature, 54
Cylindrical coordinates, 169

Differential, 80
Directional derivatives, 94
Distance traveled, 52
Divergence Theorem, 281
Double integrals, 133, 140
    double integrals in polar coordinates, 146
    Fubini's Theorem, 136
    Riemann sums, 134
Doubly connected region, 265

Flux across a curve, 270
Flux integrals, 249
    differential form notation, 255

Gauss' Theorem, 280
Gradient, 97
    chain rule and the gradient, 101
    gradient and level curves, 101
    gradient and level surfaces, 104
Green's Theorem, 261

Implicit differentiation, 90
Incompressible flow, 271
Irrotational flow, 271

Jacobian, 183

Lagrange multipliers, 124

Level curves, 62
Level surface, 64
Limit, 66
Line integral of a vector field, 200
Line integrals, 196
    differential form notation, 204
    fundamental theorem of line integrals, 212
    independence of path, 213
    line integrals of conservative fields, 212
    Line integrals of scalar functions, 196
Linear approximations, 75

Mass, 154
    mass density, 154
Maxima and minima, 107
    discriminant, 112
    Second derivative test, 112
Mobius band, 230
Moments, 155
    center of mass, 155
Moving frame, 48

Normal distribution, 158

Orientation of a curve, 197

Parametrized curves, 35
    derivative of a vector-valued function, 40
    tangent line, 42
    tangent vectors, 40
    unit tangent, 42
Parametrized surfaces, 222
    normal vectors, 227
    orientation, 230
    tangent planes, 227
Partial derivatives, 65
    higher-order partial derivaives, 71
Planes
    normal vactor, 31
Potential, 215
Potential function
    existence of a potential, 271
Principal normal, 49
Probability, 157